Hindenburg: The Wrong Paint

Hydrogen: The Right Fuel

By Richard G. Van Treuren

with the thesis of
Dr. Addison Bain, Ph.D.

Foreword By Eric Brothers

Front Cover: LZ-129 photo from the Hepburn Walker, Jr. collection
Orbiter *Columbia* photo courtesy NASA
BMW Hydrogen power engine photo courtesy BMW

Author wishes to gratefully acknowledge the help of DCH Technologies, Air Products, Inc. and BMW for their support in the making of this book.
Thanks to Cartridge Source America, Inc. for printing support.

"There are always those who know all the reasons why something can't be done.
Fortunately there are those who know when not to listen. "

Hindenburg, The Wrong Paint; Hydrogen, The Right Fuel
ISBN 0-9639743-7-8 Soft Cover

ISBN 0-9639743-6-X e-book CD ROM

Atlantis Productions
Edgewater, Florida 32132

Table of Contents

ON JOURNAL

URSDAY EVENING, FEBRUARY 20, 1930 (PAGES 21 TO 34)

Test Lightning Bolts On Dirigibles

Tests of lightning on a model airship provided by the Goodyear Zeppelin Corp. at the Ohio Insulator Co.'s outdoor laboratory at Barberton indicated that the chief, if not the only, danger facing the airship from this source is the possibility of the fabric catching fire.

Several tests on a scale model of one of the proposed commercial Zeppelins were made at the laboratory under the direction of Arthur O. Austin, chief engineer.

In the majority of these tests there was no observable damage except a pinhole in the fabric at the point at which the bolt struck. In one case the fabric near the tail surfaces caught fire.

Last line reads, "In one case the fabric near the tail surfaces caught fire."
Karl Arnstein Collection, University of Akron Archives

Foreword

No one can reasonably dispute that the German luxury liner of the air, the Zeppelin *Hindenburg*, burned while landing at Lakehurst, New Jersey on the evening of 6 May 1937. But what caused the accident has been a source of mystery and controversy ever since that time. Thanks to radio, motion pictures, and now television and the Internet, the *Hindenburg* accident has become a familiar part of the world's audio and visual memory of the 20th century. Over time, repeated viewings of the newsreel film footage have etched an indelible image in the public's consciousness. And radio newsman Herbert Morrison's emotional account of the airship's fiery demise has become an instantly recognizable milestone of broadcast journalism. Yet none of the dramatic sound and visual record explains why or how the airship became a raging inferno.

Of all the alleged causes of the accident — static electricity, sabotage, gunshots — no one ever questioned the seemingly obvious fact that the *Hindenburg* burned quickly because it was inflated with inflammable hydrogen. At least not until NASA hydrogen fuels expert Addison Bain wondered why the *Hindenburg* fire did not behave the way hydrogen usually does when it burns. This observation, based on his personal and professional experience, prompted him to re-examine what happened to the ill-fated airship on a rainy spring evening six decades earlier. Bain was able to test his hypotheses with new techniques unavailable to the original accident investigators, and he examined some clues previously not considered by them. His research also uncovered some startling information. Bain discovered some old, "hidden" information that had never been acknowledged before or made accessible to researchers.

During that inquiry, this book's author, Richard Van Treuren, sought out Bain, and was enthusiastic about what he had learned. To Van Treuren, Bain's new interpretation was a salvation. It offered a solution to an old problem — and a means to dispel years of prejudice against what some people have considered to be a "criminal" element: hydrogen. But you'll be reading more about that shortly. Once Van Treuren had found the answer, he set about to publicize Bain's work to a wider audience. Airship enthusiasts were among the first to read about the findings in The Lighter-Than-Air Society's bulletin, *Buoyant Flight*. Then came an article in *Popular Science* and appearances in televised airship documentaries.

It may be that there is nothing really new here, that it is just old knowledge being rediscovered and seen with new and greater appreciation. An example of this possibility is demonstrated in the clipping reproduced opposite this foreword. Maybe in our haste to pass judgment we have overlooked or forgotten a tiny bit of information that in the right context makes the new interpretation more plausible. Whether or not you accept the conclusions, I hope you will agree that anything that challenges us to look at accepted wisdom in a new way is worthwhile.

Eric Brothers

Editor, *Buoyant Flight*

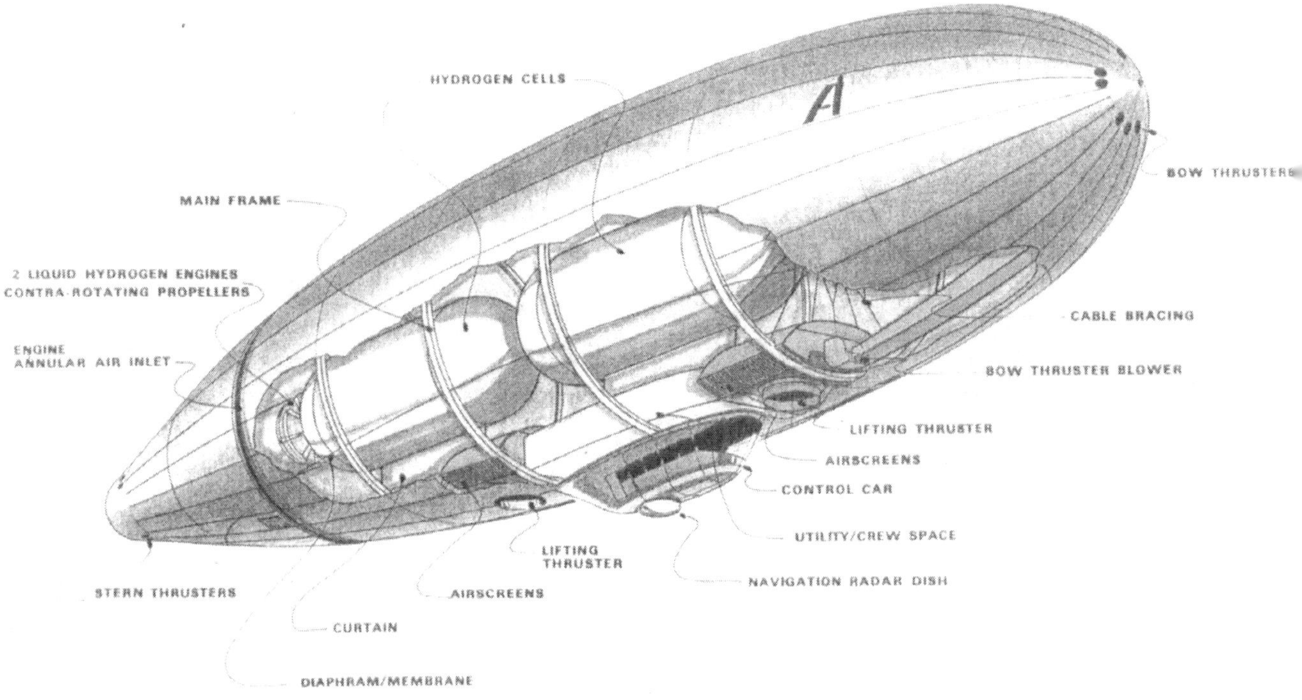

Painting by John Mellburg, courtesy Airships International, Inc.

"What a horse they are losing for want of skill and spirit to manage him."
– Alexander the Great

Introduction and Acknowledgements

James Burke's excellent television history series "The Day The Universe Changed" provided banquets of food for thought, repeatedly serving samples of how a given era's predominant thinking was gospel until suddenly upended by new discoveries and applications. Elsewhere Burke wrote, "But in the absence of more coherent explanations, they often gained devoted followings, making it difficult to substitute new explanations. Apologists have built careers and even religions around many particulars of such theories. In earliest times, such individuals were priests; later, we began calling them scientists. Their closeness to society's stores of knowledge has always separated them from the average individual. It took centuries, but the printing press helped diminish those class distinctions. Today, we're experiencing a similar but much accelerated phenomenon as inexpensive personal computers provide us with virtually unlimited access to information..." A decade after those words were published, this writer happened into an example of the former, and found application of the later that, in addition to the traditional archive research, would promote debate after, and even before, traditional publishing. Perception of all that resulted being positive or negative will polarize in the mind of the reader. The universe doesn't care what we believe, or if we poison our planet and ourselves along with it.

What began as a simple effort to list some historical examples supporting an idea, then the thesis reinforcing the idea's validity from a safety standpoint, followed by the patent application and supporting technology, dragged on for years as we gained access to a never-ending stream of historical material and new developments. On TV, Burke was on camera; but it took a team of professionals behind the scenes to provide such a wondrous educational experience. Likewise, the author of this book was an organizer of references, wrenched from obscurity by hardworking enthusiasts. Instead of credits rolling at the end, the references and their authors/researchers are given on the page without having to flip to an appendix. This author apologies to those he mistakenly omitted owing to code names, e-bumbling, or just plain CRS disease. The author gratefully recognizes even those who loathed the very idea, because they also provided valuable input, if unwittingly. Likewise, the help of family, friends, co-workers and even our Volkswagen Club cannot be overemphasized even as, like the older crowd that makes up us "helium heads," many passed on before we could finish this thing. Not wanting to play favorites, I would still be remiss without mentioning the outstanding efforts of those who tirelessly worked for the good of this project following Dr. Addison Bain's opening our eyes. They are, at least in part: Eric Brothers and Marc de Piolenc who helped even while also tirelessly editing LTA projects of their own; Herman Van Dyk, who reviewed the early manuscript while ceaselessly adding to LTA history; our strongholds of LTA knowledge and archive, James Shock and Hepburn Walker, Jr., and of course my ever tolerant, supportive, proofreading, and loving wife, Deborah. If we are successful in our writing and organizing, then irrational fear, based on ignorance, will lose this round, and our immediate universe will change for the better.

Richard G. Van Treuren

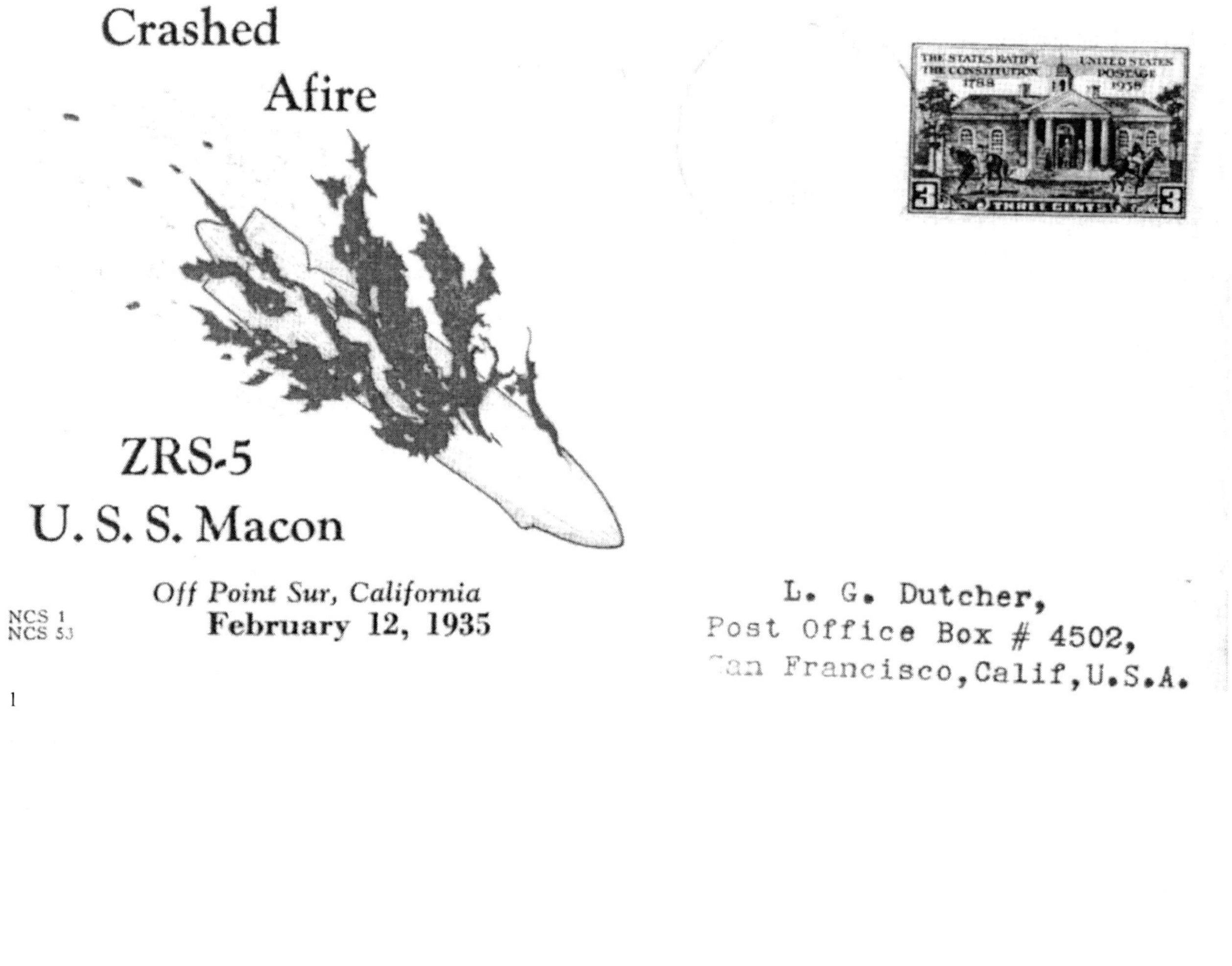

[1]

---

[1] Envelope courtesy Cheryl Ganz, Editor, *The Zeppelin Collector.* Of course the *Macon*'s covering did not burn aloft, but only after flares on the bridge contacted salt water and ignited the hundreds of gallons of gasoline floating around the wreck. That fire spread to the covering, which from then on was self-sustaining. In a newspaper item dated April 4, 1933, from Berlin, Dr. Hugo "...Eckener suggested as a possibility that the destruction of the (USS *Akron*) airship had been caused by a fire on board which destroyed a number of gas cells... 'meteorlogical reports always are so thorough and regular that *Akron's* crew must have been constantly informed of the conditions'... Dr. Eckener inclined to doubt the the *Akron* was destroyed by a storm forcing it down upon the sea."

Chapter One: Hydrogen Hysteria

"The fire was reddish-orange in color," remembered one rigid airship vet who later stood guard duty over the wrecked remains of the infamous German airship *Hindenburg,* and who described the burns on one man's hands.[2] Another remembered of his rigid airship that "...a fire started and at last the cover burned off as she went down." [3] One victim jumped to his death; another likely perished in the flames.[4] A heroic Chief Petty Officer was commended for keeping a cool head "...and remaining close to the hull even when fire broke out in the bow as it was sinking."[5] Was this a famous hydrogen airship fire?

No. On the night of February 12, 1935, the helium-inflated USS *Macon's* acetate-aluminum fabric covering burned as her control car slipped beneath the waves of the Pacific. Newsreels exclaimed, "*MACON* EXPLODES!"[6] "In that disaster, which took place ten miles off Big Sur, south of Monterey, two of *Macon's* gas cells suddenly exploded in a heavy squall, causing the airship to nose upward sharply, much like *Akron* had two years before. At an altitude of just 1,250 ft. and a speed of 63 mph., there wasn't time to make corrections, drop or shift ballast. With elevator control totally crippled by the explosion, *Macon* rapidly lost way."[7] A newspaper proclaimed, "...reports have persisted that he [the messboy] was in fact a Jap. The government is not overlooking the possibility that the destruction might have been caused by a fanatic of foreign-birth. Young patriots in certain countries are schooled from birth with the idea that a sacrifice of their lives for their nation insures their immortality... Friends of lighter-than-air cannot escape the feeling that there was an uncanny timeliness about the disaster."[8]

Consider how different our perception of this accident would be today if reports like those were constantly repeated. Mercifully, without pictures, newsreels or audio recordings, or a larger number of fatalities, the tragedy of *Macon* was quickly forgotten. The same lack of damning audio-visual records allowed many helium-filled airships to burn to their skeletons, some with total crew incineration, without public outcry over the explosively dangerous petroleum fuels they used, or the flammable materials employed in their construction.[9] Mostly, there was a war on; a 10-man blimp accident was so similar to a commonplace airplane wreck, who would care?

---

[2] Audio interview with William Clarke, rigid airship rigger, by author, December 1998. The burns to his shipmate's hands were actually *rope* burns, as the man hastily slid down a hawser without having gloves.

[3] US Naval Institute – published oral history of RADM "Min" Miller includes his own typewritten account set down shortly after the USS *Macon's* loss.

[4] Radioman Ernest Dailey, having lost his eyeglasses, jumped from too great a height, according to Clyde Padget, ZRS-5 engineman, whom the author met in 1991. Steward Florentino Edquiba, terrified of the water even wearing his life vest, was last seen climbing on the structure (probably before the fire broke out), according to LCDR William Carroll, photographer at NAS Moffett. Another version has the smallish Phillipino going back into the ship to "retrieve belongings."

[5] Commendation letter to CBM Robert J. "Shaky" Davis from Commander, Aircraft Battle Force, 5 SEP 35. Courtesy his son, Bruce Davis, WWII airshipman.

[6] "Hot Shots/Cool Cuts" stock footage reel CC 2422 also says "...250 miles off California coast!" when USS *Macon* actually floundered in view of the Point Sur Lighthouse.

[7] J. Dean, "Airships," *Airpower*, Vol.30 No. 4, July 2000, pg.16. This article states all too common misconceptions about the ZRS program, and goes on to suggest a big problem with the rigid airships was their "immovable object" heavy metal gondola not being "pliable" like the gas cells *and* framework! Of course there was time, since even with the wrong reactions the airship was aloft for half an hour following the upper vertical fin casualty.

[8] The clipping found by Eric Brothers was likely from the Akron *Times-Press*.

[9] In NASA Materials Testing Brach memo "Flammability Test on Blimp Materials" 10 NOV 71, Federal Test Method Standard 191 applied to hull and fin material averaging .016 in. thick showed they burned at about 12 in/min., fortunately much slower than Zeppelin fabric.

Thankfully, very little has been written or publicized about helium-filled airships being destroyed by fire. No matter, all airships are painted with the same broad brush. "Just once, [Zeppelin NT pilot] Scott Danneker would like to see a TV documentary or magazine story about his employer that doesn't feature the airship *Hindenburg's* bursting like a lava-filled egg over Lakehurst, N. J. on May 6, 1937.'What would happen if people felt compelled to mention Pan Am Flight 103 every time they talked about airplanes?'" [10]

Indeed, we step aboard a 747 airliner without even considering the ghastly carnage wrought when two of these jumbos collided in the Canary Islands in 1977, slaughtering 583 souls. Part of the reason was our becoming hardened by the previous fatality record, the DC-10 accident of March, 1974 that spreading the remains of 346 passengers and crew across a forest and glen. That single plane accident killed considerably *more* people than died in the famous battle of the Little Big Horn, but we have become so callous to airplane accidents that even those that kill hundreds remain headline news for but a day or two. The average airliner crash kills more Americans than died in the Persian Gulf War. Some mysteries, like the Pakastani airliner whose hard landing turned into a funeral pyre for 301 Muslim pilgrims in August of 1980, have *never* been solved. By contrast, in the sum total of *all* airship accidents (including all ground casualties) from the beginning of time, fatalities totaled less than that *one* collision.[11]

Incredibly, a single aircraft accident in which a grand total of *thirteen* passengers died seems to stand out, from all other aeronautical disasters, even after a lifetime of more than sixty years. "During the landing operation, the Airship *Hindenburg* burst into flame at an altitude of about 200 feet and was burned to destruction by hydrogen fire originating at or near the stern. The ship fell to the ground on the north side of number one mooring out circle approximately 232 yards from the mooring mast."[12] The watch officer who made that log entry noted the fire's origin location by eyewitness, but took it upon himself to condemn the lifting gas without a trial - or even a hearing. We members of the human race tend to form an opinion based on first impression or initial suggestion. If a person were arrested on suspicion of a crime, human nature would tend to agree with the first accusation. Therefore, our Founding Fathers wisely instituted a Constitutional guarantee that a person is presumed innocent until proven guilty.

Hydrogen was blamed for the 35 deaths aboard *and* the end of commercial rigid airship flight as a result of the accident at Lakehurst, New Jersey, on May 6th, 1937. With endless repetition of the words and picture accompaniment for this accusation, we've set aside the Constitution for the element accused in this case. "35 died. Not the greatest disaster in aviation history, but she was the biggest thing that ever flew; that's why the newsreel crews were there, why Herb Morrison was there. Their pictures, his words: that's why we will never forget what happened, 50 years ago today."[13]

We have long since forgotten or never knew that nothing more than rumor and hearsay, *not a scintilla of evidence* to convict hydrogen, was *ever* presented. In fact, a great deal of evidence was presented to *exonerate* hydrogen in the fire's instigation. Few realize the latter, and the former seems to make no difference. The argument is outcome-based: There was a fire; there *must* have been a leak, right? Then, there was some spark – wasn't it in a storm or lightning? An

[10] S. Zwick, "More Than Hot Air," *Time*, 17 July 2000, pg. B15

[11] B. Gordon, "Zeppelins Are Safe!", *Modern Mechanix*, August 1937, states: "In the entire world, since the first dirigible was flown, only 318 lives have been lost in peace time operations of lighter-than-air craft."

[12] Lakehurst NAS log, quoted by W. Althoff, SKY SHIPS, Orion Books, 1990 pg. 138

[13] Joel Seigel during the GMA TV program that date, courtesy Alan Gross. (total of 36 died)

Associated Press article's photo on the 50th anniversary has a caption, "Dirigible's hydrogen exploded just as airship was about to land." No mystery, there, surely, what's to explain?

Having no experience with hydrogen, people have nonetheless been conditioned from childhood that nature's simplest element is something to be afraid of. To some people, it is *mysteriously* dangerous, i.e. "they never figured out how the hydrogen caught fire" so we dare not use it least the mysterious demon also consume us. To others, it is *supernaturally* dangerous, i.e. "the airship passengers were consumed in horrifying flames akin to a small H-bomb." Human nature makes us uncomfortable with things we don't understand, and likewise frightened by potential explosives we have not become accustomed to in our daily lives. Yet, if one can restore the Constitutional guarantee for just long enough to read this book, one may come to accept that, regardless of its being all but imperceptible, hydrogen's spooky-dangerous reputation is not deserved. As you read this book, try to remember Lord Thomas Dewar, who said, "Minds are like parachutes, they work best when open."

The history of buoyant flight is the history of hydrogen, and it all came together about the time the United States of America did. The discovery of hydrogen was first made by alchemist Theophratus Paracelsus just after the year Christopher Columbus made his first voyage to the new world. James Dangerfield wrote, "Placing a drop of sulfuric acid on iron filings, he noticed 'inflammable air' was created. Isolated without the flammable air (oxygen) a flame could not exist." Over the next 500 years this basic fact would be repeatedly set aside. Still, the first recorded accidental fire involving a buoyant vehicle cannot be blamed on hydrogen. It actually predated the acknowledgment of hydrogen. In 1709, a Brazilian Jesuit launched a hot-air balloon model for the court of King Joao V of Portugal.[14] "Flight was achieved, but during its decent the balloon ignited some draperies, causing considerable damage to the royal hall."[15]

Henry Cavendish recognized the gas as a complete substance the year Americans declared their independence from his native England.[16] Also in 1776, J. Warlite found that water was produced when hydrogen was combusted with oxygen. Frenchman Antoine Lavoisier gave the element the name from the Greek *hudor* (water) and *gennan* (generate). Another Frenchman, Professor Jaques Alexandre C. Charles, was the first to employ the gas to displace air, building balloons the year Americans won their independence. The first case of hysteria was not from what was in the balloon, but the unmanned sphere itself. Yet there is no record of fire as frightened peasants tied it to a horse and dragged it across the French countryside. Charles ascended from Paris on December 1st, 1783; the following year, a French hydrogen balloon lifted off in America and was witnessed by five US presidents.

Two years later that French pioneer, Jean-Pierre Blanchard, flew American Dr. John Jefferies across the English Channel. Difficult to pinpoint exactly when it started, the association of hydrogen and fire in buoyant vehicles probably dates back to the 1785 fire that killed balloonists Pilatre de Rozier and Pierre Romain.[17] Lighting a wood fire under a silk gasbag doesn't sound very bright, but represents the lengths the men would go to in order to have more control over their vehicle's lift. Fires were also associated with gasless, hot air balloons in the early days

[14] The lavish and well-equipped German South American base, the *Aeropuerto Bartolomaeo Gusmao*, was named for the Brazilian priest and LTA experimenter.
[15] Legacy of Flight, *Aviation History*, July, 1996
[16] Others put the date at 1766 and cite the eccentric and shy Cavendish as merely the first to collect gases over water. Cavendish himself never claimed to have discovered the gas; Robert Boyle mentions it in 1671.
[17] THE AERONAUTS, Time-Life Books, Alexandria, VA.

owing to the basic incompatibility of fire with the paper, silk and other lightweight but flammable materials of the day. At the time, the fact that pure hydrogen gas cannot burn was not generally understood. That misunderstanding is still a problem today.

The relationship of hydrogen and electricity dates to 1800 when Allesandro Volta announced his 'pile.' Shortly thereafter, William Nicholson and Anthony Carlisle, using minor alterations, were the first to collect and recognize hydrogen and oxygen in the decomposition of water. Sir William Grove discovered he could split water electrically and then recombine its two elements to produce electricity in his 1839 *fuel cell.*

The more efficient lift of lighter-than-air gasses seemed to outweigh the danger associated with their flammability (or toxicity, with ammonia) for the first 150 years of "the sport of kings." There were of course problems associated with the relatively crude gas production technology of those days, meaning the lifting gas was not always pure. The mechanical details of the free balloon introduced more problems. As Walter S. Diehl wrote in BALLOON AND AIRSHIP GASES, "Whenever a free balloon descends, some air enters the appendix, thereby increasing the proportion of oxygen to hydrogen... A mixture of 36% air in hydrogen is probably the smallest concentration of air within a balloon that could be ignited…" In a vehicle in which a mere handful of sand is used to change static condition, the loss of almost 40 per cent of lift obviously means the balloon would not be floating about looking for an ignition source. Still, that's not the real problem with balloons.

Just as your ears "pop" when air in your head tries to equalize as you ascend into ever-decreasing air pressure, so would a gas balloon. The open appendix of a balloon allows the gas to escape as the balloon ascends. When ammonia was used for lift, the escaping gas had thereby sickened the balloonists. If a flammable gas is used, gas escaping and mixing with air can fall victim to an ignition source even though its overall energy value would be likewise diluted. For hydrogen, real world conditions have shown about 8 to 9 percent gas-in-air concentrations can possibly, if not easily, be ignited. If all other materials around it are also flammable, the normally short-lived fire could cause a secondary ignition if the easily inflamed materials are *directly above* the heat path. Tying off the appendix and venting gas from the top of the balloon solves that problem, as Diehl observed, "A flame once started at the bottom of a balloon containing 15 percent air would probably be serious, although a flame started at the top would not…" Indeed, the ballooning craze spread rapidly, but fires were rare. By 1878, for the French International Exposition, a gas balloon was built that could carry 50 people at a time. The 880,000 cubic foot sphere safely lifted more than 35,000 people aloft in just four months.

Contemporary wisdom begs the question why vented or leaking hydrogen cannot be ignited and follow a flame path down to the main supply in the balloon. The concept would be obvious to contemporary city dwellers. Coal gas, also known as 'city' or 'town' gas, piped into buildings, was ignited in lamps for illumination and furnaces for heating. The flame could not burn *backwards*, down the pipe, to blow up the city gas supply. Likewise, coal gas filled airships were basically safe if not contaminated with air. Rich in hydrogen, the gas was a convenient way to fill a buoyant vehicle's chamber, and many were so filled. "Balloons of this type were generally inflated with household illuminating gas, which, although no safer than hydrogen and lacking much of the latter's lift, could be readily obtained at relatively low cost."[18] We must come to understand that hydrogen filled balloons obey the same laws; pure hydrogen cannot burn or

[18] H. R. Palmer, Jr. THIS WAS AIR TRAVEL, Superior Publishing Co., Seattle.

explode. Contrary to present day "wisdom," an air-displacement vehicle utilizing hydrogen is by its nature safe, since aeronauts always sought to use the most pure gas obtainable, for the maximum lift.

The perfection of the *dirigible* [19] balloon progressed through human, steam and electric powered air vehicles. Frenchman Henri Giffard created a trend-setting design with his 1852 powered airship. It featured a three horsepower steam engine suspended twenty feet below the gasbag. With our present day thinking, we say it was a wise precaution to isolate the hydrogen from the stream of red-hot cinders streaming aft. Looking at it another way, keeping the silk bag away from sparks made a lot of sense regardless of what was in it. "A real airship fever broke out, especially in the years 1850 [through] 1900, and many tried their luck with the construction of aircraft that were 'lighter than air.' Names such as Dupui de Lome [sic], Renard, Leaudi… German counterparts were Hamein,[sic] Woelfert, Baumgarten, Parseval, Schwarz, and others." [20] Georg Baumgarten built and floated his "paddle wheel" airship in 1879 proving one man could not provide needed power. His second airship featured three-man power, but it took an unexpected trip aloft when two 'motors' jumped out. The envelope tore, the hydrogen leaked out rapidly and the machine fell from about 4000 feet. Shaken but unhurt, Baumgarten also suffered hydrogen gasbag punctures in later airships, but without fire.[21] "In 1860 Lenoir had run the first internal combustion gas engine, and now in Germany Paul Haenlein applied it to airship propulsion at six hp, a four cylinder horizontally opposed unit burning coal gas from the airship's envelope with a propeller drive at 40 rpm propelling the airship to 16 kph."[22] This first internal combustion engine for an airship burned coal gas, about twice as heavy as hydrogen, but still LTA, lifting the airship.

The definitive history of early powered airships has not yet been compiled in one volume. In America, Dr. Tom Crouch's THE EAGLE ALOFT[23] leaves off about the time performers were getting the knack of making some progress against the wind with pedal power. First in America appears to be Charles Ritchel whose 1878 dirigible had both vertical and horizontal hand-cranked propellers. According to Volume 16 of the 'Modern Aviation Library,' New York balloonist Carl Myers, of 'Balloon Farm' fame, worked for Peter C. Campbell of Brooklyn to build a hand-powered balloon. Upon losing its propeller, it was last seen heading out over the Atlantic taking its hapless pilot, Canadian E. D. Hogan, along with it. Researcher Maureen Lynch explained Myers applied for a patent for his own "Sky Cycle" in 1889 and operated various versions of it for more than a decade. Myers looked for a suitable engine for ten years. Brazilian Augusto Severo built his first dirigible in 1882. Alva Reynolds built an ovoid gasbag that he propelled and steered with "air oars."[24]

There is no record of fire destroying any of these obviously crude early hydrogen filled vehicles in spite of their being made of flammable varnished silks. Keeping the bag away from sources of ignition appears to have worked. A later example would expose the more common problem of

---

[19] *Dirigeable*, the noun form of the French word *diriger* which in one sense means 'to steer.' One can easily understand why "scout dirigible" was rejected by the British in favor of the onomatopoeic word formulated when an Army officer thumped the tight fabric of a hydrogen-filled envelope. Many English aeronautic terms – *ballonet, empennage, canard, fuselage*, etc. are from aeronautic's home; but so is the word "bureaucracy," for good reason!

[20] V.W. Geiling & M. A. Sauter, ZEPPELIN'S HEIRS, Stadler, Konstanz, 2000

[21] H. Van Dyk, "Ernst Georg Baumgarten: The Forgotten Pioneer," *The Zeppelin Collector*, Jan 2001

[22] P.R. Stokes, "Aspects of the History of Airship Power Plant Development," 3rd Airship Association Conference

[23] T.D. Crouch, THE EAGLE ALOFT: TWO CENTURIES OF THE BALLOON IN AMERICA, Smithsonian Institution Press, Washington, DC, 1983

[24] *Popular Mechanics*, April, 1906

flammable fabrics used in conjunction with open flames. The hydrogen balloon *FRNS* twice set altitude records, but instead of retiring, years later it was converted to *hot air* inflation. "The burner became dislodged and a small fire started near the appendix. This rapidly spread and in a short time the whole envelope was destroyed, the rubberized fabric giving off dense smoke." [25]

Progress in making pure gas was as important as the invention of the *ballonet*, a bag inside the envelope that could be pumped up with air to make up for the volume lost when the airship valved gas while gaining altitude. Before the turn of the century there were several schools of thought in powered airship design. Of curious note are the creations of Slovenian David Schwarz, both of which leaked great quantities of hydrogen. The first leaked so badly it never flew, the second crashed and fell over, crumpled and lost its gas. Constructed of metal, these airships did not catch fire even when crushed and their gas was leaking profusely.

The first problem with airship fire had not come with the open-flame, cinder-belching steam engines employed by Giffard from 1852 onward. The next year Michael Faraday discovered he could separate water into hydrogen and oxygen by a process he dubbed "electrolysis;" there was no fire in his electrical lab. Nor was there fire in the first airship to make a round trip, the all-electric *La France* in 1884. The Tissandier brothers flew at 5 kph with power provided from 24 potassium bichromate cells each weighing 8 kg, or about 375 lbs per hp. The motor power was varied by the simple expedient of a rope varying the immersion the cell's carbon zinc plates in the electrolyte.

There is no mention of fire in the non-rigid airship's history until *petroleum fuel* enters the picture. The fourth airship Karl Woelfert was associated with, *Deutschland* of 1897, was powered by a gasoline motor depending on a platinum filled tube fed with gasoline, heated by an open-flame burner.[26] In his 1911 book THE AERIAL AGE, Walter Wellman criticized Woelfert's operation, saying "...there can be no doubt that the gas, escaping from the valve with too great a force, reached the furnace and caught fire." Yet, another writer *did* cast doubt, stating "...flames shot from the engine and the balloon envelope caught fire. The fuel tank exploded. The airship became a mass of fire and plunged straight to the ground..." That second viewpoint, that of a visible gasoline flame shooting upward, did not have to involve the lifting gas in the ship's ignition; *the engine caught fire first*, then the bag; only then did the lifting gas become involved. A nearly identical engine was employed that same year by an all-metal airship that crashed – without fire.

In France, a young playboy from a Brazilian coffee plantation was spending his fortunes building motorized hydrogen gasbags that thrilled all of France. P.R. Stokes wrote, "In 1898 the Brazilian, Santos-Dumont, flew his No. 1 airship powered by an ingenious adaptation of two single cylinder de Dion tricycle engines into a single unit. Above the conventional arrangement he mounted a second cylinder and attached the two pistons together via a stiff connecting rod passing through a gas tight gland, producing 3.5 hp for a weight of 30 kg (66 lb)." Eventually the young Santos-Dumont won a time-distance prize circling the Eiffel Tower in 1900. He also suffered various misadventures, with none involving fire.

Count Zeppelin could be seen as a latecomer who had the new idea of putting several gasbags inside an encompassing framework. It was not taken seriously, though neither was it perceived as

---

[25] *Gasbag Journal*, No. 10, December 1991

[26] Such an automobile engine is on display at the *Verkehrsmuseum* in Dresden, Germany, under portraits of the German engine pioneers Nickolaus Otto, Gotlieb Daimler, Wilhelm Maybach, Karl Benz, and Rudolph Diesel.

being particularly flammable. "One thing is very certain, and that is that no airship of the Zeppelin type will ever carry many people. The enormous expense incurred in building such airships would be serious obstacle."[27] In England, Stanley Spenser made an impressive 30-mile flight over London in his hydrogen dirigible powered by a 3.5 hp Simms engine.

That same year, New Yorker Albert Leo Stevens, after wasting money on other French engines, acquired a more useable de Dion as well, to beat Meyers in completing America's first powered airship. He called her the *Pegasus*.[28] Though they were unable to return to their starting point, Meyers and the young upstart Stevens had some success with their motorized airships in 1901 and '02. There was even a race between Edward Boyce, an American flying a Santos-Dumont craft by then owned by George Kerr, and Leo Steven's *Pegasus* in September of 1902.[29] Both of the hydrogen filled craft suffered engine trouble and made successful landings without incident. The envelope of the French ship somehow survived until it was much later donated to the Smithsonian, where Paul Garber ordered the fabric destroyed "because it was highly combustible."[30]

Sadly, two of the contemporary French teams lacked Santos-Dumont's expertise. After an October 1902 flight over Paris, Baron Ottakar de Bradsky Laboun and his mechanic were reclaimed by gravity when their airship came apart without fire. The other fatal accident has long been blamed on hydrogen; in fact, this might be the one in which the gas *actually was* part of the bag's ignition sequence. M. Augusto Severo also hired balloon maker Lachambre, but his semi-rigid design found the bamboo frame and twin-engine drive train largely enclosed by the lifting bag. The *Pax* (Peace) airship scheme was based on the desire to use more efficient center axis thrust, and featured a propeller at the center of both bow and stern. The weight and thrust were distributed via a frame that necessarily fit up to the center of the envelope, thereby creating a slot, which also contained the gas valves. Worse, the channel was sealed at one end by a windshield. Enclosed by material designed to contain gas, all the way up to the vehicle's central axis, and with very limited ventilation, the valves could not exhaust into atmosphere. (One report says at least one valve may have been sealed with wax.) An engine just happened to quit as the airship was rising sharply, and confusion may have caused the aeronauts to drop even more ballast. It is likely a gas valve was open and venting in the tight confines of the slot. A spark or backfire from mechanic George Sache probably trying to restart the engine could have ignited vented gas trapped in the slot. If present, hydrogen flames were likely *below* the flammable bag. Even with its low radiant heat, the lifting gas could have been part of a chain of fire that eventually resulted in the flames seen bursting from the bag.[31] The lesson of *Pax* has been lost to history, but it is an important lesson – and one we should bear in mind in the discussion of airship fires. 1902 also saw the successful Lebaudy airship fly in France powered by a 40 hp four-cylinder Daimler engine.

The following year saw Dr. August Greth's *California Eagle* lifted by 50,000 cubic feet of hydrogen in a pongee silk envelope yellowed with 14 coats of linseed oil. He bragged to the San Francisco *Examiner* that his vectored-thrust ship was superior to Santos-Dumont's owing to Greth's leaving out the ballonet. "But the balonet [sic] device is cumbersome. It is heavy...and it is expensive because the air gets into the gas and spoils it, so that the balloon must be frequently

[27] *Scientific American*, August, 1900
[28] Maureen Lynch, Editor, *Balloon Post*, in a 1994 letter to the author, cited a 1909 *Fly* magazine article
[29] D. D. Eklund, "Captain Thomas Baldwin: Pioneer American Aeronaut," Univ. of Colorado, 1970
[30] Garber was interviewed in 1967. At least the airship's engine is still in the collection.
[31] H. Van Dyk, Severo's Dirigible *Pax* and his Airmail," *The Zeppelin Collector*, Vol. 57 No. 3, July 2000

emptied and refilled." Unfortunately he also lacked a lightweight and reliable engine, so during several flights his airship was never able to return to its starting point. Notable was the lack of fire, even though historian Dr. Douglas Robinson suggested the crude Brenan automobile motor was, at seven feet, too close to the hydrogen vent valve. [32] Greth had told a reporter from the San Francisco *Call* "By my experiments I proved the proximity of the gasoline motor to the balloon was not a dangerous factor." Its last recorded flight, on May 2, 1904, was cut short by a stuck-open hydrogen valve. There was no fire, and the *California Eagle* survived to be wiped out in the great San Francisco earthquake of 1906.

The builder and pilot colleague of Greth was balloonist "Captain" Thomas Baldwin. "Captain Tom," after failing with a heavy auto engine, went on to build a more practical ship supposedly after a chance sighting of a New York motorcycle built by Glenn Curtiss. Baldwin eventually patented several new features, and with his Curtiss engine-equipped *California Arrow*, joined other pioneers to compete for the $100,000 prize being offered in St. Louis. The Aeronautical Journal in October 1902 promised free hydrogen for all contestants, though its quality was soon shown to be less than pure.

Before the Great War, American aeronauts thrilled crowds with their homespun hydrogen gasbags, the first airship race taking place a year before the Wright brothers' first powered heavier-than-air flight. Curtiss Museum photo.

Explained in the book CITY OF FLIGHT, what would seem today to be a comedy of errors found.the would-be Santos-Dumonts competing for that prize money at The Louisiana Purchase Exposition of 1904. "John Berry began inflating his airship on November 19... after 10,000 cubic feet of gas had been pumped into the balloon, about 75 small punctures were discovered from slivers which had penetrated the silk and rubber envelope as it was being inflated. A few days later, a second inflation was attempted from Berry's vitriolic plant. 12,000 cubic feet of gas had been used before a strong wind sprang up and blew several holes in the envelope, spilling out the precious gas... The third and last inflation... the wooden rim to which the envelope was connected at its equator broke... and an eight-foot rent was opened in the gas bag.... On November 15, Francois had completed minor repairs... saw the framework hit the fence, and the balloon sank to the ground. As if this were not enough, while the dirigible was being taken back to its shed, a nail in the doorway caught the top of the bag and tore a large hole in it, spilling out all 65,000 cubic feet of hydrogen and instantly asphyxiating a small number of sparrows which had nested in the eves."

Benbow's *Montana Butterfly* (or *Meteor*) would not lift his bulk with the supplied gas. Horace Wild was also at the 1904 St. Louis Exposition with Baldwin. Wild reported Santos-Dumont's "...airship was built like a watch. Ours [Baldwin's *California Arrow*] looked as though it had

---

[32] Dr. D. H. Robinson, "Dr. August Greth and the First Airship Flight In The United States," *JOURNAL of the American Aviation Historical Society*, Summer, 1976

been turned out with a butcher knife and a bucksaw... They all let go. Nothing happened. The airship did not budge. For several days, it had been standing around with the bag filled. The gas, which probably wasn't any too good in the beginning, had lost much of its buoyancy." Smaller balloonist Roy Knabenshue was substituted for the oversize Baldwin. "The whole country read of the flight. Knabenshue was the Lindberg of 1904.... In one year I made $47,000 and I spent it all on dirigibles. The first of a fleet of fourteen airships, which I built at various times, was finished in the fall of 1905." [33] Knabenshue was thrilling crowds that year with Baldwin's dirigible in competition with others at the Oregon Exposition. "With it Roy Knabenshue described circles in the air, moved over the city of Portland, and returned to the place where the air-ship had risen from the exposition grounds."[34]

If the average ticket buyer hoped to see a disaster, even the crude rule-of-thumb engineered powered gasbags had an enviable safety record. James Shock has compiled a list of inventors, builders and pilots of these "rubber cows" in America prior to World War I, and it contains some 37 men who assembled and/or operated hydrogen filled airships. Several of these aeronauts flew numerous airships in many locations, with various misadventures that in large part do not appear to have been fatal in spite of impure hydrogen and flammable materials. Wild continues about an accident "...at Tampa, Florida, in February, 1907. I was flying my 'rubber cow,' as we called the early dirigibles, at 200 feet when the gasoline connection broke, the fuel ignited, and flames shot toward the gas bag... A circle of fire shot up the bag. In a flash, the envelope was a mass of roaring, shooting flames." Wild fell into a shallow pond and survived the flaming motor and framework landing nearby. In an early reference to the bag's flammability, Wild said, "Carried by the wind, pieces of the flaming gas bag fell among the barns of a large race track near by, starting a fire in which twenty horses perished. When I finished paying all the bills, I was cleaned out."[35]

Without an ignition source, leaking hydrogen was more of a buoyancy problem than a fire hazard. In front of a summer Seattle crowd in 1908, L.G. Mecklem did an improper weigh-off and rose so fast his silk envelope burst open leaving a five-foot tear. In spite of the leak being below the bag, there was no fire. Mecklem was able to reach the hole and hold it together with his teeth while he sewed it together. Settling gently to a lake, a launch retrieved Mecklem and his craft. [36] Wild told of a similar experience. Other early exploits included Baldwin and Dr. George Tomlinson piloting their airships and circling the Statue of Liberty during the 1909 Hudson-Fulton Celebration.

Most of these early aeronauts lived to ripe old ages, the main exception were a handful that set aside airships for newfangled aeroplanes – many of whom were killed in them.[37] Many have heard of the famed Lincoln Beachy, who made the switch to HTA and wound up dead in San Francisco Bay. Fewer are familiar with the touching story of young Cromwell Dixon, who pedaled a hydrogen-filled bag sewn by his own mother about the sky as he grew up on the show circuit. Later adopting a newfangled aeroplane, the young man was killed when it crashed. Army

[33] Horace B. Wild, "My Forty Years of Flying" *Popular Science Monthly*, December, 1930.
[34] D. A. Wiley, "Air-Ships," 1908
[35] Horace B. Wild, "My Forty Years of Flying" (conclusion) *Popular Science Monthly*, February 1931.
[36] "Saved By His Teeth," *Modern Maturity*, Feb/Mar 1972.
[37] First appears to be Daniel Maloney, who abandoned the safety of the balloon to drift down in a glider. He died shortly after being found unconscious in the wreckage. "Professor" John Montgomery, the heavy gliding craft's designer, died after his own glider flight in 1911. (More famous was Otto Lilienthal, who after thorough theoretical study and examination, simply jumped off a cliff for a glide to his death.) Such deaths included the first airship-lofting airplanes experimenters in England almost a decade later.

Lt. Thomas Selfridge, who had worked with Curtiss on the SC-1 and was trained by Baldwin to fly the airship, met his end when the Wright aeroplane crashed.[38] Yet as late as 1910, the dirigible was still competing with the aeroplane for spectators, sometimes flying in winds too dangerous for winged machines. The safety record was near perfect. One exception to the record of early American airships may have been notable, though it was evidently not well publicized outside California. One "Professor" Moore died following a fiery accident to his airship near Redlands on July 5, 1909. "When it had risen to a height of about 40 ft. the propeller ripped a hole in the envelope, the gas caught fire from the exhaust pipe of the motor, and the bag exploded. "Prof." Moore was so badly burned that he died four hours later."[39] The accompanying Eastman View Company photo shows the smoky fire of the balloon fabric flaming furiously – and the streak caused by the burning aeronaut's body falling from the height, an image that makes an impression. Years later a man named Tolliver started an unsuccessful airship in central California and might have been dogged by the memory of this accident.

Walter Wellman had the idea to avoid the failures of his 1889 dogsled effort and make it to the North Pole by air. He hired a French team and created the airship *America*. According to Volume 16 of the MODERN AVIATION LIBRARY, "He set plans for a summer expedition by chartering the *Frithjof* and loading it with hydrogen generators from Paris, 110 tons of sulfuric acid from Hamburg, and 70 tons of iron turnings from Norway. On the 2nd of September 1907 the attempt was slowed by headwinds. "Encountering a glacier, the crew was forced to pull their knives and slit the gas bag open so it would collapse where they could get out." Cold and dry are the best conditions for hydrogen ignition, but if the gas burned there no one noted it. Enlarged for a 1909 second attempt, the ballast system had rotted and fell off shortly after leaving the Spitzbergen hangar. Massive amounts of hydrogen had to be vented to compensate for the 1200 pound loss. On the way back they were able to drop a line and be towed by the Norwegian steamer *Farm*. "As the *Farm's* crew started to haul the *America* in they were shocked to see Wellman light a big cigar – right in the face of the discharging hydrogen gas."

Since Commander Robert Perry had claimed he reached the Pole on land, Wellman rebuilt *America* for an attempt to cross the Atlantic in 1910. There was gasoline motor trouble only a few hours after liftoff on October 15th. "What a night it was! Soon the exhaust pipe heated red hot and began to belch sparks – not an occasional flash of fire, but thick showers flying aft along the varnished cotton enclosure of the steel car, up against the under part of the balloon itself, against the canvas rudder, and the ship's color floating above the helm. Nor were these mere evanescent flashes of filmy flame, but great constellations of living coals, many of them glowing for a few seconds after they had fallen into the sea. It seemed only a question of time until one of these fiery, incandescent masses would lodge in some nook or cranny, *set fire to the canvas*, and bring our little world to an end." (Emphasis added.) The crew removed and tossed the auxiliary motor over the side, but eventually had to give up. Lowering their lifeboat, they were rescued by a nearby steamer, their hydrogen ship intact but drifting away out of sight.

That year, southern California neighbors of Charles Toliver might have been the first Americans to be uncomfortable with something they'd heard was a problem. After all, this was not part of a traveling show, but rather being constructed in their town. Now second-time airship builder Charles Toliver ridiculed the ignorant San Diego natives who feared a catastrophe as he built his

---

[38] J. E. Lacouture, "CDR George Cook Sweet: First U.S. Navy Air Advocate," *Foundation*, Vol. 10 No. 2, Fall 1989, explains one of the special props cracked and fouled in the rudder wires. Selfridge became the first of many whose death would result in airfields being named after them.

[39] SEE THEM FLYING, Houston Peterson's Air-Age Scrapbook, 1909-1910. Baron Pub., Inc. New York, 1969

Toliver I. We have reason to believe his hydrogen was far from pure, owing to the lack of lift in relation to its size and weight.[40] After a storm wrecked his unflown ship, Toliver said "...the escaping gas went up into the heavens in as straight a path as it could find. There was no possibility of it ever doing anything else. The idea of it... coming into contact with fire and exploding was utterly ridiculous. Why, we had open lights burning in our building not thirty feet away at the very time the gas was escaping."[41] There was good reason for Toliver's confidence.

As Wellman with his cigar, Toliver with his open flames, Greth and tens of others repeatedly proved in practice, hydrogen, being the least massive thing in the universe, basically heads for the stratosphere unless vigorously restrained. It will gain velocity by colliding with atmospheric gases and actually escape the earth's gravity. Hydrogen has been used to displace air in containers harnessed for buoyant flight since before the year Americans signed that Constitution guaranteeing innocence until proof of guilt. "...since it is the lightest of all gases, it has an ascensional force per unit volume superior to any other," wrote Burgess. Only trace amounts of the fast moving molecules, detectable only via sensitive instruments (such as mass spectrometers), tend to head in all directions when first released. Nevil Shute put it another way: "In an airship, there is no great danger of death from violent impact with the ground. Although both ships [R.100 & R.101] were filled with hydrogen there was no great danger of fire in the air, for any gas escaping would go upwards and out of the top of the ship, remote from any likely source of ignition."[42] The gas cannot linger about in flammable concentrations in the open air for another reason. Hydrogen is also natures' most *rapidly dispersing* element, disappearing into the air 3.8 times faster than methane. While urban legend warns of supernatural dangers, the actual record shows that, if for no other reason that it ascends and dissolves so quickly, nature's most abundant element is actually no less safe than other fuels.

This principle was given a dramatic demonstration near San Francisco in May of 1908. The 485-foot long gasbag of John Morrell was filled with about 500,000 feet of hydrogen-rich "city gas" filled from Berkeley's city gas pipe system. The bow end was not released at the right time and the bag tilted upward at 45 degrees, causing a rush of gas to balloon into the high end. "The oiled cloth, of which the envelope was constructed, could not withstand the pressure, and it burst, where upon the machine fell rapidly to the ground, carrying with it its nineteen passengers, who were tangled in a mass of broken machinery, flapping cloth, and network."[43] The settle rate was so slow everyone survived.

Contrary to popular belief, the history of even more complicated early hydrogen-filled airships is *not* a chronology of fires automatically attributable to their lifting gas. The common misconception dates back to at least the 1912 accident of the Vaniman *Akron*. Wellman's mechanic, Melvin Vaniman, went on to draw up plans for a 400,000 cubic foot airship. Not following his predecessor's French envelope purchasing, Vaniman went to balloon maker Goodyear Tire and Rubber Company. A team built his new trans-Atlantic attempt ship *Akron,* which first flew in 1911. Goodyear produced a booklet advertising the new business, saying "Balloonists will be interested in knowing that you can send in an order, and the balloon – fully equipped with engine and parts already assembled – will be delivered without further bother to you." Goodyear's own history, TRAIL BLAZING IN THE SKIES, tells of the fatal end to the *Akron*. "Apparently the valves were too tight, making it impossible to valve out gas as the ship

---

[40] H. Van Dyk, "Charles H. Toliver and His Airships," *Buoyant Flight,* Mar-Apr, May-Jun, Jul-Aug, 1998
[41] San Diego *Union*, 22 DEC 1911, courtesy San Diego Aerospace Museum
[42] N. Shute, SLIDE RULE, Ballantine Books, NY, 1954. Courtesy Kent O'Grady
[43] *Scientific American*, June 13, 1908, quoted by Payne in LIGHTER THAN AIR

ascended, causing the bag to burst. The escaping hydrogen ignited, either from static electricity or a spark from the engine, no one ever knew." Here we see the first demonstrable example of the lifting gas taking the rap, with the "mystery" twist added for effect, when the engineers *actually knew* it was not the problem.

The ill-fated *Akron* of 1912, featuring Goodyear's reinforced envelope. Courtesy LTAS.

Ralph Upson's unpublished 1913 report on the accident[44] concluded, owing to no evidence of fire on the keel car or the bag, that there was *no hydrogen fire*. Suspected causes of the crash, consistent with witness accounts, included over-pressure of the gasbag or a broken suspension line. "Eyewitnesses of the disaster speak of a wreath of smoke that appeared, followed by caving in of the gasbag."[45] *Everybody's Magazine* reported, "His craft was a mistake, and in his methods he violated principles the soundness of which have been thoroughly demonstrated. There is good reason for believing the accident to the *Akron* followed an attempt to inflate the air ballonets inside the big gas-bag, by which means Vaniman erroneously believed he could compress the gas and prevent it from expanding under the influence of the sun or the rarefied atmosphere of higher altitude. This strain burst the envelope. The disaster was all the more deplorable because it was unnecessary." The addition of strong springs to the valves appears to have been the method by which he thought he could compress the gas,[46] and is another example of how far men have gone to resist aerostatic law. Perhaps Vaniman thought he was improving on the British 'Nulli Secundus' of 1907,which featured an expansive envelope of gold beaters skin rather than the well-established ballonet system.

Meanwhile, Zeppelins were becoming the world's benchmark in aeronautic progress in the century's first decade, in spite of frightful teething problems. LZ-2 was ground-pounded during a storm, broke away and without fire, was simply dismantled in the field. Demonstrating LZ-3 to the German Government, Count Zeppelin had to fly his creation 480 miles, through a storm that had wrecked eight aeroplanes participating in the Berlin-Vienna race. Passenger hauling was not the original idea, but eventually Zeppelin's businessmen set up *Deutsche Luftschiffahrts Aktein-Gessellschaft*, commonly known as DELAG. There was no reputation of high flammability in these rather unsophisticated vehicles, whose operational manuals were being written pretty much as they went. Nor did their passengers have some sense of impending doom. Out of gasoline on

---

[44] Found in the Goodyear archives by Eric Brothers in 1999

[45] *Scientific American*, July, 1912. Courtesy Jeff Cook

[46] H. Van Dyk, "The Evolution of the Wellman/Vaniman Airships," *The Zeppelin Collector*, Jan-Oct, 1996

the 29th of May 1909, LZ-5 made an emergency landing during a storm and a pear tree demolished 30 meters of the bow and cells, or about one quarter of her length. In the storm there was probably a huge hydrogen fire, but no one noticed. Without gasoline to explode, the airship was secured, makeshift-repaired, fueled, and flown back some 450 miles. It was later smashed around a hillside, again without fire. Leaving with an inadequate fuel supply, LZ-7 became a pile of wreckage in the *Teutoberger* Forest near *Bad Iburg* without fatality or fire; it was out of gasoline. LZ-8 was bashed beyond repair backing out of her hangar; passengers and crew suffered no fatalities – or burns.

Most common are pictures of the bare silver skeleton of LZ-4, whose fabric covering burned away in a storm. Commonly thought of as the first hydrogen fire to destroy a Zeppelin, a contemporary account does not say so, explaining "... '[A] not inconsiderable wind' suddenly kicked up dust and dirt, which irritated his eyes and ears. A kilometer away, the same wind ripped the anchored LZ-4 from its moorings. Spectators watched in horror as soldiers gripping the anchor ropes were lifted up into the air before they jumped to safety at the last moment. The huge airship rose higher and higher and raked its anchor through the crowd. At the same time, it began to keel over, stern to bow. After a few moments the dangerously inclined ship slipped to the ground. A dull thud sounded and then 'flames shot up from the hull, a second, a third detonation... a column of fire rose into the sky, immense, horrible, as if the earth had opened up releasing the flames from hell. Huge flames ate their way up the balloon, piece by piece... an enormous cloud of black smoke marked the spot where the elements had sacrificed the creation of man. It was all over in three minutes.'" [47] In our time, nothing more than another hydrogen fire; then, " In 1907, at *Echterdingen*, the Zeppelin IV, while on the ground, was set afire by a charge of electricity that had collected in her metallic frame during a flight that had just ended. Instruments now detect these charges before they become dangerous, and *a fabric has just been made for covering the balloons that will obviate frictional electricity*." The optimistic 1912 article went on to say, "It is worthy of special note, too, that, in all of these disasters, no Zeppelin has ever lost a life among either crews or passengers. Indeed, in Germany the airship is regarded as a vehicle of greater safety than the ocean steamship."[48] Sadly, once hydrogen was blamed, there was little incentive to try and publish a retraction once the evidence to convict the gas was not found.

No matter that aeroplanes were falling, breaking up, crashing and burning all the time in those days. "Since then breaks in the mechanism have furnished the cause for a greater number of accidents. And such accidents have happened to well-known aviators – [E.] LaFebvre... met death at Juvissy, and Captain Ferber... was killed near Boulogue. The death of [Leon] Delagrange, which occurred near Bordeaux on January 4, 1910, removed one of the most brilliant figures..."[49] "There seems to be also a lesson to be drawn from the accident to Charlie Hoxsey, who was killed in Los Angeles... the wind was seen to turn it completely over in the air, after which it dashed to the ground before it could be righted, instantly killing its daring pilot."[50] Kaempffert reported "Le Blon was killed at San Sebastian on April 2, 1910, as a result of similar motor trouble."

---

[47] P. Fritzsche, A NATION OF FLIERS, quoting K. Clausberg, *Zeppelin: Die Geschichte eines unwahrscheinlichen Enfolges* (Munich, 1979), Harvard Univ. Press, Mass., 1992
[48] Emphasis added. T. R. MacMechen & C. Dienstbach, "The Greyhounds of the Air," *Everybody's Magazine*, Vol. XXVII No. 3, September, 1912
[49] Augustus Post, "The Bird-Man and His Wings," *Cosmopolitan Magazine*, May 1910
[50] Augustus Post, "Learning to Fly," *The Outing Magazine*, May 1911

In all this time, these early airships were leaking hydrogen at what today would be considered a foolhardy rate, but so much progress had been made a trans-Atlantic Air Service was envisioned by 1910. The newest ships had "... tougher gas-chambers, they leak only thirty to forty pounds of gas in twenty-four hours, instead of a hundred, as a year ago."[51] Few thought aeroplanes would ever be anything but thrill rides. "Accidents were commonplace for exhibition pilots during the day, and the night only compounded the dangers of stunt flying. At a meet in New Jersey, several Wright exhibition team members ascended into the darkness, equipped with port and starboard lights, as well as automobile horns to 'honk each other out of the way.' By 1912 so many of these pilots had been killed or injured in 'trick' flying that Walter Brookins, a well known daredevil, formed the 'Safe and Sane Club' in which pilots pledged not to stunt... Howard Gill, who had been seriously injured while attempting a night landing, joined the club... [nonetheless] Gill was killed in a mid-air collision over Chicago five months later."[52] Nor did air-planing machines discriminate by pilot gender. Harriet Quimby, first female to fly the English Channel, was killed at the Boston-Harvard air meet in 1912. Another aviatrix, Julia Clark, met her end in the same year the *Titanic* sank. Bessie Coleman, first black female aeroplane pilot in 1912, later died in an aeroplane accident. When the War To End All Wars started two years later, the carnage from airplane accidents quickly outgrew our historian's ability to tabulate them.

Thrilled Chicago crowds mob the *Wingfoot Air Express* in 1919. Jack Boettner, pilot in the Great War, is standing tallest in the car. Note height of the gasoline engines. Boettner would go on to fly many Goodyear airships, and train a WWII generation of blimp pilots. LTAS photo.

Between 1909 and 1914, the hydrogen-inflated DELAG airships flew more than 107,000 miles carrying 34,288 guests – without the loss of a single *fahrgast*. As the nation prepared itself for what was thought would be a post-war boom in aeronautics, there was an accident. "On July 21st, 1919, *Wingfoot [Air] Express* departed from its base at the old White City amusement park on Chicago's South side never to return. Loss of this airship was due to ignition of the highly inflammable hydrogen with which it was inflated, and resulted in Goodyear's subsequent pioneering in the use of helium as lifting gas." [53] Yet again, this is not the complete story.

In FLAMMABLE GASES, Don Overs wrote: "Augie O'Neil...spoke of hydrogen fire tests conducted at Wingfoot Lake after the loss of Goodyear's *Wingfoot [Air] Express* ... J. F. Cooper... conducted tests such as punching a hole in a hydrogen filled barrage balloon and lighting the escaping hydrogen. Cooper noted that the fire would not progress closer than a few inches to the balloon...." and that a slight disruption, in this case passing a hat in between, would snuff the fire. This action illustrated a simple fact: hydrogen fire cannot "back up." It also showed a puncture leak, even if ignited at the point the pure gas mixes with air without

[51] T. R. MacMechen & C. Dienstbach, "Over Sea By Air-Ship," *The Century Magazine*, 1910

[52] W.E. Fischer, Jr. "Into the Night: The Evolution of Flight Under Cover of Darkness Before the First World War," *Air Power History*, Vol. 40 No. 1, Spring 1993, citing *New York Times* from 1910 and 1912.

[53] Z. Hansen, THE GOODYEAR AIRSHIPS, Airship International Press, Bloomington Illinois, 1977

disruption, and then ignited, did not radiate enough heat to start a secondary ignition of even the flammable fabrics of the day. And that's not mentioning windstream snuffing!

Sadly, in all of these early accidents, level headed investigations always lagged behind newspaper headlines, fact-filled reports were back-page small print filler items in contrast to yellow-journalistic front pages. As airships increased in size, taxpayer investment, and number of souls on board, sensationalist media continued pandering to an eager market.

James Shock has compiled a list of seventy-four Dirigible Balloon Pilots as qualified by the Aero Club of America as of 1919. Again, most seem to have lived to see jet airplanes, unless they tried heavier-than-air flight themselves. Sadly, the efforts of these pioneers are not published in a history book. Airplane accidents were by far more common. "... I ran out of the building and saw the machine fall and strike. A wing had broken. Both bodies were considerably messed up." Aeroplane designers preferred the more efficient and well-balanced pusher configuration, but the pilots were not fond of what happened during the usual crash. Of course moving the engine out front caused a different sickness – gut wrenching caused by the ingestion of castor oil thrown out by the engine."[Saufley] was making an endurance flight... when the flipper which had been improperly attached came loose and he fell and the motor fell on him crushing him."[54] The ghastly aeroplane accidents attracted little attention since they often killed small numbers of people, wasting fewer taxpayer dollars and causing politicians less embarrassment. In the final total, many more pilots were killed in Great War *accidents* than died in combat! The astonishing toll of fatal Heavier-Than-Air training accidents would have to be re-learned in the next war, whose practicing richly earned such sayings as "a plane a day in Tampa Bay." In several airplane models, the accident loss rate actually did exceed the combat loss rate – in *both* World Wars. The WWII airship's most common search and rescue mission was to try and locate downed airplanes. Bad design or faulty materials in aeroplanes usually killed only their pilots or perhaps some hapless people on the ground. Sadly, the airships' enormity did not make them immune to the similar folly of smaller air vehicles.

Early military Zeppelin accidents did carry dozens to their deaths. It has been said that the German Navy had not learned to operate its first rigid, the L1, as it was lost in a fierce storm in September, 1913, killing fourteen, seemingly similar circumstances to the US Navy's first rigid airship. Yet the loss of the second a month later has always been blamed on hydrogen. Supposedly the L2 (build number 18) had been modified with windscreens that made it difficult for vented hydrogen to escape. As the story goes hydrogen somehow became trapped by these screens and then found an ignition source. The accident, shortly after 'up ship' at Johannistahl, on October 17, 1913, established a new record of twenty-eight people killed. Once again witnesses do not describe a hydrogen fire. AIR CRASHES reports, "But no sooner had the engines started than a tongue of flame shot outward from the engine gondola. Within seconds, the flame had doubled back on itself – blocked, fatally, by the windscreen – and both gondolas and the hull were consumed by a roaring, cataclysmic fire." Strange how many people assumed lifting gas was in the cars, even after two crewmen on L53 were overcome by gasoline *fumes* alone, without their ignition, fives years and fifty one Naval Zeps later.

The British R.38 (which became the American ZR-2) suffered from Whitehall's heavy power-to-weight demands and its frail, already corroding structure not having been stressed for in-flight, dynamic loads. Emilio Herrera observed, "...the airship showed various original and daring

[54] 1916 letters to his sweetheart from Marine Aviator Francis Evans, in "Love, Tom E," *Air & Space Smithsonian*, September 2000

innovations, which were in no way sanctioned by experience."[55] In ICARUS OVER THE HUMBER: THE LAST FLIGHT OF THE AIRSHIP R.38/ZR/2, author Tom Jamison explains the design ignored aerodynamic loads *entirely*! Long publicized as being air-polluted and explosive, ZR-2 was subjected to an extensive accident investigation also not included in general histories. Herera continued, "We see therefore that the dynamic effect of a violent rudder displacement, which was easily produced on this airship with its balanced rudders, could be more than four times as large as the combined static and aerodynamic loads during ordinary flight. It is not strange, therefore, that a rigid airship, whose frame had already been strained by the two latter forces and was then within a few minutes, while still under maximum engine power, put to the fearful test of having its rudder thrown hard over, should break in two." The official report relates how the broken airship "...descended comparatively slowly, but did not catch fire... The forward portion caught fire at the fracture..." when severed electrical cables sparked atomized gasoline pouring from broken fuel lines. "The fire spread rapidly, due to the presence of escaping petrol in the keel. An explosion followed, which led to the collapse of the structure and the ignition of the *liberated* hydrogen gas." [56] In spite of this report, "helium correct" historians basically ignore the fact the aft half of the ZR-2 was also filled with hydrogen, but merely settled into the river without fire. It was unfortunate chief designer Charles Ivor Rae Campbell had so little knowledge of dynamic loads, those stresses found in flight for which the lightweight structure had been unprepared to handle. Worse, even less was known about corrosion and fatigue. "The failure of the R-38 was traced to a materials problem. Unvarnished three-year-old Duralumin, known by those who had experience with it to be highly vulnerable to corrosion, had been used in the structure."[57] Other airships would break up, but as P. R. Stokes wrote, "The accident's high fatality rate, ascribed primarily to *fuel fire on the water*, strengthened incentive that flight worthy heavy oil engines utilizing high flash point fuel be developed…" (emphasis added).

One would be hard pressed today to find a single person that did not think the *Wingfoot Air Express* victims, the L2 crew and the R.38 crew were killed by hydrogen, presumably burning from the cell outward. Gasoline has a lower flammability limit and would tend to hang around potential ignition sources, but in airship fires, gasoline escapes attention. Others imagine the hydrogen gas *exploding,* braking windows in the town of Hull, even though the relative energy of explosion of vapor for H_2 is 2.02, while gasoline is 44.2. In spite of partially empty gasoline tanks having 22 times the explosive energy, and containment of the lifting gas not being sufficient to permit an explosion, hydrogen was blamed. The gas *had* to be blamed; hydrogen was the *only thing that could be changed* - or so it seemed, after the isolation of helium.

The British R.34 crisscrossed the Atlantic in 1919, but later was bashed badly, puncturing several hydrogen cells. There was no fire. Whole Zeppelins were taken as war booty by the major powers; in all, eleven Zeppelins were awarded. Three were disassembled, but the others flew with inexperienced crews and had various misadventures. Italy got two, including the proven liner *Bodensee* (LZ-120) which it renamed *Esperia.* It served in the Italian Navy until dismantled in 1928. England got a couple and used them mainly for study. France got the finest example built to date, the LZ-121 *Nordstern.* In spite of banging them up a bit and even ground-

---

[55] E. Herera, Chief of Engineers, Madrid, "The R-38 Catastrophe and the Mechanics of Rigid Airship Construction," *Memorial do Ingenieros,* 1922.

[56] (Emphasis added.) TECHNICAL REPORT OF THE AERONAUTICAL RESEARCH COMMITTEE FOR THE YEAR 1921-22, VOL. III, AIRSHIPS, His Majesty's Stationery Office, 1924.

[57] M. B. W. Graham, "R&D and Competition in England and the United States: The case of the Aluminum Dirigible," courtesy Alcoa.

handling them to destruction without fire, the new owners operated them safely. The last of the wartime Zeps, LZ-114, with 1560 horsepower mounted below 68,500 cubic meters of hydrogen, had been demilitarized with hopes of an Atlantic-crossing attempt. Under its new French operators, it instead made a 118-hour voyage across the Mediterranean. Not surprisingly, you only hear about the single fiery death in all these reparations ships. Naval Zeppelin L72 was renamed *Dixmude* by the French commemorating the Flemish town where a pivotal battle of the Great War had taken place.[58] At war's end, "Germany has delivered to us some of their best hangars and they are to be erected at Marseille, Tunis, Algiers, and Casablanca."[59] Yet the hangar at Tunis was never started, inviting disaster. Not designed for four-day long runs in stormy weather, *Dixmude* likely suffered engine failure and may have been free ballooned for 435 miles before impacting the water off Sicily.[60] Evidence suggests a violent explosion, and right away the replacement Astra-Nieuport cells were accused of leaking air and causing the explosion. If anyone doubted that, a backup explanation of lifting gas being vented into the lightning storm was offered. The basic impossibility of hydrogen cells exploding is not included in these assessments. Dozens of empty bomb-like gasoline tanks also escaped attention. A British review saw the lightweight structure as the problem. "The loss of the *Dixmude*, a Zeppelin handed over to the French as reparations, was due to the same [structural weakness] in all probability." In an unpublished manuscript, Zeppelin Captain Hans von Schiller wrote, "...in all probability the ship ran into a mountain in the Atlas range."

In this and other accidents, helium proponents happily quote hydrogen fuel's explosive potential, highest of any gas on a mass basis. This is misleading, because the whole idea of an airship cell is *lack* of mass. Hydrogen actually has the *lowest* explosive hazard of any fuel when measured by volume. An explosive yield is a complex interaction among factors such as composition, level of ignition energy, geometry, and degree of confinement. The danger, or yield, in case of an explosion, is measured by comparing the relative energy of explosion of fuel vapor (EEF) in equivalent mass of TNT per volume. Measured, gasoline vapor is rated 44.2 and hydrogen only 2.02. That makes gasoline 22 times more "dangerous" than hydrogen when both were adequately contained. More important, we must remember a Zeppelin gas cell does not constitute a chamber capable of permitting a hydrogen explosion, because it does not constitute sufficient containment. As well, the minimum air pollution capable of creating an explosive mix (the lower limit is about 18% in air if actually contained) would represent an unacceptable loss of lift for any airship. The ullage of largely fume-filled gasoline tanks, on the other hand, is ripe for window-smashing explosions, but was rarely indicted.[61]

This popular thinking would predispose that Great War airships of all types were hydrogen fires waiting to happen. The technology of the day would seem to insist on it. As Americans were introduced to European airship flying, reliable hydrogen valves continued to be a problem. The *Capitaine Caussin* was under the command of future ZR-2 skipper Lieutenant Lewis Maxfield. On its way to convoy escort, the envelope pressure began to drop. "Upon looking at the gas valves and the cover was hanging loose, allowing the gas to escape freely. After several futile

---

[58] N. Walmsley, "DIXMUDE", THE AIRSHIP AND HER COMMANDER FROM CONTEMPORARY SOURCES, Airship Heritage Trust, 1999

[59] *The Writing of Lt. Cdr. Jean du Plessis, French Navy*, translated from the French by Anton Heinen.

[60] LOSS OF THE *DIXMUDE*, Hugo Eckener, *Luftfahrt*, 23 JAN 24, translated by NACA, tech memo #256

[61] *Ullage* is defined as "the amount that a container (as a tank or cask) lacks of being full." Hardly a record, a gasoline tank explosion in Texas City smashed windows in Galveston, some 15 miles away. Investigators concluded the TWA 747 that exploded in pieces over the North Atlantic was caused by an ignition source reaching fumes in the nearly empty center fuel tank. Flammable temperature range decreases with altitude. So fuel with a lower flammability limit of 100° at sea level will have a lower flammability limit of just 55° at 40,000 feet.

attempts to close the valve by pulling the valve cord, Lieutenant Culbert went aft and attempted to close the valve by operating the wire controls by hand on the stern of the nacelle. This also proved useless…the gas was escaping faster than air could be put in." Bombs were dumped and the ship made for land, but settled into the water. Drifting ashore, it was beached, hand-ripped and disassembled without further incident. [62]

Today, it is generally believed that hydrogen escaping from valves would spell the fiery end of an airship. Not one, but two Zeppelin losses have been blamed on their crews valving gas in thunderstorms. Of course there is not proof either way, but we must understand how unlikely these accidents found some exception to the rule that hydrogen fire does not "back up." The Bunsen-burner principle that pure gas cannot burn allows us to light our kitchen stove without the flame following backwards, down the gas pipe, to blow the supply tank to smithereens. Small propane burners fed by slight pressure can simply be blown out. This is not unlike what winds aloft in a Zep slipstream might do to a hydrogen fire fed with less than 0.10 PSI of pressure.

The official history of US Navy LTA does mention one fire. "The O-SS was destroyed by a fire in the summer of 1918 shortly after it had been received at NAS Hampton Roads."[63] It does not say hydrogen was part of the ignition. Often called a hydrogen accident, the SSZ-23 was consumed when fumes from *cleaning* gasoline were ignited by activating the tube set radio.[64] Somehow, though it was soaked into the wooden car, and its fire spread to the other flammable components first, the hydrogen above was still blamed – via the usual excuse the fire surely was much worse as a result of no helium.

James Shock's tabulation[65] of early US Navy airships is revealing. Not counting the hangar fires, we find the hydrogen filled blimp B-5 had an engine fire that spread to the envelope. Another B-type airship kept a single inflation charge for 9 months, during which it logged 743 hours of flight time. The other B-type airships, B-1 through 20, survived long enough to be made obsolete by helium. A few were lost to operational accidents; one was supposedly flown into the ground by its pilot deliberately,[66] but no fires. The airships flew 13,600 hours covering approximately 400,000 miles without a fatality.

Twin-engine C-type airships were famous in US Naval service. These include C-1's record distance flight and airplane loft and launch, and C-5's attempt at crossing the Atlantic. C-2, following its transfer to the Army Air Service and record-breaking 1922 cross-country flight, was smashed against its hangar during docking. Ruptured gasoline tanks spread fire to the car, then the bag, which began burning. Last to join the conflagration, hydrogen was first to be blamed. The C-6 slammed into a foggy mountaintop, but the crew survived; no fire is recorded.[67] Following C-3's establishment of in-air refueling, it was lost to an "engine spark." A report states that LT R. R. "Rip" Paunack was awarded the Distinguished Flying Cross for saving the hydrogen inflated C-8 and six crewmen. Straddling a support beam and pulling himself away

---

[62] Bulletin No. 2, U.S. Naval Aviation Forces Operating In European Waters, DIRIGIBLES, May 18, 1918 "The following statement is quoted regarding the "Capitaine Caussin".
[63] KITE BALLOONS TO AIRSHIPS…the Navy's Lighter-Than-Air Experience, Edited by Roy Grossnick, US Government Printing Office, 1986(?) There was actually no OSS airship, rather these were British –built SSZ airships purchased from England after WWI. This was possibly the SSZ-23 and the accident likely destroyed the airship before it made a single flight at Hampton Roads, VA.
[64] Author's interview of Jarvis Frith, Airship Heritage Trust, in Scottsdale, AZ, 1993.
[65] J. R. Shock, US NAVY PRESSURE AIRSHIPS 1951-1962, Atlantis Productions 1994
[66] From an interview with the last surviving B-ship pilot, 103 years old, courtesy Joe Panzarino.
[67] JACKRABBITS TO JETS, The History of North Island Naval Air Station, San Diego Publishing, 1992.

about 10 feet from the control car, with nothing below him but space, "Rip" extinguished a blazing carburetor. When the C-8 was later lost to fire, you already know what was blamed. Likely in the few fires that did happen, gasoline was ignited, spreading fire to the bag, then *the bag went first, before the gas;* then and only then could the gas inside escape to mix with air and have a chance to get involved in a fire overhead.

Not throttled by the heavier-than-air pilot's distaste for pushers, the D-type airship also showed the innovation of suspending the gasoline's weight along the envelope's equator via small tanks. No D-type airships, or for that matter E, F, H or J ships were actually lost to operational fires. Their safety (or lack of) would be obvious to anyone who knew the characteristics of the materials involved, yet little was published as such.

In an unpublished paper, British airship design worker P. L. Teed spelled out "The Disadvantages of Helium For Airships."[68] (Teed was co-creator of the Scott-Teed airship slide rule. Its "A" scale gives lift in pounds per 1000 cubic feet of gas; the "B" scale is graduated in percent of purity; the "C" scale in degrees F and the "D" scale gives barometric pressure. With this ingenious calculator, airshipmen could set the temperature opposite the barometer and read the lift opposite the purity.) He ridiculed the idea that airships up to his 1925 writing were prone to fire from their lifting gas. "...such fires as have occurred in airships in the past have been due in the first place to petrol vapor, which, becoming ignited ultimately destroys the ship and would, without a doubt, also destroy a helium-filled vessel." Teed was right, almost exactly ten years before USS *Macon's* outer bow cover burned away. And, Teed never even complained about the rigid airship's weak outer defense, its volatile covering, in his anti-helium paper.

In the face of such seemingly obvious facts, how did hydrogen get such a fatal reputation? Could the doomsday rep have been developed because of what happened when there was *shooting* at hydrogen filled ships?

"In military balloons or airships the fire hazard is, of course, greatly increased, a single well-placed *incendiary* bullet being sufficient to transform the whole costly structure into a mass of flames; and in this even the fate of the crew is practically sealed."[69] Even Goodyear would repeat this conception. "British aviators knew that the gasbags were vulnerable; a single incendiary bullet or tracer entering one hydrogen gas cell would cause a fire that would burn a Zeppelin in seconds."[70] In the Michael York – Elke Sommer movie ZEPPELIN, the British are shown firing a special bullet to get a hydrogen balloon not just to burn, but to explode – in a mass of curiously yellow and orange flames. Later, one did not even need a glowing bullet. Viewers of the motion picture THE ROCKETTER[71] were kept on the edge of their seats as the FBI agent character shouted "Watch out for that Zeppelin! It's full of hydrogen! One bad shot and we'll all fry!" In HOPE & GLORY, rifle shots completely destroy a runaway barrage balloon. Not surprisingly, gunfire has been often blamed either directly or indirectly in the LZ-129 fire.[72] Oh, that it were that simple, the defenders would lament.

---

[68] The paper is dated 19/3/25 and is courtesy of Charles P. Hall.

[69] Emphasis added. G. S. Rogers, "Helium, The New Balloon Gas," *National Geographic*, May 1919

[70] G. Larson, BLIMP!, Van Notrand Reinhold Co.,1981, pg.26.

[71] Screenplay by Danny Bilson & Paul De Meo, The Walt Disney Company

[72] A resident of Barnegat Light, New Jersey, for many years bragged he hated the Nazis so much he had shot LZ-129 with his .306 rifle and his bullet lead to the disaster.

The Italians were actually the first to use a hydrogen airship in a shooting war, starting in their 1912 war with the Turks. "Greyhounds of the Air" carried the Italian commander's reports: "Two baby Italian dirigibles cruised day after day a mile above the ground, surveying every Turkish camp within twenty miles of the Italian camp. While these small airships were at work, swift aeroplanes were also in service, but they did well if they explored one camp in a flight and got safely back to headquarters... On the other hand, the airships reported constantly by wireless to the troops and to the war-ships... When the enemy opened fire on them and they discovered they were within possible reach of the bullets, they simply took a higher altitude, from which they deliberately continued their operations."

These seemingly fragile vehicles were surprisingly hardy in practice. "While walking the ship out, [L22 from Tondern's *Toska* double shed] a violent gust broke the port forward tackle and the door smashed the nose of the ship. Remarkably there was no fire."[73] Remarkable, so it would seem, because the Zeppelin was a flammable container *itself*, while also containing volatile compounds. Yet there are many examples of survivable accidents. Baltic fir trees cushioned L38's final fall, and it did not burn. Hans von Schiller wrote, "In December, 1916, a group of airships returning from an attacking expedition ran into heavy fog, with the barometer falling. One of the ships, miscalculating its altitude because of the reduced atmospheric pressure, struck a building head on and was so badly damaged that it had to be dismantled." Hans von Schiller was aboard the L6 when it encountered a flotilla of minelayers and two British cruisers near Heliograd. "We were placed under steady shrapnel fire from the cruisers and rifle fire from the mine-layers, to which we replied with our machine guns. On the way home the ship became heavier and heavier, and everything moveable had to be thrown overboard before we reached the landing field. Once in the hangar, we discovered over a hundred holes." The same crew and ship took a direct hit from an artillery shell, which did not explode. "We were at a height of 20,000 feet, and although our largest cell, the one in the middle, was out of commission, we managed to get home safe and sound." Time and again hydrogen's smothering and cooling effect saved airships even after massive hits.

Airships are displacement vessels, so they enclose mostly emptiness; something vital would have to be hit to cause serious damage. Petroleum fuel tanks were relatively small, but plentiful. The L7 suffered anti-aircraft fire from HMS *Galatea* and *Phaeton*. Robinson wrote, "The warships had already turned back when they saw the Zeppelin stand on her nose and plunge blazing into the sea. One of the last shells had got home in a fuel tank and started a fire that quickly enveloped the whole airship... A British submarine, E 31, surfaced nearby and rescued seven survivors." Again, the engine fuel released the greatest energy. Robinson continued, "The L31 fire, fed by its petroleum fuel, flared and smoldered for hours in spite of a weeping drizzle."

This stands in contrast to what Winston Churchill would later optimistically suppose, that the sky monsters were quite vulnerable since they were filled with flammable gas. Surely, it seemed obvious, glowing tracer bullets could be used to set them afire. The defender's tracer rounds, designed to assist in aiming machine gun fire, were made with the base of the shell containing phosphorus mixed with potassium chlorate, which would react in air and leave a visible light-streak. Trouble was, the glowing bullet did not contact the flammable fabric upon entry for long enough to transfer enough heat energy to produce a secondary ignition. Inside the gas cell, pure gas won't support the reaction. Best hope was the exit point, but fleeting tracer rounds were not

[73] Dr. D. H. Robinson, THE ZEPPELIN IN COMBAT, A History of the German Naval Airship Division, 1912-1918, Schiffler Publishing, Ltd. Revised & reprinted 1994. The Tondern site is now part of Denmark and the city is named Tønder.

effective against gasbags even as the bullets left a ragged exit wound. L12 was so badly shot up she settled into the sea, but in spite of her structure failing, she was photographed with covering intact. A similar case was the L19, which was so severely shot up over Holland she ditched in the North Sea. Her crew huddled atop the buoyant wreck expecting to be rescued by a passing British trawler. There was no fire, nor rescue; the L19 crew all drowned.

Maritime airship L15, *werft* number 48, after repeated hits by British defenders, settled into the Thames. One crewman was lost, seventeen were taken prisoner. One Zeppelin battle illustrates what the defenders were up against early in the war. A total of twenty-one fighter planes were sent out to intercept the airship; five crashed just trying to get to it. LTAS photo.

Lesser known are other theaters of battle, where the same physical laws applied without regard to national origin or airship types. Anti-aircraft artillery holed the Italian V.1 *Citta di Jesi* causing it to ditch in August 1915; there was no fire. German ace Oswald Boelcke shot down 40 some French and British airplanes in flames, but in September 1915 he was unable to bring down an English Astra-Torres non-rigid airship. As a matter of fact the British operated more than 200 hydrogen airships, but the short-ranging enemy was only able to bring down the C.27 and the second C.23 during the entire war. That C.23(a) fell victim to a U-boats' deck gun – without catching fire. "The airship broke in two and fell into the bay. Her pilot, Captain A. J. Elliot, and the engineer were saved but the W/T operator was drowned… The wreckage of the airship was recovered and towed into Newquay harbour… [Elsewhere, in the case of C.9] On examination, both the envelope and the car were found to have received bullet holes, mementos of C.9's time over Jersey that day… repairs were quickly effected and C.9 was soon back in harness…"[74]

The French launched the first air raid of the war when on August 9, 1914, the *Adjudant Vincenol* dropped four 155-mm shells in German territory on Konz-Kartaus. Neither the Germans, nor the uneducated French soldiers who thought it a Zeppelin, were able to shoot it down in the 17 flights it made from August to October 1914. The *Montgolfier* carried out a series of raids topped by one lasting twelve hours. "The *Alcase* was one of the new ships of 500,000 cubic feet and had already carried out several raids. On the night of October 2, 1916, she was struck by a German

[74] P. London, U-BOAT HUNTERS; CORNWALL'S AIR WAR, 1916-1919, Truro, Dyllansow Truran 1999, ISBN 1850221359, courtesy H. Van Dyk

shell and forced to land in the enemy's lines; the crew were taken prisoner."[75] Even as late as November 1918 the Italian semi-rigid airship F5 landed safely after one of its hydrogen cells was deflated in action.

The tracer's burning phosphorus was denied oxygen by the extremely low air pressure at the base of the rapidly moving shell. Robinson wrote, "All the while Culley was firing he could see the incendiary bullets striking the back flanks of the airship, which showed no signs of igniting…" Yet even after iron filings were added to the tracer rounds to help create ignition sources outside their flight path, tracers were ineffective in setting gas cells and covering alight at the exit point. Too little heat was available for too short a time to get the material to ignite. What of the supposedly highly explosive gas leaking from the exit wounds? With cell pressure little more than the atmosphere surrounding them, no significant amount of hydrogen would escape to mix with air while the ignition source was present. The crew of the German raider L49 wasted so much time firing *flare pistols* into the hydrogen cells in a vain attempt to scuttle their downed ship that French soldiers were able to capture them with their Zeppelin intact.

Others recognized there was a more reliable, repeatable way to start a Zeppelin on fire. The German crews had actually shown the way to do it themselves.Von Schiller wrote, "…L-33 was struck squarely by an artillery shell which detonated in the interior of the ship. Several cells were torn open by the explosion but no fire started. Too heavy now to reach base, the ship was landed near London..." A witness reported, owing to the loss of gas, "The Zeppelin struck to ground heavily." The Zep's crew used flare pistols on her, least she fall into enemy hands. "The riddled gas cells were almost empty and the resulting fire did little more than burn off the outer fabric covering…"[76] Likewise, also shot up and exhausted, the L45 ditched on an island in the river Bueche. Robinson wrote, "Kolle… fired a signal ball into the top of the hull. A puff of smoke, a roar of flame, and L45's duraluminum bones were quickly bared of their fabric integument."

Surely potential attackers would apply this lesson and attack the Zeppelin's weak outer defense. "Others had a strange incendiary bomb covered with fish hooks which would theoretically catch on the Zeppelin's outer cover while the bomb set the hydrogen on fire."[77] Again, failure, the fishhook bomb could not ignite the gas. The answer, it seemed, would be flechettes, or burning steel darts. Engineer Lieutenant-Commander Francis Ranken developed a complex projectile featuring pop-out vanes that would stick in the Zeppelin's outer covering. (Rimell:) "On leaving the tubes… the arms swung outwards serving as a grapnel, and at once the body of the dart entered the fabric."

The Imperial War Museum in London has one of these clever inventions on display with a caption. "Developed for use against Zeppelins by engineer Lt.-Commander Francis Ranken, this type of dart was initially an RNAS weapon but was also adopted by the RFC in 1916. The dart contained high explosive black powder and phosphorus, which were ignited when the head of the dart penetrated the airship's outer skin. The tail incorporated spring-loaded vanes which opened and unlocked into position to ensure the dart remained stuck fast in the Zeppelin's fabric." [78] The

---

[75] WONDERS OF WORLD AVIATION, Ed. By C. Winchester, Waverly Book Co. London, 1938. Pg. 1009. Courtesy the late Francis Scales.

[76] R.L. Rimell, ZEPPELIN: A BATTLE FOR AIR SUPREMANCY IN WW I, Conway Maritime Press, London, 1984, courtesy Hepburn Walker, Jr.

[77] Leroy Thompson, "Zeppelins Over London," *Aces & Aircraft of the Great War*, 1994.

[78] The display also reads: "There was marking on this dart. '32 – 78' This could be an acquisition number. It follows the same numbering format on exhibit 2. of the German Airship SL11 made into souvenirs and sold to raise

Ranken dart had to be dropped from *above* its intended target. An opportunity came when the L15 was hit by AA fire from Purfelt. "Three gas cells amidships are shattered... by some miracle the highly volatile hydrogen does not ignite..." Losing gas, L15 sank so low, "Brandon overhauls and manages to get above the ship, a rare opportunity for home defence airmen... hurls out a batch of explosive Ranken darts... L15 got away but finally settled into the water an hour later, losing a crewman." Why, if they struck, were the darts not effective? It's likely the design, which was supposed to lower a flaming insert into the gas cell, found the pure hydrogen wouldn't burn. Regardless, aircraft of the day rarely had the upper hand on the high-flying Zeppelins, and no other opportunities seem to have come. Dropping something downwards wasn't the answer; something that could be *fired upwards* was needed, even if it was not yet known what the enemy's weakness really was.

"By June 1915 trials of a new tracing composition of one part magnesium to eight parts barium peroxide" was rolled out, official designation SPK MK VIIT, but nicknamed 'sparklet.'" But even these improved shells did not ignite even their exit point on the Zep bombers. The British defenders hoped that their upward-firing machine guns loaded with the new ammo would cause the sky monsters to explode in hydrogen fire. These dreams disappeared with their seemingly unharmed Zeppelin raiders sailing away. How could one miss such a huge target? They weren't missing their targets: "L65, returning, found bullet holes – three hundred forty in all –in the after gas cells. There was no fire of course." Unable to easily see or even hear their attackers some of the time, the airshipmen were not always sure they had been attacked until post flight inspections revealed the damage inflicted. At the *Aeronauticum*, the museum located on the grounds of the WWI Nordholz Zeppelin base in northern Germany, there are displays of outer cover material and gas cell material. One cell section framed and wall mounted, saved from the L35, prominently displays its bullet hole. As the Zeppeliner had said, "There was no fire of course." The British defenders reached the seemingly logical conclusion the Germans must have been using an inert gas in between the hydrogen cells and outer cover! Thompson wrote, "...it was widely believed in Great Britain that the gas bags in the German airships were surrounded by inert gas piped in from the engine exhaust which would render an incendiary bullet ineffective against the hydrogen."

If it sounds unreasonable now, it was not so to the defenders who pumped their combination ammo into the gasbags with no ill effects. So, if they contained some inert gasses, the defenders reasoned, then *explosive charges* could be used to compromise the gas containment. Commander Fredrick Aurthur Brock developed a bullet that would *burst* in the interhull area, like shooting small hand grenades in between the cells and covering. Then the Pomeroy or Sparklet bullets would seem to have something to work on. Rimell wrote, "Later experiments led to aircraft being armed with machine guns carrying a mixture of armor-piercing, explosive, incendiary and tracer bullets." Yet with cell pressures of near barometric, the Zeps weren't losing gas fast enough even when big holes were blown in the cells. Rimmel continued, "Over Clemsford a home defense aircraft of 39 HDS from Hainhault Farm attacked Bocker's ship for nearly 20 minutes, and although a great many of the Pomeroy and Brock bullets struck several fuel tanks, the Zeppelin stubbornly refused to catch fire."

The Germans had reason to believe the British had not perfected a flaming bullet. Kapitanleutnant Heinrich Mathy, most experienced of the Zep commanders, reported being attacked in L13: "Because no incendiary shells were used by the enemy, the only danger came

---

money for the Red Cross. After its destruction, the SL11 was initially wrongly identified as the L21 by some British authorities." Courtesy Anne MacGregor, Moondance Films.

from fragmentation shells which punctured the gas cells or damaged the structure of the ship."[79] Little did he realize the defenders were pumping incendiary bullets in between little 'grenades' into the ship as fast as they could!

British Lt. A. deB. Brandon attacked L15, later telling Robinson: "The Brock ammunition seemed to be bursting all along it but the Zeppelin did not catch fire. I was using Brock, Pomeroy and Sparklet. I turned again and put in a fresh drum." L15 slipped away into the clouds! "In 1915 alone, 87 British planes had attempted to intercept zeppelin raiders, all in vain," wrote Eric Niderost. When artillery pieces were hastily reconfigured to fire upwards, they had more negative effect by their shells falling back on British citizens. "Anti-aircraft fire opened up, and a shell ripped open three of the airship's gas cells. No fires were ignited, but it was plain the ship [L6] could not continue on this course."[80] The artillery at least turned L6 away that day.

US Army airman William "Billy" Mitchell believed the rigid airship was an excellent bombing platform, perhaps without realizing when the Zeppelins' bombs "stabbed at the cowardly heart" of their enemies, they vented off a corresponding amount of lifting gas or they did not make it back to base. Photos courtesy Stefhan Bauer and John Miering.

Finally, chemist James Francis Buckingham developed a version of his cupro-nickel 'envelope' bullet that would fit the standard .303 cal. machine gun. The Buckingham design mixed *powdered aluminum* with the phosphorous in a round featuring a number of holes plugged with antimony solder. When the round was fired, the heat of the round passing down the barrel melted the solder and allowed the phosphorus to contact air and ignite the aluminum.[81] Rimell wrote, "Subsequent events proved that the most effective weapon against the Zeppelins was the standard American-designed Lewis .303 machine gun armed with a mixture of incendiary, tracer and explosive bullets."

As impossible as it is to believe, even this formidable cache was not always effective. Rimell wrote, "We fired four trays of Brock, Pomeroy and Buckingham at her from 11,500 feet, which was our maximum obtainable height, but without apparent effect." As mentioned, The L17 survived two attacks. The L40 was twice attacked by Curtiss H12 flying boats and escaped. The

[79] M. Griehl and J. Dressel, THE GERMAN AIRSHIP STORY, Arms & Armour, London, 1990.
[80] E. Niderost, "Terror by Night," *Military Heritage*, Feb. 2001 , pg.48
[81] I. V. Hogg, THE ILLUSTRATED ENCYCLOPEDIA OF AMMUNITION

L56 escaped after being shot up by an F2a. L62 fought off an attacking FE2b. L64 escaped no less than six British planes. Next raid, as the Airship Heritage trust puts it, "The [L64] was then involved in another raid on 2nd August 1918 but as the ship was crossing the English Channel, it was intercepted by a British Naval Flying Boat and attacked. The ship was damaged by fire but survived and struggled back to base at Alhorn, Northern Germany. It was damaged upon landing as it landed 'heavy' due to the loss of gas from the attack. The ship was put into the shed and repairs undertaken." Small wonder, as Thompson recorded: "Primarily ...a policy of attacking airship bases from Belgium and destroying them on the ground was considered most effective."

In spite of these challenges, the airships gave the British cause to invent the aircraft carrier warship. Admiral Jellicoe would write, "On August 18, 1916, the British Fleet decided to undertake a long-range reconnaissance of the North Sea... At 1400 the Fleet believed itself so near to the German Fleet that it deployed in battle formation, but the tie passed and the German ships did not appear. The zeppelins had followed all the English movements. They had advised the German admirals of the formidable attack which threatened them and permitted them to escape. At 1800 the Harwich squadron coming from the south made out in the distance an important group of enemy ships convoyed by two airships, but it was too late to attack them and the pursuit had to be abandoned. During the whole day, the British vessels, notably the cruisers, never ceased to be under the observation of the airships." England's vast maritime superiority was checkmated until a waterborne airplane carrier could be developed.

Observation balloons, the style called 'kite' by the US Navy, were nothing more than flammable bags containing hydrogen. Yet, according to Rosendahl, "Enemy planes attacked 89 (hydrogen filled) American Observation balloons and brought down 35 of them in fames." That's less than a 50-50-success rate against a sitting duck, supposedly explosive target. In the entire war, there was only *one* US observer lost to fire.

Other defensive technology along with the three-part ammo eventually did catch up with the Zeppelins. Marconi radio direction-finding apparatus was employed to locate the raider's approaches and vector fighters to struggle to altitude and meet them. Thompson wrote, "Sgt. A. E. Hutton of 39 Squadron, for example, developed special night sights for the Lewis gun by putting a small red light bulb in the foresight and illuminating the backsight with a white bulb powered by a two volt flashlight battery. This improvement in aim allowed pilots to shoot for a 'Zepp's' most vulnerable points like the engine nacelles where the gasoline tanks were located [sic]." Finally, the tide was turning. Lt. William Robinson was quoted in THE ZEPPELIN FIGHTERS: "I flew to 800 feet below it from bow to stern and distributed one drum (alternate new Brock and Pomeroy). It seemed to have no effect." Robinson expended another 47-round drum along the side of the Zep with no effect, but with a third drum he concentrated his fire *on one part* and got a fire started.

We have to read carefully to realize just what the three-part ammo was setting afire at the exit point. Second Lt. F. Sowrey attacked L32. (Robinson:) "I fired at it. The first two drums of ammunition had apparently no effect, but the third one caused *the envelope* to catch fire in several places; in the center and the front." The emphasis is added, for reasons we shall discuss in Chapter Four. As we consider the record of hydrogen in the next chapter, we should bear in mind words from one reference:

"The adamant refusal to think, so often viciously defended, is indeed a dangerous privilege."

Chapter Two: The Lost Art

Before natural gas, lamps provided light in American cities fueled by "town (also called city) gas," manufactured from coal and typically containing 50% hydrogen, 50% carbon monoxide and trace amounts of other gases like methane. This readily available and cheap pipeline gas was often used to inflate balloons and even airships. Like this forgotten enterprise and its good safety record, it will be difficult for modern readers to imagine an industry that safely produced, transported and utilized gaseous hydrogen in thousands of balloons and airships. If hydrogen were unmanageably dangerous, it would be so from the moment of conception. Besides dozens of traveling-show balloon gas makers in the US since 1784, several hydrogen generating plants from White City (Chicago), to Toledo, to California were in use over many early years and there seems to be no record of even this primitive equipment exploding or burning.

Ft. Logan's hydrogen generating plant (right) and gasometer (center) is seen in this Air Force Academy photo. The earliest permanent Army gas balloon facility was established here in 1897; this is the later steel "balloon house." Courtesy James Shock.

From the outbreak of the Great War in 1914, Germany, Italy, England and France employed an increasing number of airships, as well as a number of observation balloon squadrons to scout for the armies. Historian Joseph Branch states that France alone built 4,170 balloons. As America built airship bases, so came with them hydrogen generators and storage gasometers. It would seem housing a hydrogen-filled gasbag inside a building would present a grave danger. The Army airship sites include Forts Logan, Meyer, Omaha, Monroe, Knox and Fort Bliss (also named Biggs); fields Langley, Ross, Phillips, Brooks, Wright, Scott, Pope, Lee Hall, Post, and Camp John Wise. The Naval sites included Wingfoot Lake, Pensacola, Chatam, Montauk, Rockaway, Norfolk, Cape May, Key West, Coco Solo, Quantico, and Yorktown. In addition, there were at least 15 coastal artillery forts with balloon hangars and hydrogen generators in the US, Panama, Hawaii and the Philippines. At first the war effort was supported from the generator and building at the White City Park, but soon the new generator and hangar at Wingfoot lake was completed and LTA men were trained in Ohio. Goodyear was producing four observation balloons a day, and by 1918 was employing 2,800 people in the manufacture of hydrogen balloons. There is not a single fatality recorded during the generation, transfer, and storage of the hydrogen gas or inflation of the balloons inside or outside any of these facilities. No accident there; Dr. Michael Swain demonstrated how simple vents placed atop and on the bottom of a garage door would prevent a safety problem even if the hydrogen car inside was leaking at 7200 liters of hydrogen per hour!

Ironically, the most commonly published photograph associated with the "danger of hydrogen" is that of a kite balloon seeming to explode in flames as its frightened soldiers scamper away. It was first publicized with the grandly optimistic *National Geographic* 1919 helium article. It is rarely captioned this accident took place while *deflating* the balloon. Kite balloons made of the finest contemporary materials were still subject to air contamination since their 1898 invention. "The kite-balloon consists essentially of an elongated gas-bag which is divided into two unequal portions, the larger of which (comprising about four-fifths the total volume) is filled with hydrogen; the remaining one-fifth constitutes the ballonet, or air-cell, and this is automatically inflated by the wind through a convenient aperture."[82] The winding down and disassembly phase of the operation was obviously most sensitive to lapses in proper procedure. In this seemingly terrible mishap, no one was seriously injured.

The safety record must also include the hazardous and fire-prone environment of waterborne vessels ever since Thaddeus Lowe took his hydrogen generating plant aboard a raft to loft balloons capable of spotting Confederate movements from the Potomac. "Construction of the *Mayfly* brought the first assignment of a Royal Navy warship to aerial service. The officers of Sueter's inspectorate were carried ... on the roster of the cruiser *Herminone*, and the vessel herself attached to the airship's building site, Barrow-in-Furness. She transported the crew for the *Mayfly* there in September, 1910, and was equipped with a hydrogen-generating apparatus to help produce gas for the airship." [83] This first British airship was caught coming out of her waterborne hangar by a freak storm and broke in half; in spite of free hydrogen and flammable materials loose in that storm, there was no fire on the airship. (Had that split not occurred, perhaps the history of the rigid airship would not be so dominated by one side. His Majesty's Airship No. 1 demonstrated many advanced features, including exhaust ballast-water recovery.)

For years navies operated hydrogen balloon scouts from their ships. While the US Navy only had one brief experiment with a purpose-built observation balloon ship, the USS *Wright* (AZ-1), navies of other major powers all had observation balloon support ships. Of course every one had to be equipped with a hydrogen storage and/or generation system. "Kite balloon units were also based in France and Ireland with specially equipped ships assigned to tow them during escort duty. These included five destroyers and the yacht *Harvard*. Five more destroyers and two more yachts were being equipped when the war came to an end."[84] The War To End All Wars also saw such innovations as anti-airplane cable curtains suspended between multiple balloons.

The United Kingdom operated the largest fleet of hydrogen-filled non-rigid powered airships known to history. After several unique prototypes dating to 1908, the Great War forced some fast paced innovation. As Airship Heritage Trust explains, "In total 158 'SS' Class ships were built.

[82] A. A. Hopkins, OUR COUNTRY AND ITS RESOURCES, Scientific American Series, Munn & Co., Inc., NY, 1918.
[83] R. D. Layman, TO ASCEND FROM A FLOATING BASE, Associate University Press, 1979.
[84] "ASW Aircraft," *Naval Aviation News*, March 1970 pg.33

The 'submarine scouts' with the prefix of 'SS', were to be so successful on coastal patrols that the Admiralty wanted bigger and better ships and fast. Three further classes 'Coastal', the 'C*' and 'North Sea' class ships were developed. Each having larger engines, envelopes and crews than the previous class ships, the patrol duration increased. Equipped with small bombs, these ships proved to be not only 'observers' but also active participants to the fleets' battles. It was common that a U-Boat on patrol, once spotted by airship, had a choice of either moving away or engaging the airship in a race. The battle was between the U-Boat surfacing and being able to mount his gun and try to bring down the airship, whilst at the same time the airship would be signaling the location of the U-Boat to the fleet, and preparing to drop it's bomb, before the U-Boat could take a shot at the ship."

Enemy planes were also a concern, and the British C-17 was shot down in flames. On December 11, 1917, a German seaplane shot down the C-27, with the total loss of her five crewmen. As Kent O'Grady wrote, "Approximately 32 Coastals were built (including the 5 which were sold to Russia), and those combats were the only two losses. Of the C Star (C*) airships, Ten were built and all were decommissioned... none destroyed. Of the much larger North Sea Class airships, some 14 were built, one was lost to lightning, the NS-11, with ten dead."[85]

Two good works have summarized the UK's Great War airship history, Ces Mowthorpe's BATTLEBAGS and Patrick Abbott's THE BRITISH AIRSHIP AT WAR. Therein we see many other opportunities for hydrogen to prove unacceptably dangerous even without all the enemy's shooting. SSZ-27 was scouting and spotted a mine near a trawler on April 12, 1918. (This is not to be confused with the SS-27, wrecked on a church steeple at Marquise; she did not catch fire either.) The trawler fired at the mine but ricochets apparently punctured the envelope of the SSZ-27 and she came down in the sea or on the trawler. In either event there was no fire, and trawler was able to salvage her almost entirely. (After repairs the SSZ-27 was flight worthy and back in service in May. She flew a total of 782 hours, and was disassembled in December of 1918.) The USS *Downes* found the partially deflated SSZ-51 floundering with a dead engine and took up a line to help. "The pilot and crew of the SSZ-51 were very much concerned on account of the heat waves from the smoke pipes, fearing that the airship would catch fire, and after some conversation, requested that we send a boat out to pick them up."[86] The rough seas thwarted efforts to recover the wreck, at one point draping over the ship, but there was no fire and the crew was saved to fight another day.

Of course the greatest concentration of hydrogen production for airship use was by Germany in the Great War, as they created and flew an air force unique in history. A Frenchman inspecting German airship hangars after the Great War noted "The gas conduits (which are generally double and 0.5 to 0.6 m in diameter) are buried in the floor of the hangar and are provided every 20 m. with strong branch valves. The connection between the stationary conduits and the ballonets is made by strong flexible sleeves from 6 to 7 cm. in diameter, terminating in dust filters. Precautions are taken to prevent the formation of dangerous sparks from the electrization [sic] due to the friction of the gas on the metal parts of the filter."

---

[85] This may have been a 'cover' story, according to historian Nick Walmsley, the airship was attempting to re-establish its endurance record, and the lightning was not in the area until after the accident. Paranormal research suggests a PBM flying boat disappeared in the Bermuda triangle; the Navy admitted some just blew up.

[86] Letter from the CO to the Force Commander, 16 AUG 1918, National Archives RG 45 Box 137. Courtesy Richard Peuser, NARA Military Reference Branch.

The French could hardly believe the capacity of the German bases. "At Alorn their capacity of production is 70,000 cu m per day. Let us recall, for the sake of comparison, that the large French works, based, it is true, on other processes, delivered during hostilities hardly 4000 cu m a day. Most of the German works employ the so-called 'iron' process, that is the decomposition of steam by red-hot iron and the regeneration of the iron oxide by a current of water gas... extensive facilities for storing the gas are necessary... In fact, for a large airship of 70,000 cu m., with a total theoretical carrying capacity of 50% the maximum unballasting (fuel, ballast, bombs, etc.) during a voyage, is 15 to 20 tons. On its return, therefore, it must receive a fresh supply of 15,000 to 20,000 cu. m. of gas for immediate replenishing... Ordinarily the 30,000 cu.m. of the gasometer would seem sufficient for the daily replenishing and for serving as a balance between the continuous production of the works and the irregular consumption of the balloons... it must be possible to fill a large airship in 10 to 15 hours (The L-72 was filled at Friedrichshafen in 12 hours). At this speed, there is the advantage of not requiring so large a personnel [contingent] and, above all, of facilitating the process of balancing, which might needlessly strain the frame of the balloon. It is, in fact, just so much easier to avoid a wrong distribution of the weight and lifting forces, as the means are more powerful for producing a rapid flow of gas into the balloon. It was probably for this reason that most of the German stations have a gasometer of 30,000 cu.m. and a battery of cylinders containing 30,000 to 50,000 cu.m. of gas under pressure, making a total of 60,000 to 80,000 cu.m., a volume practically equal to that of the largest airship... it would seem more economical to resort to the exclusive use of gasometers, and thus avoid the expense of compressing, which absorbs about 0.5 hp per cubic meter... the containers are enormous cylinders 15 to 20 m. long by 1 m. in diameter, each holding about 1000 cu. m. of gas at 100 kg... in order to avoid explosive mixtures during the first filling, vacuum pumps are provided for removing the air from the reservoirs... Since there is little chance for the introduction of air into the cylinders, the purity of the gas is excellent. (The lifting force of the L-72 was about 1188.) At Nordholz the hourly output (1200 cu.m.) of the compressors differs but little from that (1000 to 1500 cu.m.) of the hydrogen producing works. The situation is about the same at Tondern (5000 cu.m.)... at Alorn, where the gas production exceeds 2500 cu.m., the hourly output of the compressors differs but little from that of Nordholz."[87] At the large Tondern base an elaborate tank and pipeline system had capacity for 56,000 cubic meters of hydrogen. A gasworks built southeast of the hangar area produced 8000 cubic meters of hydrogen every 24 hours, but this was not enough. A rail line directed special wagons carrying additional hydrogen into the base. In spite of many similar facilities spread throughout the Continent, there seems to be no report of even the most crude and hard-pressed facilities falling victim to accident. The Tondern gas works building stands to this day, now in modern Denmark.

The US Navy was just learning how to design, build and operate hydrogen filled blimps late in the Great War. The US also purchased airships abroad and brought them home for study. Though small in number, inexperience would have offered ample opportunity for 'Murphy' to strike. A comedy of errors found the British import SSZ-24 wind-slammed against the Wingfoot Lake hangar entrance, tearing it and discharging its gas. A letter from the Commanding Officer noted the various contractors and military people in and about the hangar did not work well as a team, but there was no fire that anyone noticed.[88] Coastal defense forces successfully deployed spotters in kite-type balloons for deadly accuracy in firing their artillery. Balloons were sometimes kept in nearby sheds or "bedded down" in the open. The manned observation balloon continued to be used by the US Army until 1942, some being real airships since they were equipped with motors.

---

[87] J. Sabatier, "Large German Airship Stations," *L'Aeronautique*, March 1921

[88] To: The Bureau of Operations (Aviation) Navy Department, Washington DC, subject 'Report on accident to the Dirigible S.S.Z. 24, April 24, 1918.' National Archives RG 45 Box 137. Courtesy Richard Peuser.

By far the largest hydrogen consumers were unmanned balloons. Not well published, the British launched a propaganda and sabotage campaign against occupied Europe using small hydrogen balloons to carry everything from leaflets to snare wires to incendiaries. Another division launched "free barrage balloons" into the swarms of oncoming Nazi bombers. "During the 2 ½ years that Operation Outward was in action, 45,599 balloons equipped with trailing wires and 53,543 balloons carrying incendiaries were launched... almost 99,000 balloons were released in 540 days: an average of 183 balloons a day."[89] The much larger, more familiar barrage balloons, sometimes with explosives attached, had cables designed to snare airplanes. Barrage balloons squadrons went ashore with the British Expeditionary Force in France. By 1940 "...the British had forty-four balloon squadrons consisting of approximately 33,000 officers and men and 1,500 balloons. The London area was protected by ten squadrons operating approximately 500 balloons with 120 officers and 5,380 enlisted personnel." During the V-1 offensive there were some 2,000 in the area. (Historian Norman Longmate credited balloons with snagging seventeen V-1 "buzz bombs" in one day, July 20th, 1944 – ironically the day of Hitler's narrow escape from an assassination attempt.) Therefore, during WWII we had hundreds of hydrogen generators inflating many thousands of balloons, with tens of thousands of hydrogen bottles in flux. Hillary St. George Saunders pointed out one squadron alone received some 19,000,000 cubic feet of hydrogen in a two-month period. Gas bottles went out on barges in the Thames Estuary and the Humber so balloons could prevent the mining of these vital waterways by seaplanes. As the October 1941 *National Geographic* revealed, much of all the work was done "by the ladies."

US Army photo showing a typical WWII hydrogen generation plant utilizing the ferro-silicon/sodium hydroxide process for inflation of barrage balloons, courtesy James Shock. Today, 194 ft^3 K-bottles used to move GH_2 are pressurized to 2015 PSI. As of this writing, Via-Tech is perfecting a third-generation generator that produces hydrogen gas through a time-release alkaline metal emulsion released into a water reservoir.

In "The Shipboard Balloon – The Beginnings of Naval Aeronautics," R. D. Layman and S. McLaughlin show two types of hydrogen generators employed by the Russians aboard ship. One is shown mounted on the transport *Kolyma.* They point out "The Second World War saw a revival of the kite balloon, but only as an unmanned form of aerial defense. Large numbers of these barrage balloons were flown from vessels in British, German and Italian harbors and from landing craft in Europe and the Pacific. A few manned balloons were employed on land early in

[89] H. Van Dyk, "The British Balloon Offensive Against Continental Europe, 1939-1944, *Buoyant Flight*, Vol. 48 No. 2, Jan/Feb 2001

the war by Poland and the Soviet Union, and later by Japan." Germany also used a few manned observation balloons, the author having seen a newsreel showing one employed on the Russian front, for example. Likewise the French employed this simple but effective defensive measure.

A 1944 intelligence report pegged the number of German barrage balloons at 1,826. "U-boat pens, building yards, torpedo testing stations and oil storage facilities at varying locations were allotted twelve barrages of 209 balloons each."[90] Submariners even took hydrogen into their cramped u-Boats to inflate radar decoy balloons they named *Aphrodite*. One balloon saved the U-566 by misleading attackers on August 8, 1943. Largely ignored are the balloons used to loft paratrooper training platforms, providing a safe and reliable method of getting the trainee soldiers high enough to test their skills.[91] Like the unmanned balloons, large numbers of women (WAAFs) quickly trained for the purpose did a lion's share of the work with the gas balloons. Here again, even with little expertise and hasty training under wartime conditions, the safety record seems to be all but perfect.

"Hydrogen plants were located at the Army training centers at Camp Davis, NC, and Camp Tyson, TN. The Marine Corps training centers were at New River, NC, and Camp Lejuene, NC. The six Marine Corps BB squadrons that served in the Pacific probably had one unit each. There were 22 Coast Artillery Barrage Balloon Battalions, and 3 Group HQ units... my guess is approximately 25 more generating plants."[92] THE U.S. ARMY BARRAGE BALLOON PROGRAM [93] shows a typical hydrogen generating plant and hydrogen storage "bottles" (high-pressure steel cylinders) hand-stacked. As America entered the war "...hydrogen gas generators and gas bottles were in short supply. Adequate gas facilities were available on the West Coast and by January of 1942 the Pennsylvania Salt Manufacturing Company and the Hooker Electro-chemical Company, both of Tacoma, Washington, and the Shell Chemical Company of Monterey, California, were able to supply the necessary gas for the West Coast barrages." The US Army eventually had some 36 barrage balloon battalions, each with about 135 balloons. Each battalion was equipped with a truck-mounted hydrogen generating plant, probably a spare, and a host of other trucks to move equipment including bottles and winches. One can only imagine the opportunities for 'Murphy' to strike amid hastily organized and quickly trained recruits, but amid many broken cables and embarrassing runaway balloons, there is *only one* recorded fatality blamed on hydrogen. (This was an asphyxiation case that would no doubt have been as fatal as helium displacing air in the lungs. Helium has unfortunately killed Navy and civilian workers when proven safety rules are relaxed.[94])

The tropics offered other opportunities for Murphy to strike, from the Italian bases around the equator to the German base described by von Schiller. "The hydrogen obtained is compressed into high-pressure containers, which reduces the time required to fill the ship. In the early days of the South American service we had to spend two days in Pernambuco for filling alone; today in

---

[90] K. O'Grady, "Barrage Balloons of WWII (Part II)," *Aerostation*, December 2000

[91] Illustrated in various historical documents and TV programs, such a system was assembled for the Alan Ladd film *The Paratrooper* – realistic even to the inclusion of women in the training process.

[92] Letter to the author from James Shock, July 2000.

[93] J. R. Shock, World War II Historical Monograph Series, WWII Historical Society, Bennington, VT, ISBN 1-57638-052-1

[94] Two AEROS workers were asphyxiated by helium in December 1999 while working inside their airship's ballonet. Some Navy fatalities have been in the helium cell during "helium dives" where masks were not sealed or equipment has malfunctioned. Any gas that displaces air in a confined space is very dangerous. Two Space Shuttle workers, Bjornstead and Cole, were killed in 1981 when their team was ordered to enter the Orbiter *Columbia's* aft compartment while it was still being purged with nitrogen.

Rio, where a large high-pressure storage tank is available, even the *Hindenburg* can renew its supply of hydrogen in a few hours."

The operation of a hydrogen airship industrial base is an unknown art today. Generation and storage of the gas was one potential safety problem, certainly an extra concern afloat. Also difficult to imagine today, hydrogen-filled air vehicles enjoyed freedom of operation modern helium airshipmen could only dream about. This is mainly due to the fact that lifting gas could be vented as often as ballast was released, if not more often. William Ballantyne, an R.34 crewman determined to overcome the oversight of not being on the roster for the British Atlantic attempt, hid in the ship's cell valve access. "Ballantyne had intended to stay hidden for the whole voyage if possible, but hydrogen from the relief valves made him sick so that he had to move to a more open place."[95]

Lost hydrogen gas, let alone the huge volumes purposely vented, would seem to be a serious safety problem. Yet the lifting gas was constantly being vented for a variety of reasons, from overcoming pressure height, to compensating for dropped payloads, and to lose lift no longer needed. "Hydrogen airships could afford to be extravagant with their lifting gas because it was replaceable, and since any airship grows lighter as it burns up fuel they would climb to high altitudes perhaps twice in an ocean crossing to force out the surplus gas no longer needed. The *Hindenburg* might 'blow off' as much as one and one-half million cubic feet of hydrogen in an Atlantic crossing."[96] If today's wisdom is correct, such constant hydrogen venting was foolhardy. The actual safety record, however, speaks otherwise, and it has from the first days passengers were carried.

"In 3 years of service the four airships *Schwaben, Viktoria Luise, Hansa* and *Sachen* made 760 flights and carried 14,000 passengers; the distance that they covered totaled almost 100,000 miles. In all these flights there was not one mishap, nor was any passenger injured."[97] In contrast, the first pilot of the first scheduled aeroplane line, Tony Janus, was killed October 12, 1916 in a plane crash. His brother Roger flew almost another two years until a DH-4 crash took his life.

The official US Navy rigid airship history film concedes that "even with hydrogen" the Zeppelin service had an enviable safety record. From 1908 through 1937 they racked up, by one estimate, 440,704,000,000 passenger miles. Only *thirteen* passengers did not arrive safely. Less well known is the French airship passenger operation, the Compagnie Generale Transaerienne. Their *Ville de Lucerne* made 267 flights, carrying 2,590 passengers without an accident. Figures on the Russian airship operations are not available, but since their V-6 once held the endurance record, it's likely the Russian program also racked up miles without unreasonable fatality rates. Yet with all of these airships, discarding a supposedly explosive gas was a way of life.

The airships thrived because of hydrogen's basic safety in operations. In a booklet introducing the R.100, the British team wrote under 'Danger of Fire', "An airship carries large quantities of two inflammable substances – hydrogen and petrol. Of these two, the petrol is probably the more dangerous. The reason for this may be easily seen. Hydrogen is a gas very much lighter than air. The engines, passage-ways, passenger quarters, etc., are all located in the bottom of an airship. Consequently, if a tear should be made in one of the gasbags of the ship the escaping hydrogen at once flows *upwards*, away from possible sources of fire, and escapes out vents provided at the

[95] P. Allen, THE 91 BEFORE LINDBERG, Flying Books, Egan, Minn. 1984
[96] P. Litchfield and H. Allen, WHY? WHY HAS AMERICA NO RIGID AIRSHIPS? 7c's Press, 1945/1976
[97] R. S. Holland, "The Zeppelins," HISTORIC AIRSHIPS, Macrae Smith Co. 1928

top of the ship." [98] Much greater quantities of the gas were released in operation, even if venting hydrogen into the air surrounding our vehicle would seem foolish today. The airship's skin would have a static charge accumulating as it flew along. As Diehl wrote, "Airships in flight acquire an electrostatic charge due to impact of dust and moisture particles in the atmosphere. This effect is noticeable at 40 km/hr, increasing approximately as the 16th power of the speed."

Furthermore, we know strong electrical energies are constantly present in the atmosphere itself. Most of us are familiar with the legend of Ben Franklin's experiment with a kite and atmospheric electrical energy in a storm. Recreating Franklin's famed experiment, *Popular Science* magazine found "...we had no trouble getting sparks in *cloudless* skies. And *voltage increased as the kite rose*." [99] Just flying along under hydrogen would seem an uncomfortable worry even with today's materials, let alone in an airship whose gas was haphazardly retained in balloons fashioned by gluing cattle guts to huge cotton sheets.

As if there was not enough to worry about in clear skies, wet *clouds* were actually *sought out* by airshipmen to collect extra ballast weight to compensate for fuel burned. "To maintain altitude, the *Hindenburg* would brush against rain clouds and collect rainwater in gutter that fed the ballast tanks. If the airship passed through a shower every few hours there was no need to valve off the expensive gas.[100] Frequent cloud rubbing would seem ill-advised when one remembers the microchip-destroying high voltages that can be generated simply by dragging one's feet walking across a carpet. Rubbing clouds in a hydrogen-filled airship seems like a crazy way of doing business with what we think we know today. It takes only a minuscule amount of electricity to ignite the perfect (stoichiometric) mixture of hydrogen gas in air, 0.02 milli-joule (mj, or thousandth of a joule) to be exact. Admittedly this increases dramatically as the mixture becomes less than perfect; for example at 10% gas in air, it takes hundreds of times more energy. Still, would it be advisable to rub clouds *on purpose*? There were hundreds of hydrogen filled airships slowly seeping and even deliberately venting hydrogen gas at altitude, regardless of their possibly charged skins, in atmosphere where scads of milli-joules of electrical energy exist continually. Given the wide range of hydrogen flammability in air, 4.1% to 74% in theory, a combustible mixture would have been *unavoidable*, even commonplace. Its ignition, given the unlimited energy available, was assured.

Hydrogen seems frightening, with its very high flame speed, from 3.24 to 4.40 meters per second. In other words, at the 40-plus per cent gas-in-air mix, hydrogen advances at *7 feet per second*. Its flame temperature reaches a withering 2050°C, if mixed just right.[101] Of everything classified as fuels, hydrogen has the very highest specific energy when defined by weight, a staggering 120 mega-joules (million joules) per kilogram. LZ-129 had more of it than any airship built to that date, 190,000 standard cubic meters (6.7 million SC Feet). The energy content of the hydrogen was equivalent to about 1,900 gigajoules (GJ), or 19 GJ *per passenger!* The word "volatile" is tossed around to heighten the phobia, since hydrogen is already a gas; it doesn't have to evaporate to be ready to mix with air. The word "explosive" congers up visions of even greater danger. Walter S. Diehl wrote "Mixtures of hydrogen in air containing between 9 and 64 percent hydrogen are explosive *in the sense* that flame will travel in any direction after

[98] AIRSHIP R.100, 1929 brochure reprinted in 1983 by Mr. Pye Books, Howden, Goole, U.K.

[99] Emphasis added. *Popular Science*, October, 1897

[100] R. Pilkington, "By Airship to America," *Air & Space,* Dec 1991, pg.24. The LZ-129's rain gutter ballast collection system was at the equator amidships, but is only visible in selected photographs.

[101] A jet of hydrogen flame is applied directly to the glass rod in the manufacture of fiber optic cable, its flame temp of 3,713° F, only a couple hundred degrees less than our old friend gasoline at 3,987° F.

ignition."[102] Gas balloons can't help but get a little air in them. "The tube hangs loose and open just above the car, and small quantities of gas constantly escape from it."[103] Early sportsmen none the less flew hydrogen balloons into storms. "The storm was so grand that in its horror it failed to terrorize. The lightning was not forked or in streaks; there would simply be an opening up of blue flame which extended on all sides and cannot be described by anything except the old-fashioned idea of hell. The electricity was so prevalent that the ropes snapped as I touched them. Yet the very fact that I was immediately in the storm with the lightning completely surrounding me prevented the gas-bag from exploding..."[104] How could balloons and multi-cell Zeppelins have routinely seeped, vented and even dumped such a "highly dangerous" gas into ever-present static electricity without blowing themselves to kingdom come?

Simple: *explosion* is more properly defined as a fast chemical reaction between a flammable gas and an oxidizer, which generates a pressure wave. The better word in this instance is 'detonation.' Popular gases propane and methane can detonate at 6% concentrations; it takes *three times* that figure for hydrogen, and only then if properly contained.[105] Lifting gas pure enough to loft an airship, and so loosely contained in a flexible membrane *cannot* detonate or explode under *any* conditions. An extreme example was seen in the Great War, when HMS *Furious* had been modified to carry 7 Sopwith 2F1 camels and successfully bombed the double hangar *Toska* at Tondern. L54 and L60 were destroyed by fire. Inflated, but containing no gasoline, they simply burned. Rimell wrote, "The fact that there were no explosions prevented the total destruction of the shed: it was soon repaired."

Discharging some gas in air does make the concoction briefly susceptible to ignition before the gas disappears. At first there was little concern, since so little energy would radiate that it, even if it somehow found an ignition source, was unlikely a secondary ignition would occur. Of course gas vented into the upper altitude range would inevitably encounter ignition sources; Zeppelin designers eventually provided isolated vent shafts that effectively dealt with the necessity of pure gas having to mix with air when it was released from the cell.[106]

"Patent No. 1,416,580 relates to venting the spaces between and surrounding the gas cells and safely discharging as escaping from the safety valves... so far as our investigation of the prior art patents have gone, we have failed to find a patent combining a flue ventilator for the space surrounding the gas cells with a discharge pipe for the safety valves."[107] During the Great War, all patents were off; L34 (*werft* number 78, or LZ-78) was the first Zeppelin to have true gas vent shafts. They were not armor-plated, since it was not possible for the mixture to explode with pressure wave at any time. Designed so the "chimney effect" would immediately dilute the gas and funnel its upward escape, the vent system was witnessed working even in the hangar. At

[102] Emphasis added. BALLOON AND AIRSHIP GASES, Part I – Hydrogen and Helium Production Processes: the compression and storage of gasses, by Charles deForest Chandler; Part II – Physics of Gases, by Walter S Diehl, Ronald Aeronautic Library, The Ronald Press Company , NY 1926.

[103] Countess Grace Di Campello Della Spina, "Ballooning By Moonlight, Narrative of a Woman's Trip over the Apennines," *The Century Magazine*, May, 1907

[104] Dr. J. P. Thomas, "The Perils and Pleasures of Ballooning," 1906

[105] Commonly used explosive limits at STP are 18% – 59% hydrogen-in-air, contained.

[106] At the Dorotheum auction of "*Zeppelin- Und Luftfahrthistorika*" on 5 OCT 96, a patent drawing showing a 1924 vent system enhancement was offered for minimum bid of 1,200 Austrian shillings. "*Patentanspruch v. 8.2.1924 betr. Einrichtung zum Beluften v. Luftschiffen mit Einzelgaszellen d. Fa. Luftschiffbau Zeppelin und Ing. K. Siller in Friedrichshafen.*"

[107] "Key To The Development Of The Super-Airship, Luftfahrzeugbau Schuette-Lanz," Collated and Edited by F. S. Hardesty, NY, 1930. This "Hardesty Report" caused great frustration for Goodyear-Zeppelin as the appearance of competition was offered, although the organization had no realistic method of building an airship in the US.

typical airspeeds of the LZ-129, the corridor flow was so powerful that crewmen reported having to pull themselves along against the current. Ordinary operational dumps would be quickly diluted beyond the mixture that typical atmospheric electricity could ignite. The description in a 1937 book explains what may be too tiny to appreciate in published plans: "Each pair of gasbags has, between the two gasbags, a gas manifold shaft into which the gas release valves are made to discharge. The gas escapes through a special discharge fitting on the outer cover of the ship. The gas manifolds serve also to ventilate the interior of the ship, not only the parts that lie above the central gangway, (above which all the gas valves are arranged) but also those lower parts of the ship down to the keel gangway."

The Material Safety Data Sheet calmly states, "Care should be taken to eliminate sources of ignition such as sparks from electrical equipment, static electrical sparks, open flames, or any hot object in excess of 900°F (482C)." If this sounds like common sense, it is; but the gasoline that runs your car demands more stringent precautions, since its lower flammability limit is smaller. Crewmen wore non-sparking overshoes and took other precautions against accidental ignition of lifting gas or petroleum fuel. If, say, pressure height was suddenly exceeded by an updraft and all of the 36-inch cell valves came open at once, a flammable concentration might be fleetingly present in the shafts. In the unlikely event all these precautions still allowed an ignition source to be present, the shafts would have protected the cells from its intermittent, quick burst of low-radiation heat. The lightweight material forming the covering of the wire-hoop vent column, equipped with a ladder for passage, had no extraordinary fire reflective materials; such weight need not be installed, it would be useless because of the nature of the gas.

The hydrogen safety handbook[108] helps the reader understand that the much ballyhooed lean flammability limit does not make hydrogen some kind of stand-out dangerous. "...the differences between hydrogen and methane for the *lean flammability* limit are minor. The lean limits of propane and gasoline are even lower."[109] What about that seemingly ultra-low ignition energy? The handbook helps put the difference in perspective. "The minimum ignition energy for hydrogen applies only at a fuel concentration of about 25 to 30% in air... so the extraordinarily low ignition energy required for hydrogen at stoichiometry may have little practical significance... At leaner or richer fuel-air ratios, the ignition energy for hydrogen-air mixtures increases sharply." As a reference, the rather heavy saturation required for best combustion is in stark contrast to other fuel's optimum fuel-air mix. It is less than 10% for methane, 4% for propane, and a frightening 1.76% for gasoline! (The next time someone screams at you for using a cell phone while gassing your car, remember it's for good reason.)

Owing to the ever-present electrical energies on the ship's exterior, fire was more likely even though hydrogen gas quickly disappears through buoyancy, convection, and diffusion. As a precaution, Zeppelin exit vents were equipped with metal escutcheons. A quickly bursting (six times faster than gasoline), upward-burning, intermittent lifting gas fire would radiate so little heat downward that the metal surround was more than adequate to prevent secondary ignition of even the flammable fabric. In fact, the *Hindenburg's* version of this proven system was actually put to the ultimate test in 1936, as she once approached *pressure height*. Pressure height is defined as that altitude at which the gas cells have expanded to their structural maximum. Operations in the tropics challenge airships, as conditions conspire to reduce pressure height. On a 1936 trip out of Rio, pressure height was down to 800 feet; automatic overpressure valves began to chatter open as the ship bumped up to that height. Cell III's automatic valve suddenly

---

[108] SAFETY STANDARD FOR HYDROGEN AND HYDROGEN SYSTEMS, NASA NSS 1740.XX 1995

[109] 4% sounds really dangerous – until one realizes propane is about 2% and gasoline is a terrifying 1.4%.

stuck wide open, and proceeded to dump 160,000 cubic feet of hydrogen into the vent shaft. Mixing with air and exiting the louvered vent, this blimp-sized volume of dangerous gas *must* have offered a flammable mix to the atmospheric electrical energy.

The hydrogen applications book[110] warns: "Consequently, in actual accidents with brief releases, the lean flammability limit is particularly important regardless of the total flammability range because possible ignition sources are almost always present when a fuel-air mixture first reaches a flammable concentration." This is even assuming the leak is at ground level, not the charged lower troposphere. With the ever-present electrical activity at altitude, there was most certainly a huge fire, with perhaps half of the released gas bubble being consumed. Remembering that such a mix gets more dangerous as concentrations increase, such a volume would have certainly passed the lower limit of *downward propagating flame*, which is ten percent gas-in-air. Some fire should have been able to burn back down toward the source. Further, radiant heat was certainly felt downward on the sheet metal surrounding the vent hood.

However, like countless other gas releases, this seemingly certain catastrophe was a colossal non-event. The huge fire went unnoticed, something to remember when later blame was placed on a *slight* leak. The cell had lost *half* of its gas, right up to the level of the valve. The cell had "mushroomed" well up in the structure, becoming noticeably different than the other cells in inflation level. Yet with skillful airmanship the crew was able to overcome the loss of four and a half tons of lift dynamically, without alarming the passengers! The inflation line was used to distribute gas back into the cell, ballast was dropped, and the entire episode did not even make Hugo Eckener, then in the passenger compartment, suspicious something was amiss.

The hydrogen applications book explains why. "The apparent or average emissivity of the flame is a more significant factor than flame temperature when considering thermal radiation damage from fires...The radiation hazard from a fire is directly proportional to exposure time, burning rate of fuel, flame emissivity, size of the burning surface as viewed by the observer, heat of combustion, and percent heat of combustion immediately appearing as radiant heat that can reach the receiver and be absorbed by it. The radiation hazard is inversely proportional to the square of the distance between emitter and traversed by the radiation. A considerable amount of the radiation emitted from a hydrogen flame is absorbed by water vapor in the surrounding air. The radiation flux decreases rapidly with the distance if the air contains as much as 2% by weight water vapor." Coming up on the earth's equator on that flight out of Rio, it was probably fairly humid at 800 feet. At Lakehurst the night of the last landing, the station recorded 98% humidity. Seemingly even more insane today, Eckener and other experienced air-sailors reveled in *stormy* weather's ability to greatly increase their ground speed. "Airshipmen like storms. They go for them... even the swift-moving air of a storm with lightning all about," bragged *The Literary Digest* in 1936. Today we'd think someone had a screw loose that would fly a hydrogen filled airship into an electrical storm. Yet *The Century Magazine* back in 1910 had stated flatly "A ship in the air is safe from lightning." When asked about storms, Captain Lehmann was emphatic: "Air-ship men like them! They go looking for them!" As the Goodyear promoters wrote, "A thunderstorm is well for any aircraft to steer clear of, since it indicates the presence of violently moving air. But lightning does not worry even the hydrogen-filled airship, because the metal structure can be efficiently bonded against electrical charge." The 1936 *Literary Digest* verified, "But for a third of a century, German airships have been operating on hydrogen without a single

[110] SOURCEBOOK FOR HYDROGEN APPLICATIONS, Hydrogen Research Institute and National Renewable Energy Laboratory, 1998

casualty directly attributable to hydrogen, though on several occasions they have been struck by lightning." This would seem to indicate no hydrogen could have been present; but it was.

Easy manufacture of *wasser stoff* anywhere on the planet promoted worldwide operation. Globetrotter *Graf* ranged from the Artic to the Brazilian rain forests. (Center) Gas consumed crossing Siberia on the way to Tokyo will be refilled in LZ-127. Courtesy Stefan Bauer.

We should be aware the *Graf* had the extra problem of rather permeable – meaning gradually seeping - hydrogen gas cells. Created by painstakingly gluing bovine gut, called goldbeaters' skin, to a cotton base, the gas tended to seep through the cell walls, on average, about one liter per square meter every 24 hours. Multiplied by the thousands of square meters of cell area, a goodly amount of hydrogen would be leaking into the air space between the cells and the outer cover *constantly*. Even given nature's fastest dispersal rate, one would have to assume flammable mixtures were sometimes present.

Ignition sources were certainly along for the ride even in clearer weather. At times atmospheric electrical activity even caused *visible* manifestations aboard Zeppelins. Lehmann had written, "Tongues of fire are licking around the muzzles of my machine guns, and around my head, too. And when I spread my hand little flames spurred out my fingertips," describing his experience aboard the airship LZ-98, a wartime Zeppelin the likes of which were typically constructed in less than two months time. Robinson wrote, "…in L17, due to a broken winch, was unable to wind in her antenna. In a hail squall, long sparks cracked from the antenna to the hull, lightning jumped from the control car to the top platform, and static electricity flickered on the fur collars of the lookouts." How did that collection of leaky gasbags survive? Wartime Zeppelins were constructed in as little as six to eight weeks; LZ-45 was built in an amazing 33 days. Shortages of bovine gut, even with every slaughterhouse under strict orders to save the entrails for the Zeppelin effort, meant the more spark-prone cells sometimes had to be substituted. "Besides the concern about fire, crews working in confined spaces between the airship gas cells were sometimes fatally overcome by leaks of the colorless, odorless gas, which could act as a euphoriant, like nitrous oxide (laughing gas) making inexperienced crewmen act *lustig* ("merry").[111] That author's rendition is not very likely, since the gas absorbs some odor from its cell, and inhaling hydrogen does not make one silly; it causes nausea. More likely this is some mistranslation of symptoms of little-understood "high altitude sickness" caused by lack of oxygen. Yet the gas *was* seeping out; it seems impossible that a single mission could have succeeded if hydrogen could somehow become trapped under the outer cover awaiting only an ignition source.

Early British Zeppelins had a reputation for leaking hydrogen to an even *greater* degree, owing to the difficulty in manufacturing gas cells. Yet the R.34 was first to crisscross the Atlantic in 1919. Just lucky, as well? Hardly; ignition energy was constantly present in the air, let alone other sources. "Experience had shown that when the [R.34] airship was descending it was

[111] F. A. Contey, "Zeppelins Bomb London", *Aviation Heritage*, Sept. 1992; only one fatality is known.

prudent to wind in the trailing vertical radio antenna as on the outward journey the antenna had touched the sea and all the static electricity in the airship was discharged, with a bang, through the antenna, frightening the wireless operators, but luckily doing no damage to the vintage Morse code radios." [112]

While the operation of such rigid airships is a lost art, in the songstory CURLY'S AIRSHIPS, Judge Smith's extensive research offered insights. "Smell that smell? That's the airship smell: The sweet and heavy scent of aircraft dope, the sour, animal odour of the gas-bags. There's a whiff of petrol and everywhere the indefinable bouquet of hydrogen. The text-books tell you that it's got no smell but that's rot. It's all around us, leaking from the bags or venting from the valves. Get too much down you and your voice goes daft, too much more and you pass out cold." Fantasy? Peter Andrews wrote, "The British R-38 was a leaky ship that used a million cubic feet of hydrogen a month – about 40 percent of its capacity." Shute reported, "During these two flights R.101 grew steadily heavier… she seemed to be losing lift at the rate of nearly a ton an hour. At the time it was thought by the Captain of the ship that this was due to the oversensitive gas valves chattering on their seatings…" (Elsewhere Shute said "The gas valves were of a novel design and reported to be very sensitive; being located on the sides of the bags a small degree of rolling would cause them to open and lose gas.") "…but more probably it was due to the chafing of the gas bags against the girders of the ship since the wiring had been let out to increase the volume, which had resulted in so many holes."

How were such operations possible when we see Diehl quoted saying "A mixture of 36 percent air in hydrogen is probably the smallest concentration of air within a balloon that could be ignited from small sparks near the surface of the envelope." Although Diehl ignored the fact that with nearly 40% of its lift diluted, no airship could fly, he suggests a fire could actually happen if ignition sources were just close! With these high leakage rates and ever-present electrical activity, how is it that hydrogen ships did not blow up on ordinary missions, or were able to complete any missions without bursting into flames?

The perception of even slight leaks being deadly is not borne out in practice. According to the 1940 manual, "At 77°F (25°C) and 30 mm. pressure the diffusion of hydrogen through balloon fabric (rubberized fabric) varies from 8 to 15 liters of hydrogen per square meter per 24 hours." Yet the LZ-127 *Graf Zeppelin* has an unmatched record of safe performance motoring along on spark-ignition engines fed by gaseous fuel kept in bags, sampling every kind of weather all over the world as she flew 1,060,000 miles, crossing the Atlantic 144 times. How is it that in today's "wisdom" she has become a ship of fools, an *uninsurable floating bomb*? She might have been called "lucky" – but her success was not by chance. Prudent precautions were always taken, especially on the ground.[113] In an unpublished manuscript[114], the *Graf's* last Captain, Hans von Schiller, put fears to rest. "We hasten to assure all who may still have fears based on deplorable mishaps in the past that there is not the remotest danger of the ship's 'going up in flames' so long as it is in the hands of experienced officers and a well trained crew. If there is any secret to explain the success of German lighter-than-air aviation, it is simply that the whole ship is constantly subjected to the most careful inspection; even the tiniest leakage of other disturbance

---

[112] D. White, "First Radio Over the Atlantic," *short wave Magazine,* Vol. 54 issue 7, July 1996

[113] A popular photo shows Navy and other officials in a car ready to take LZ-129 dignitaries away from their Mines Field landing site. Not as popular is a photo run in FIRST TO THE FLAMES: THE HISTORY OF FIRE CHIEF VEHICLES (E. L. Peterson, Krause Publications, 1999) showing an armed fire chief enforcing a strict no-smoking zone on August 26, 1929.

[114] H. von Schiller, A MILLION MILES IN A ZEPPELIN, kindly provided by Wilbur Cross.

is immediately eliminated, and the men are sufficiently rested and alert." "On 8 September 1915, and without having been hit itself, the airship [SL-2] returned to Germany; it had been in the air for 13 hours. During the 980-km trip to London and back it had consumed 2,500 liters of fuel and 45 liters of oil. When it touched down in Germany, the airship had excessive buoyancy. The first pilot pulled the valve at a height of 2,000 m to release gas. The airship went down to a height of 70 m. Suddenly it descended rapidly. The crew didn't have time to release enough ballast. The airship crashed onto a house. Its smoking chimney was the first thing to pierce the airship's hull. Within a short time the gangway was filled with smoke. There was a great danger of explosion. Captain Schramm, the engineer Nordhorst and some other crew members jumped from the nacelle, ran to the house and put out the fire." [115]

Consistency of operation and the experience that comes with trained crews has become the lost art of the hydrogen airship. However, the problem of leaky cells and ever-present electricity was solved by the later rigid airship's basic design. The need to allow the air pressure outside an aircraft to equalize with that inside is not familiar today, in the age of even smaller airplanes being pressurized. Yet the Space Shuttle Orbiters still use such a system. On accent, vents in the unpressurized aft, mid, interhull forward and wing sections are closed to protect the interior from the punishing noise of lift-off. They are then opened to allow air to escape as the vehicle climbs into the stratosphere, least the enclosed air is trapped and would fracture the structure. The doors are also opened following the blistering re-entry, least the pressure of our atmospheric ocean crush the structure like an aluminum soda can. The principle is familiar to anyone who has ridden in a fast elevator. They have noticed their ears "popping," as air pressure behind the eardrum tries to equalize with the air pressure outside.

The experience of Naval Airship L51 revealed the problem that showed up after the topside spine was varnished, as her skipper later recalled in Karl Arnstein's biography: "We were descending from high altitude and, with the outer cover completely and tightly Cellonized, there was no way for the pressure to equalize. First thing we knew, there was a shriek of tearing metal amidship and then a grinding crash. The pressure outside had caused the longitudinal girders in the lower midships portion to bend inward between the main frames, and they fractured in many places..." From the point L51 limped back to base, Arnstein's later Zeps were equipped with a robust interhull vent system that insured adequate air exchange. "There are however, openings in the hull so that air may flow readily into or out of the ship to maintain the inner and outer airship pressure at the same value."[116] While this necessity had the side benefit of diluting seeped gas, the outer cover represented little barrier to hydrogen compared to the cells. Typical losses due to cell permeability naturally sailed right through the outer cover "in as straight a pathway as it could find." In the case of unusually large leaks, such as those caused when cells were rubbed through against the structure (enough to need holes patched, as recorded) the air exchange setup would have easily diluted the leak on its way out the vent.

If during that time a flammable mixture was available, it would need only an ignition source to instigate combustion. Here again, basic design prevented atmospheric electricity from initiating ignition, even when struck by the lightning bolts that had been so powerful as to have damaged girders. Von Schiller had first hand experience. "During the war the L-11 was caught in a violent storm and struck by lightning many times, mostly on the nose, where it was found afterwards that the metal had fused. The framework of the ship offered such a large metal surface that no

[115] Karl-Heinz Hoffmann, "The Airship Constructor Schuette-Lanz and his Rigid Airships," 3rd Int'l Airship Conference.
[116] AIRSHIPS TRANSATLANTIC TRANSPORTATION, Summer 1936 Lakehurst Souvenir book

further damage was done, and the electric charges left the ship immediately through the exhaust of the motors, which consists of pure water-vapor." Later improvements provided more protection, as we shall see in Chapter Ten.

Said the *Literary Digest*: "If you want complete immunity from lightning, sail on a Zeppelin, where lightning cannot enter. The 803-foot long hull is composed of huge rings made of aluminum alloy. Every section of these rings and the longitudinal girders are connected so that the whole outer structure of the hull makes a single electric conductor." Even the most powerful lightning bolt would pass harmlessly through the structure on its way from one potential to another. While the main idea was to prevent structural damage, it also had the benefit of not allowing the powerful electrical forces of nature anywhere near the hydrogen gas.

In the *Graf Zeppelin's* near decade of operations from the Artic to the Amazon, its interhull vent setup was challenged with violent up and down drafts as well as outer-cover shredding weather fronts. Its ramie cord netting failed to protect its cells on occasion, necessitating cell patching in these rub-though areas. If that wasn't dangerous enough, the ship's engines ran on propane-like fuel gas also kept in fabric-based cells. *Graf* was struck by lightning countless times, causing cover maintenance headaches as holes had to be patched. Flying with storms was aggressively pursued. "At one point, the Zeppelin took advantage of the south-side fringe of a typhoon, which boosted her [ground] speed to a temporary 98 mph."[117]

It cannot be overemphasized that the basic nature of an air-displacement vessel demands that its displacement medium be free of contaminants. A small percentage on such a large scale would have a measurable reduction in the ship's payload. So, keeping the gas free of oxygen came part and parcel with purity checks. The 1940 manual explains, "Balloon gas must be tested to determine its lifting power and its purity. The lifting power is determined directly by the determination of the specific gravity… Therefore if the specific gravity is within allowable limits the determination of the oxygen content would be unnecessary to determine whether or not the gas was safe for handling."[118] If there were some flaw in the scheme of the hydrogen rigid airship, one would be hard pressed to name parameters not explored by the *Graf Zeppelin*. Yet, yesterday's marvel of the age was somehow transformed into today's ignorant folly. In the most outrageous case of apparent mental illness, *how did men bearing scars from the LZ-129 fire calmly board the next hydrogen filled Zeppelin and fly it?*

All these seemingly foolhardy men had experience and knowledge we lack in our modern age. Also not aware of the hype, crews flew the ex-US Navy L-19 airship for ten years around Europe, freely smoking cigars and cigarettes![119] A single engine hydrogen filled airship was built late in 1956, but was storm-torn off its mast. Then, a twin-engine hydrogen non-rigid was actually built by Zeppelin *Metalwerke* with its envelope constructed by *Augsburger-Ballonfabrik.* It began flying for Trump candies in August, 1958. Despite various mishaps, even deflations, it flew 3,400 safe hours under hydrogen until decommissioned in 1963. Of course free balloons, naturally more prone to air contamination, have had mishaps. Anthony Smith told of the demise of his balloon *Jambo* in the book THE DANGEROUS SORT. Yet in hydrogen balloon flying since 1950, with some 28,000 flights in Germany alone, there have been six

---

[117] E. Niderost, "Globe Trotting Leviathan," *Aviation*, July 1993, pg. 23

[118] The manual describes the various absorption methods of determining specific gravity, including the use of alkaline pyrogallol, or ammonia-copper solution, and the effusion method.

[119] Author's phonecons with *UNDERBURG* pilot, the late Charles Probst and ex-USN training pilot Lou Prost. Ironically, this airship became a total loss following its being re-filled with helium.

accidents resulting in fire due to electrostatic ignition; one of these involved fatalities. Since that one fatal accident, fabric for gas balloons must meet minimum conductance standards. If the pilot refuses to use the rip panel, the valve has been known to strike the ground and make sparks. Most care has to be taken upon deflation, although the gas escaping from the top rising at 60 meters per second has little time to be a flammable mix.[120]

What about hydrogen *explosion*, the words so often used but no one seems to have actually witnessed? On July 28, 1934, Army LTA men Kepner, Anderson and Stevens lifted off for an altitude attempt only to have their three million cubic foot hydrogen balloon tear at 60,000 feet. Valving and preparing to abandon ship as the vehicle dropped to 20,000 feet, the entire bottom of the balloon ripped away. The top retained some hydrogen, but racing downward it became impregnated with air, and of course electrical energy at those altitudes provided ample ignition sources. "The final disintegration of the upper portion of the bag was caused by the explosion of the hydrogen-air mixture which it contained... A hydrogen flame is almost colorless and could not have been seen under the circumstances. Finally, it would not set the balloon on fire or even scorch the fabric. None of the fragments in the tests at the Bureau of Standards showed any signs of scorching."[121] The three men helped each other out, then parachuted to safety, probably wondering why people later wrung their hands about the danger of 'hydrogen explosions.'

So how did we come to villainize nature's most simple and abundant substance? How did it become the nemesis of all things lighter-than-air? How did it become the "criminal element?" Surely not for so shallow an excuse as the yellow journalism of the newsreels, like Movietone News, proclaiming, "DIRIGIBLE HINDENBURG EXPLODES! 50 KILLED!" [122]

"Airships of World War I had to keep the car well away from the envelope to prevent sparks from igniting the highly inflammable hydrogen used as their buoyant gas," claimed *The Airship Car* training manual. Another booklet in that series stated the expected leakage rate of the type of envelope shown here was half a cubic foot per square yard of envelope every day. The ex-L-19, here being assembled, enjoyed ten years of service with greatly improved performance, under hydrogen. Courtesy pilot Lou Prost.

---

[120] Courtesy of Markus Haggeney of Germany.

[121] "Report of the Joint Board of Review Appointed to Investigate the Cause of the Accident to the Balloon "Explorer," 1934 courtesy Eric Brothers.

[122] NARA footage, listing the photographer as 'Al Gold,' likely the senior man. Author interviewed Charles White, who worked at Paramount on 44th Street in NYC in 1937. When film of the crash was brought in, Charlie recorded the narrator on a Vitaphone record, and was asked to come up with some background sound for the edit. Charlie had another Vitaphone record of a fire in Philadelphia, so he used that. Other sound stripers were not so levelheaded, adding horrifying screams and explosions. Like most of the film in the newsreels showing LZ-129 in clear skies actually from 1936, the sound is not real either. Charles White later worked from Grumman on the Lunar Lander.

Chapter Three: The Plot Thickens by Twice

The semi-rigid airship, which can be defined as a blimp-like envelope attached to a rigid keel, offers a different set of parameters distinct from the pressure or rigid airship. To proponents of this design, it is the only way to fly, the perfect mating of the rubbery bag and the unforgiving hardware of engines and cargo. A study of many early airships reveals 'keels' spreading the weight across the length of the net. There were 27 Parseval semi-rigid hydrogen-filled airships, one having been purchased by England and on her side when the Great War started. French Lt. Stapher wrote in 1921 the obvious solution to the blimp's battle of balancing the opposing forces of buoyancy and gravity was to decrease the shearing stresses and bending moments inherent in the pressure airship design. "Thus we arrive at the semi-rigid type, which accordingly appears to be the result of improvements applied to the non-rigid type and which offers, moreover, along with other advantages, increased facility of maneuvering."[123] When he turned his complete attention to blimps, master airship designer Dr. Karl Arnstein created the M-ship, incorporating many advantages of the semi-rigid design. It featured a 117-foot long, three-part articulating car that spread its load gently across the catenary of its some 300-foot long envelope. Four were built before the US Navy lost interest, which also canceled a scaled-up next generation proposal, the O-ship, that would have carried a scout/bomber airplane. The semi-rigid philosophy was adopted by a new generation for the Zeppelin NT and the giant *Cargolifter* airships in Germany as this book is written.

Meanwhile, helium had been discovered via solar spectrograph by a British scientist serving in India back in 1868. He logically named it for the sun god "Helios." It was little more than a laboratory plaything after being isolated in 1895, and an annoyance to townspeople trying to light their natural gas found laced with helium in 1905. Ironically, for the way things turned out, it was a 1916 British request that set Canadians and Americans in motion to isolate and concentrate the so-called "bullet-proof" gas. The concentrations in Canada were too weak to allow practical extraction, and helium has yet to be harvested from the Turner Valley Field in Alberta. After scientists figured out how to extract it from the richer American fields, a large infrastructure was built. The first shipment of helium was being loaded on a steamer when the Armistice was signed, so it was not tested in that conflict. The accident of a British airship might not have by itself lead to the widespread belief that the lifting gas must be changed, but a second catastrophe, this to a semi-rigid airship, had a more profound effect.

The great bulk of early semi-rigid development took place in Italy, and they have a proud history spanning many continents. "The inventor of the characteristic Italian semi-rigid airship was the engineer, Enrico Forlanini, who constructed his first ship, the *Leonardo da Vinci*, at Baggio near Milan in 1909."[124] The Italian semi-rigid *Norge* is still the only airship to have flown across the top of the world. The UK purchased one of their semi-rigids, dubbing it S.R.1, and a British crew reported to Rome to fly it home of the 28th of October 1918. As the Airship Heritage Trust tells it, "It was not an easy flight, the crew were too busy to sleep as they had to battle strong headwinds and rain for much of the trip. The ship had mechanical failures, [including] a fractured oil line which spewed hot engine oil throughout the control car. The crew worked frantically to repair this damage but [the oil] turned the control car floor in to slippery platform resembling a skating rink. Comical though it may sound, in an undulating control car in the air the crew had to take great precautions to avoid anybody falling overboard. To further the

[123] *Premier Congrès International de la Navigation Aérienne*, Nov. 1922

[124] C. Dean in the Italian section of HOUSING THE AIRSHIP, The Architectural Association, London 1989

discomfort on the flight, at one point the exhaust manifold fell off the midships engine above the control car, burnt through on to the petrol tanks below. The red-hot manifold lay in close proximity to hundreds of gallons of high-octane aviation fuel and beneath almost half a million cubic feet of equally inflammable hydrogen. Captain Williams and Petty Officer Leach lept up a ladder and pushed the burning mass overboard, at the same time brushing out glowing sparks with their hands." Nonetheless, S.R.1 became the first aircraft to fly from Italy to England.

US War Department officials sought to purchase what seemed like high-tech imports whose expertise dated to the first use of airships in war in 1912. Historian James Shock explained the only semi-rigid the US Navy operated was the Italian O-1, built by the Stabilimento Construzioni Dirigibili de Aerostati of Rome. It had a short but interesting operational life with the Americans from its first flight in September 1919, including launching towed gliders from 3,000 feet in July 1921. Suffering dual motor failure on a return trip to Cape May, hydrogen was valved and the ship was successfully landed amid peach trees. When requested help was refused, the entire gasbag of hydrogen was discharged; it probably caught fire, but no one noted it. Affectionately known as "The Wop Ship," it was not re-erected or flown. Yet, another Italian semi-rigid in US hands would have a profound effect on all buoyant vehicles that followed.

The *Roma's* rather questionable tail assembly in happier days, and what was left after it flipped over on adjacent power lines at the Army Quartermaster Depot near Hampton Roads, Virginia. Courtesy LTAS.

The US Army purchased the *Roma*, the huge Italian semi-rigid, when Army Col. William Hensley's mission failed to buy a Zeppelin. *Roma* arrived from Italy very moldy, with the last layer of paint peeling off the envelope. Repair crews applied hundreds of patches, but actually repaired only the most obvious leaks. On the first American flight a small door flew off into a prop, resulting in envelope-puncturing shards piercing the bag. The envelope shredding was so bad the in-flight repair crew had several men *overcome by hydrogen gas*, but they recovered and saved the ship.[125] There were other misadventures before her unreliable engines were replaced,[126] but *Roma* continued to suffer from mechanical problems before her fatal plunge into power lines on February 23, 1921. The official investigation produced three possible causes of the accident, and the word "hydrogen" does not appear in any of them. Neither explanation of the most probable causes of the loss of control and subsequent impact had anything to do with the lifting gas. General Mason Patrick had written "Billy" Mitchell that the fabric "was getting pretty rotten" and (later Lt General) W. E. Kepner believed that "a seriously deteriorated bag" caused the crash. When John Mitchell interviewed participants, their recollections of the accident were still vivid. Major Reardon said, "I jumped for this opening and with my bare hands tore away the flaming fabric." Master Sergeant Harry Chapman calmly took his knife from his pocket and cut

[125] H. Van Dyk, "The U.S. Army Dirigible *Roma*," *Gasbag* No. 42, Dec 1999.
[126] At the USAF Museum in Dayton, Ohio, the small LTA display features a *Roma* engine.

away at the burning fabric, and suffered the most extensive burns of any of the survivors.[127] Because the *Roma* had come down in the path of high voltage wires held up by telephone poles, it was assumed the arcing cable end set the nearly pure hydrogen afire. Yet, having only been flying for an hour, tons of gasoline was available to splatter from the impact, ignite from the broken high-tension line, and turn the victims into what looked like typical air and automobile fire crashes of today. The grisly task of removing what witnesses called "blackened charred chunks of humanity" was unforgettable. Kepner's recollection of "Arms, legs, heads burned away... a head shrunk in heat and burned to the size of a fist..." is consistent with gasoline conflagrations. Why was there little mention of gasoline in the accident's history?

Edward Wagner, Aeronautics Professor at MIT, urged readers[128] to be calm in spite of the accident's arrival just after the cream of the Navy's airship corps had been wiped out with the ZR-2. "The catastrophe at Hampton Roads, however, even following that at Hull, should not blind us to the real potentialities of the airship and to its real usefulness in the future... the preceding 20 years show hardly a single accident of the same sort." However, the funeral procession and the bandaged burn victims were heart wrenching, demanding some sort of action. Never one to miss a media opportunity, Congress held an open-door hearing, inviting the press.

A member of the NLHS posted an interesting comment to the airship e-mail list. "In his book HELIUM: CHILD OF THE SUN,[129] C.W. Seibel relates how a demonstration was conducted before a Congressional Committee in December, 1922, using a small red toy balloon inflated with hydrogen, a small yellow toy balloon inflated with helium and a burning taper. The taper merely seared the helium balloon, but the red hydrogen balloon exploded violently, rattling the windows and making a few of the attendees jump from their seats. Seibel, himself, then only a young man at the start of a brilliant career with the Bureau of Mines, merely grinned and shrugged when later confronted by one of his superiors about the possibility of whether (maybe, just maybe) somebody might have added some *air* to the red hydrogen balloon!"

The message continued: "Some of Harold Dick's writings indicate what I tend to believe is [G-Z President Paul] Litchfield & Goodyear-generated hype about the "dangers" of hydrogen; I again take much of this as a concerted "push" to make sure that America... with it's "monopoly" of helium gas... and Goodyear... with its heavy investment in an airship building plant, design staff, ... would be seen as "logical" components of any world airship enterprise."

All the propaganda had the desired effect on Congress. The message closed with: "One Congressman said 'We can't make the Army and Navy use helium, but we can say that none of the money we appropriate can be used for airships filled with hydrogen...' " So with no trial, nature's simplest element was made the scapegoat. From that day forward it would be forever ostracized. "After the loss of the Italian-built *Roma* by hydrogen fire resulting from striking a high tension line near Langley Field in February, 1922, a policy of "helium only" was fixed for U.S. Army and U.S. Navy airships."[130] No one thought the action would forever change the course of an entire technology. It sounded like a great advance of science over nature. In a 1919 *National Geographic* article, G. Sherburne Rogers proclaimed "Helium, the new incombustible gas which promises to revolutionize the science of ballooning... and this country thus has a powerful advantage in the competition for supremacy in the air which the next decade is bound

---

[127] J. B. Mitchell, THE ARMY AIRSHIP ROMA, Syms-Eaton Museum, Hampton VA 1973, courtesy John Miering
[128] *The Christian Science Monitor*, July, 1922, courtesy Eric Brothers.
[129] C.W. Siebel, HELIUM, CHILD OF THE SUN, The University Press of Kansas, 1968.
[130] THE AIRSHIP CAR, Manual No. 4, Lithographed at USNAS Moffett Field, CA, 3-11-44

to witness." Dr. Richard K. Smith offered a more sobering view, "...it was the *Shenandoah* which taught the Navy and indirectly the world, that however "safe" helium was by virtue of its nonflammability, it was no panacea. Indeed, helium created its own peculiar problems to afflict airship operations."[131] Peter Andrews would detail, "She had been designed for hydrogen, and although helium produced 92 percent of the lifting power of hydrogen, its use reduced the *Shenandoah's* range by almost 40 percent, which effectively robbed the ship of its long-range reconnaissance capability." Perhaps we can understand why that airship's last Captain, who had trained on hydrogen-filled ships, held private reservations that the helium experiment would work, even while publicly supporting the Navy's position.

The first problem was cost. One of Rogers' photo captions admits that the typical 35,000 cu ft observation balloon, which had only cost $300 dollars to inflate, would then cost $4,000 to inflate. The caption also conceded the R.33, then the largest airship in the world, had only cost about $23,000 to inflate with hydrogen. Small wonder only four rigid airships would ever be inflated with helium – and of them, but for a very brief period with two, only one flying at a time. According to A. F. Simpson's research, Army LTA expenditures for helium in 1921 were $668,500,00, when the previous year Army LTA *entire appropriation* had been $692,548.54! Simply put, helium lift cost *70 times* the equivalent hydrogen lift.

Helium's enormous cost vs. the cheapness of hydrogen cannot be understood until one appreciates their origins. Pure gaseous hydrogen is not common on Earth, but it can be quickly created with easily manufactured machinery fed by abundant raw materials. Making it even less expensive is its bounty as a by-product of other manufacturing process, such as steel making. Its cost is therefore only reflective of the expense of gathering, storing and moving it. Gas for Zeppelin replenishment at Lakehurst was purchased from Esso, the same company that sold them the LZ-129 engines' diesel fuel. The gas was a by-product of oil refinement and was shipped down by railroad from the Bayway refinery near Linden, New Jersey. The *distribution* of energy is more highly visible, has a greater environmental impact, and usually costs more than the energy itself. The cost, which includes intermediate storage and transportation, will always be kept in check by the ease by which it can be manufactured at the point of consumption. History shows us oftentimes it was more practical to make the gas at the point of use, meaning the supply was virtually unlimited. In other words, there was rarely a question of not being able to get the gas, only the cheapest option for obtaining it. The War Department's 1940 manual explains, "Hydrogen is prepared almost exclusively for practical commercial purposes by the decomposition of water [one-ninth of water's weight is H_2] either directly or indirectly." In those days, the more efficient and more portable was the silicon method. "The process in use by the Army for the manufacture of hydrogen for balloon and airship work is the silicon process, in which ferro-silicon (an alloy of iron and silicon) reacts with a strong, hot solution of caustic soda (sodium hydroxide, NaOH)... [it] depends on the reaction between silicon, sodium hydroxide and water with the formation of sodium silicate, called water glass, and hydrogen."

"A popular delusion has been built up over a period of years concerning the avialability of helium."[132] "The cost and availability of helium both suffer from the problems inherent in any by-product, in that both the price and the availability of helium depend on the demand for the primary product, natural gas. In times of high natural gas demand...production may exceed the extractive capacity of the helium plant, forcing well operators to bypass the helium plant and

---

[131] Dr. R. K. Smith, "The Airship in America 1904-1976," TWO HUNDRED YEARS OF FLIGHT IN AMERICA, AAS History Series, Vol. I, Univelt, San Diego, CA 1977

[132] F. Buckley & R. Barkley portion of *The Rigid Airship in Peace & War*, privately printed, 1942

send precious helium up the chimney...[or] helium must be stored or dumped at a low price to make room. If natural gas demand is low, helium is scarce and of course expensive."[133] This price is unlikely to be kept in check by any alternative. The Rogers article claims one Kansas field was "over 2 percent," but later says the largest field, Petrolia, 100 miles north of Fort Worth, features gas at the 1,400 to 1,600 ft. levels that "...carries practically 1 percent of helium, which is considered a good working proposition." Again a reality check is in order. "Nearly all the world's helium supply is found within a 250-mile radius of Amarillo, Texas (the helium capital of the world). A byproduct of billions of years of decay, helium is distilled from natural gas that has accumulated in the presence of radioactive uranium and thorium deposits...The Kennedy Space Center uses 75 million cubic feet annually. Liquid helium, which has the lowest melting point of any element (-452 degrees Fahrenheit) cools infrared detectors, nuclear reactors, wind tunnels, and the superconducting magnets in MRI equipment. At our current rate of consumption, Cliffside will likely be empty in 10 to 25 years, and the Earth will likely will be virtually helium-free by the end of the 21st century."[134] Since later on Texas helium was mainly found in useable quantities in the Cliffside site, Congress had to be lobbied to pay for an elaborate liquid air extraction plant. This was only the beginning of a strange governmental functionary that is still the subject of regular ridicule – and political mudslinging.

A 1995 TV report is typical, as is this news item for sale by NBC: "The Fleecing Of America: Amarillo Helium Plant Is Costing Taxpayers 120 Million Dollars A Year. Man opens chamber in field showing helium plant pipes. Helium plant manager Moore says it's space age material but a lot of people think of it as World War II. B&W clip of US Navy blimp floating over aircraft carrier seen. Moore tells Cummins plant has 31 million cubic feet under the ground. Borrowed 250 million never paid back; 1.4 billion dollars; annual interest 34 million. Congressman Cox says 'For so long it's been losing money. We don't have any substantive reason for government to be doing this any longer.'" "In 1996, Congress ordered the Bureau of Land Management to stop recovering helium and to sell the reserve. As a precaution, it also asked the research council to keep tabs on helium use and pricing. 'If it's so strategic, then why allow it on the novelty market? That should be a clue it's not strategic,' [Air Products & Chemicals helium business manager Curt] Maier said."[135]

Second, all the billions of dollars expended on the impossible dream of total fire safety came at a considerable price in performance. No more simply topping off with fresh gas; *repurification* was now necessary for efficient lift. As Buckley and Barkley wrote, "Helium operation necessitates the frequent return of airships to bases equipped with purification facilities – in one case severly restricting the availability of the airship, in the other increasing the expense of the base equipment." Optimistically, G. Sherburne Rogers conceded: "The only apparent disadvantage of helium is the fact that it is about twice as heavy as hydrogen, 100 cubic feet of hydrogen weighing 17.8 ounces and the same volume of hydrogen only 9 ounces... this is of little practical importance, the buoyancy, or lifting power, of helium being 93 per cent that of hydrogen." Indeed, the only directly admitted change during the transition seems innocent enough. James Shock found a Goodyear record of a type UA, 190,000 ft^3 envelope, D-72, ordered for Navy airship C-7 noted to be "for helium inflation" though her normal envelope was 172,000 ft^3. In metric terms common figures state hydrogen weighs 88 grams per cubic meter. At STP, air weighs 1275 g/m^3, subtraction gives us 1187 g/m^3 lift per cubic meter. One would

[133] F.M. de Piolenc, "Why Bring Up Hydrogen Again?" *Aerostation* March 1999
[134] E. Jenkins, "A Helium Shortage?" *Wired*, August 2000, pg. 162
[135] G. Terrell, "Up Up and Away," *Photonics Spectra*, August 2000, pg. 58

assume using helium weighing in at 180 g/m^3 would deliver 1095 g/m^3 lift, or only a 7.4 to 8% lift penalty.[136] Oh, that it were that simple.

As Robert Wilcsek wrote, "While helium has 93% of the lift of hydrogen, *it weighs twice as much*. Therefore, the first obligation of helium over hydrogen is to lift itself." To put another way, "Lift depends on the weight of air which is displaced, less the weight of gas displacing it. Air weighs around 75 pounds (34 kg) for 1000 cubic feet (28 cubic meters), while a similar amount of hydrogen weighs only about 5 pounds (2 kg), so giving some 70 pounds (32 kg) of lift. The same amount of helium, by contrast, weighs about 10 pounds (4 kg) and so gives 65 pounds (30 kg) of lift."[137] However, the problem is far more complex than just the simple difference between the lift of one gas versus another.

With gasoline tonnage spread along the lower keel, a large percentage of the hydrogen gas kept in the cells (being filled at right) was needed just to lift the motive power. A 1-inch brass main fill line fed the system, with intake points on the nose and lower keel. LZ-126 photos courtesy NAA.

It must be remembered the weight of an airship is not subject to drastic change. So in an airship designed for hydrogen, substituting helium greatly reduced the only thing that could be changed, its *payload*. P. L. Teed's paper comparison was for a five million cubic foot airship in which helium would lose the airship 12.1 tons of gross lift, or a reduction of 7.45%. Useful lift showed a 12.1-ton reduction, but that was a 13.85% drop from hydrogen. Smaller ships could not be helped. As James Shock explains in his forthcoming Army Airships book, "The OB-1 was originally inflated with hydrogen, however on August 31, 1923 the Chief of the Air Service Engineering notified the commanding officer at Scott Field that the OB-1 was to be helium inflated. Test flights were made by Lt. Robert Robillard which proved unfavorable due to loss of lift (helium being heavier than hydrogen) and the control car had to be repositioned twenty inches forward to overcome tail heaviness. The airship also experienced reduced ceiling and endurance with helium. It was apparent a complete redesign of the airship would be necessary..." Now difficult to control, there were two deflation landings. "The OB-1 was never reinflated and was declared surplus at the end of 1923." Not surprisingly, small airships were sent to the scrap

[136] After some discussion the Airship Association published H_2 lifts 1.139 kg/m^3, He lifts 1.055 kg/m^3.

[137] P. Abbott, AIRSHIPS, Shire publications, Ltd., Buckinghamshire, 1991

heap. "...the policy adopted in 1923 to inflate all airships with helium gas instead of hydrogen, which necessitated increase of volume of all existing ships in order to have the same useful lift as originally designed with hydrogen... This policy made impracticable the further production of any airships of the A-4 or Class C types..."[138] Everyday leaky components and permeability losses generally called for a complete volume of replacement gas over a 12 month period. Shortly the U S Navy was all but out of the blimp business, keeping only the J-3 and J-4.

Teed also said the 7.45% decrease in gross lift comes with a reduction in the airship's maximum altitude by 2,200 feet. Indeed, when the US Army ordered spherical balloons to be switched to helium, those stationed at the Pope and Post Fields could no longer reach the required altitude.[139] Nor was *that* the worst part.

Airship guru Charles P. Burgess explained the harsh realities of the problem in his 1927 book AIRSHIP DESIGN.[140] F. Marc de Piolenc detailed the book section covering lifting gas effect on performance in *Aerostation*, March 1999. [141] "It is common practice in America to take the unit lifts of hydrogen and helium as .068 and .060 lb/ft^3 in the standard atmosphere at sea level.[142] Both units are conservative. From these figures, 11.8% of the gross lift is lost by the lift of helium instead of hydrogen; but the percentage of losses of the useful and military or commercial load are much greater because the weight of the ship empty is a fixed quantity, and the absolute losses of gross and useful lifts are therefore equal." In the example given, a mathematical formula proves that the airship's useful load is decreased by 31%, and the loss of commercial load is an astonishing 59%!

This seems a very high, seemingly unacceptable figure, and it comes as a result of totaling all the individual performance decreases. The LZ-129 was flying when he started writing his two hundred and sixty-fifth design memo, but its wreckage was still lying at NASL when Burgess released the memo comparing the performance of commercial airships inflated with hydrogen vs. helium. Ignoring water recovery equipment weight, the memo's chart showed how a *seven* million ft^3 hydrogen ship would actually exceed the ton-mile payload performance of a *nine* million ft^3 helium ship. How? Remembering the airship was at first an air displacement vessel, the more gas providing lift, the greater the performance. As von Schiller wrote, "Consequently, the hydrogen cells are filled to capacity in order to create the maximum lifting power. It is really "lighter than air" and rises from the ground when released. But since the gas tends to expand under the reduced atmospheric pressure of the higher altitude, the gas cells would burst were it not for the over-pressure valves which let out the requisite amount of gas. Thus when the *Graf Zeppelin* rises to a height of 200 meters (the height usually chosen for reasons of safety), it must "blow off" 2% of its hydrogen, or 2000 cubic meters." A helium airship loosing 2% of its gas every launch was then, and is now, completely out of the question. The solution, while not very elegant or particularly clever, was simple enough: don't fill the balloons all the way. So with helium, the air-displacement vessels simply did not displace as much air, further decimating their payload capabilities. Burgess wrote "A further loss in performance from the use of helium follows from the necessity of starting flight with only partial inflation in order to avoid valving the costly gas as the altitude is increased." If only a displacement vessel that could no longer displace a full complement of air were the last of the problems; helium caused more yet.

---

[138] T. Blakemore, W. Pagon, PRESSURE AIRSHIPS, Roland Aeronautical Press, NY 1927, courtesy Hep Walker.
[139] Courtesy of James R. Shock by letter of 19 July 2000.
[140] C. P. Burgess, AIRSHIP DESIGN, Ronald Aeronautic Library, The Ronald Press Company, NY 1927.
[141] Inside of *Gasbag Journal* #39
[142] Loose Metric rule of thumb: One cubic meter of H_2 lifts 1.1 kilogram. One m^3 of H_e lifts 1 kg.

R. K. Smith continued, "Airships were never permitted to be as free with helium as they had with hydrogen. Airships were not permitted to take off with their cells 100 percent full; a 10 percent slackness was mandatory to leave space for expansion at altitude so helium would not be valved off automatically...Helium was [fire] safe. But it also possessed substantially less lift than hydrogen." Nor could you use as much of it in the same sized vessel.

The new lack of performance was hard to swallow. One could easily point to the LZ-126, which flew from Friedrichshafen to Lakehurst with hydrogen, but after borrowing helium from her sister ZR-1, could not re-cross the Atlantic. "Unfortunately, the helium that lifted the *Shenandoah* also kept it from going to the North Pole."[143] In a 1925 proposal by Goodyear to take over the operation of the LZ-126 to establish a sort of airborne merchant marine, options included "...the use of hydrogen as the lifting gas, and emphasizes the success and safety of the German Zeppelin-Rederei in commercial operations with hydrogen." [144] Master designer Karl Arnstein suggested adding 10% hydrogen to the ZRS-4 helium, to add 4600 lb. of badly needed lift.[145] The Government publicly backed away from this idea, coming so soon after the crash of the R.101, but even a high school chemistry student would realize this dual gas mixture would still be fireproof even if slightly air-contaminated. Why not do it?

There was a body of opinion within the naval airship organization in the 1920s that, particularly during wartime operations, the ships should be re-inflated with hydrogen and the water recovery units be put into storage to have much better performance and versatility of operations. Ironically, when the shooting started, the rigid airships might likely have been refilled with hydrogen, negating the millions spent to develop a so-called "bullet-proof" gas!

National Geographic writer Rogers was sure with helium "...the speed and cruising radius of the craft may be materially increased." But Teed's calculation showed just the opposite, since a helium filled airship could not also burn at least some hydrogen in its engines, "the maximum length of full-speed flight of the airship is reduced 40.6%." To understand why this huge difference – *more than 40%* - is factual beyond just what energy is lost by not using surplus lift energy as fuel, we must realize that hydrogen airships always had the luxury of flying *at equilibrium*. Flying nose up for lift, or nose down to prevent an upward stall was the *exception* rather than the rule. In his Design Memo #390, C. P. Burgess explained 'Since the drag of an airship at any given speed is greater when flying heavy than when at static equilibrium, the gain in endurance by heavy take-off is somewhat less than the extra load of fuel divided by the hourly consumption when in equilibrium...from a combination of increased air resistance and diminished propeller efficiency when flying heavy... It appears reasonable to expect about 60% greater consumption when [required attitude gives a lift coefficient of 0.10 rather than at no lift at EQ] at any given speed. Furthermore, the increase in consumption varies approximately as the square of the heaviness." He concluded, "Fuel consumption is a minimum when an airship is in static equilibrium." The NACA reached the same conclusion. "Any flight angle other than zero angle of pitch results in an increase in drag; hence more power and greater fuel consumption are required for an equivalent speed."

So not only was a helium ship's larger size less fuel-efficient, but in operation a helium ship could never enjoy the lowest drag of level flight. Flying heavy was tougher on the airship as

143 C. Roberts, "Ship's Vain Quest," *Aviation Heritage*, March 1991, pg. 43.
144 *Aerostation* 23/2, in *Gasbag Journal* #44, June, 2000
145 Eric Brothers located this clipping dated Dec 27, 1930, titled "Moffett Denies US Navy Plans Diluting Helium," while researching the master's biography.

well. "The *Sparrowhawks* were ferried cross-country, fortunately on the eastbound trip, since *Macon* suffered some structural damage in the area of her port fin while flying heavy in hot, turbulent weather over west Texas."[146] NACA had warned, "...the hazards of handling an airship are greater when it is not in a state of static equilibrium." However, driving the bag into the wind at an inefficient angle was not the most serious problem.

The only option was to design larger airships to hide the helium penalty. Cost aside, this does not seem at first to be a great disadvantage. As far back as 1909 *McClure's Magazine* pointed out that enlargement was common in waterborne ships. "This is due to the mathematical fact, similar to all shipbuilders, that the lifting power of ships increases according to the cube of their dimensions, while the resistance on their surfaces increases only as the square of their dimensions, and the weight demanded by their structure remains always at about the same proportion to the lifting power. Consequently, as the craft grows, there is a constantly increasing margin of lifting power for cargo." Part of that cargo is fuel, and that weight gradually disappears. If the helium airship flies long enough, it may not be practical to come down!

Rare photos of experimental water recovery apparatus on ZR-1 (left) and running up the side of ZR-3 (right). In his 1921 writings, Frenchman du Plessis used the problem of ballast water freezing as an illustration of how the French bureaucracy stifled airship development. The non-freezing solution expert would recommend chloride of calcium as anti-freeze without knowledge of the airship's problem of it corroding the tanks. Robinson and Keller observed, "While the theory was simple, the practice was not, and some of the problems were never solved." Tupper photos from the collection of Hepburn Walker, Jr.

Smith went on, "As an airship burns off fuel it becomes lighter and wants to climb higher. Ordinarily, gas would be valved to correct the situation. But the Americans were obliged to devise a system to generate ballast to replace the weight of the fuel. The Navy developed a means of condensing water vapor from the exhaust gases of the airship's engines, which more than replaced the weight of the fuel consumed. However, the weight of this water recovery apparatus was substantial, its aerodynamic drag was phenomenal, and its maintenance was a constant headache." Teed had only seen such equipment in use in the ZR-1 at the time he wrote "...this means a *double* reduction in the efficiency of an airship." (Emphasis added.) Bob Wilcsek added, "The diminished payload was also due to the enormous expense of helium, which required heavy additional hardware such as a water recovery system...and a flight technology that abhorred venting the gas..." The power-robbing hardware hanging out in the slipstream increased aerodynamic drag. Units added to ZR-3 were found to increase drag

[146] H. Andrews, "Curtiss F9C-2 Sparrowhawk", *Historical Aviation Album*, Vol. XVIII, P. R. Matt, 1987

anywhere from 5% to an astonishing 10%. Just how much all this penalized the airship was not appreciated until Garlon Fulton detailed it out. The ZRS water recovery equipment itself weighed 16,000 pounds. Typical New Jersey weather required an additional 12,000 pounds of anti-freeze.[147] (At 2300 pounds each, that added up to a *dozen* Waco UBF two-seat hook-on airplanes.) For a 6-day mission the additional drag would increase fuel consumption by 9,000 lbs. Naturally the engines lost efficiency due to back pressure, and the radiators had to be bypassed for full power – just when fuel weight was disappearing at the greatest rate. But the biggest problem was not just the heavy plumbing that had to be designed, purchased and subtracted from the payload capacity owing to the no- practical-venting requirement, or the lesser engine efficiency. Worse yet, as Robinson and Keller reported, "…extensive corrosion of the keel girders caused by the use of calcium cloride antifreeze in the ballast bags… Subsequentley, alcohol was the preferred anti-freeze despite the greater expense."

Now there would be no more "up-ship" takeoffs. Blimps, for which water recovery was forgotten, would forevermore have to be equipped with heavy *landing struts and wheels* and invent the technique of *taking off heavy*. Just how heavy a liftoff could be tolerated was therefore the limit on endurance. Max gross takeoff weight would dictate how much additional payload would have to be left behind. That's not to mention the weight of the landing gear capable of supporting such loads over uneven operating fields, or the drag it produced. "The airship cannot then maintain height below a certain speed, and must take off and land like an aeroplane. This requires space: the military Goodyear blimps generally had extensive bases and could get off with enough extra weight to ensure residual 'heaviness' at the end of a long mission, but the operation of small advertising blimps from restricted fields would not permit such a high initial superload and called for a certain amount of finesse. An unforeseen headwind or a blaze of sunshine could have the ship returning to base 'light,' requiring a ferocious headlong dive at the ground with throttles cut at the last moment in the hope that the ground crew might get a grip on the beast before it lost momentum and soared out of reach."[148] The solution was already in work.

The J-type airship (top, J-3 before her loss searching for USS *Akron* survivors) was adapted for use with helium. Not surprisingly the Government never published specifications that compared their performance difference under the two gasses. From their first flights after the Great War until the "neutrality patrols" of the next war, J-ships served 18 years. The Naval Aircraft Factory built a custom car for a J-type envelope, thereby creating the K-1 (below, early in its life). Fuel gas was kept in a bladder inside the helium envelope. The combination's unique capabilities were not appreciated by the US Navy. It was not popular with crews but saw limited service at Lakehurst, also serving as a jumping platform for the parachute rigger's school. Courtesy Naval Airship Association.

Without a chance to get a new commissioned vessel through the depression-era Congress, the Navy cobbled together a "replacement" J-type envelope and designed and built a landing strut and wheel equipped car at NAF, which became the K-1. (In addition to a gaseous fuel bladder inspired by the *Graf Zeppelin*, K-1's cost was justified to make "fuel gas experiments." With the

[147] G. Fulton, "Improving Airship Performance'" ASME Transactions, Vol. 56#5, May 1934. ZR-3's units had frozen and become useless during winter flights.
[148] E. Mowforth, INTRODUCTION TO THE AIRSHIP, Airship Association publication No. 3, 1991

very real possibility of "light" landings being a fact of life with helium, K-1 also ushered in the study of mechanical ground handling equipment for small Navy airships.[149]) The US Army TC-13 and TC-14 were the first government designed-for-helium airships, pioneering the five-fin empanage that provided a split lower fin and rudder assembly. This allowed the nose to be pulled up into extremely steep angles during heavy takeoffs without dragging the tail in the dirt. Sadly this innovation was missed with the Navy's K-2, prototype for the 134 K-type airships built for WWII, and knocking off their lower fins was an all too common occurrence.

When it seemed that at least a majority helium fill was inevitable, the LZ-130, already a wonder of weight efficiency, was extensively reworked, saving 23,582 pounds. This was not even a good start, since the difference between hydrogen and helium first lost 55,125 pounds of lift – and the required water recovery system subtracted an additional 11,466 pounds of payload. They eliminated more than 12 tons of peace-of-mind fuel reserve, dropped 17,640 pounds of operating flexibility of water ballast, and denied a backup of 4,400 pounds of lube oil. Even after all that, ten crew members have to be dropped off, and they still lost the fares of ten passengers![150]

However, even cost is secondary when one remembers the old Libertarian saying "nobody gets none if there ain't any." Teed pointed out the gas would also cause operational headaches not envisioned even when the only practical field purification technique was to use liquid air. Since hydrogen was constantly valved during operations, it was regularly replaced by fresh, pure gas. But the stagnant helium charge would naturally be more likely to become impure via air diffusing into the cells. The heavier lifting gas was naturally more sensitive to impurities' decimation of its lifting power.

Smith continued, "As bad if not worse, it was expensive; so the operator was not only getting less for his money but paying more for less." Teed echoed "...helium is 29.8 times as costly as an equal volume of hydrogen and only gives about 92.6% of the latter's lift." The 1975 prices of Chemtron Corporation in Houston showed a thousand cubic meters of helium cost $1,695.00 vs. $230 for the same amount of hydrogen. To put that in budget terms, you'd have had to fork out $78.68 to lift each 100 pounds with helium, but the same payload would only cost you less than ten bucks worth of hydrogen. Since the helium bag has to be much larger, the gasoline cans it lifts also much be larger and heavier to push the big lummox through the air; less paying payload, less distance transported, less revenue taken in. But again, *regardless of any costs or income*, nobody flies if there isn't any. Hydrogen of acceptable flying and fuel quality could be made anywhere. Ironically, the Germans did not trust *propane* they might get in South America: von Schiller wrote, "The propane is manufactured in Germany, shipped to Brazil in liquid form, and then restored to gaseous state and mixed with hydrogen in such proportion that its specific gravity approximates that of air." When the US Navy operated airships in Brazil, crewmen remember "helium being issued like from a medicine dropper."

Smith continued "Flying with helium assumed an adequate supply of the gas. In the mid-1920s there occurred a nagging helium shortage as the gas fields around Forth Worth lost pressure and the Congress debated the wisdom of funds for a new extraction plant near Amarillo which would tap the helium-rich Cliffside gas site...This 'unusual richness' nevertheless amounted to something less than one percent of volume...conservation was the order of the day and severely inhibited flying operations. More than a few American airship operators came to regard "safe"

---

[149] Information supplied by the late Zenon Hansen.

[150] H. Dick and D. Robinson, THE GOLDEN AGE OF THE GREAT PASSENGER AIRSHIPS *GRAF ZEPPELIN* AND *HINDENBURG*, Smithsonian Institute Press, 1985, pg. 157.

helium as less a blessing than a dark curse." When the Goodyear *Pilgrim* completed trial flights with hydrogen and was ready for re-inflation with helium, technically they could not purchase, but only 'rent' the gas, with the understanding it was to be returned! Putting on a brave public face, a helium balloon was entered into the 1922 Gordon Bennett race. 150,000 cubic feet of the precious gas was thereby trashed, just as no one could be sure if it would be possible to make the first gas fill of ZR-1 with helium alone. The scarcity lead to the airship becoming more vulnerable to nature's whim.

Hot air balloons owe their existence to the principle of cooler air tending to push warmer, lighter air skyward. *Superheat* became a bigger problem with the adoption of helium. Actually not so super, the term describes the effect of nature, mostly sunlight, in heating the airship's gases to a higher temperature than the surrounding air. The 'superheat meter' on an airship instrument panel is nothing more than a differential thermometer, registering the difference between one bulb inside the envelope, and a second outside. In a hydrogen pressure airship, superheat was an irritation. If sunlight expanded the gas in the bag until all the ballonet air had been pushed out, then the airship would have to be flown at an inefficient nose-down attitude to prevent upward stall. If this continued, some gas would have to be vented. If this was followed by entry into a dark cloud and the gas cooled by radiating its heat away, air would be pumped back in the ballonets, flight would be nose up, and then if necessary, ballast would have to be dropped. This could go on only so long as there was a surplus of gas and ballast. In a helium ship there could be no venting, and there was already far less ballast. Hotter gas expanding forced out more ballonet air, reducing pressure height, but with helium superheat was no longer just an irritation, it became an injurious and even deadly problem.

A lovely Macon belle models a 110 gal. gasoline tank that would be installed with a suspension system that could be cut in an emergency, punching right through the outer covering. Some featured remote dumping valves. A G-Z man was sent to NASL in 1932 to investigate complaints of fuel tank bulging. Smith wrote that on the USS *Akron's* trip to California, "...a fuel tank burst and filled the port keel with gasoline fumes, a fire hazard which lingered on board for more than an hour." LTAS photos.

R. K. Smith wrote, "The *Akron* left Lakehurst with 55 tons of fuel and ballast; arrival at San Diego she had only five tons of fuel, ten tons of ballast, and was 40 tons lighter. The launching of her planes removed a further two and one half tons."[151] The combined effect of using up her fuel weight was tough enough, but when the Navy's untrained recruits just stared at *Akron* instead of grabbing the lines, she had to go around again – adding more superheat from the San

[151] Dr. R. K. Smith, THE AIRSHIPS *AKRON* AND *MACON*, Naval Institute Press, Annapolis, MD 1965

Diego sun while the ground handling chiefs screamed commands to make damn sure they held on the next time. Even with Rosendahl's venting helium he knew could not be replaced, in the black comedy of errors that followed, *Akron's* mooring at Camp Kearny on 11 MAY 32 was not complete until two line handlers were dead. The men fell from altitude, arms and legs flailing in stark terror while the motion picture cameras recorded it all. The resulting newsreels ran coast to coast and overseas, sometimes with the men's actual impact bounce that would be censored in later endless TV programs about airships. The tragedy had many fathers, but no one would admit Harold Edsal and Nigel Henton were, in the final analysis, victims of the helium policy.[152]

No replacement helium was available at any cost. Not only did *Akron* miss the fleet exercise planned to demonstrate its capability to the doubtful brass; its planes and part of its crew had to be sent on to the Bay area on their own. Smith: "The public took these necessary actions at face value. No one was about to consult a book on aerostatics, and there was only one conclusion to be drawn: the *Akron* was not only uncontrollable, but she could not even carry the load she was supposed to carry." What might be called 'The Airship That Can't Come Down' would become the second most publicized event in airship history, and would become rooted in the public consciousness for eternity. HERE COMES THE NAVY was the first Hollywood version of 'The Airship That Can't Come Down,' featuring Jimmy Cagney rescuing the stranded line handler Pat O' Brian. Even as late as the 1950s the 'origin' episode of THE ADVENTURES OF SUPERMAN had Lois Lane (then played by Phyllis Coates) lament an airship had pulled a lineman aloft and was "unable to land." Dabbs Greer's character fell to certain death, only to be caught in mid-air by the Man of Steel. "It is an undeniable fact that they were huge, fragile and cumbersome machines virtually uncontrollable in fierce winds."[153]

All but unknown was the fact that a similar incident with the German Army Zeppelin LZ-79 in which 3 line men were carried aloft resulted in no injuries.[154] Not well publicized was the charming story of how a mascot dog, accidentally entangled in an early blimp's lines and unwittingly carried aloft, had been saved when the pilot, LCDR Maxfield, deposited the dog safely into gentle hands. After helium made it impractical to pull the vent toggle, there would be other close calls (K-57 at Houma) and deaths. "Squadron 32 at Moffett Field, California, reported a fatality in April 1942, when a member of the ground crew was caught in the handling lines of the former US Army blimp TC-13, carried aloft, and fell seventy-five feet. A similar accident at Lakehurst in October 1943 when a man was also carried aloft and he, too, fell to his death." [155] Sadly, the accidents continued, including the two men lost with a WDL ship in Giessen, Germany in 1994. Yet the "blessing" of helium would not just result in the deaths of crewmen, but eventually the loss of airships *and* crewmen.

ZR-1 was actually a copy of "maritime airship L49", company number LZ-96, a lightweight height-climber captured in France in 1917. The Zeppelin was designed to lose a cell and continue its mission. Indeed, the ZR-1 lost two forward cells, ripped during her mast breakaway incident. It was successfully recovered because trained, experienced men were aboard. If another accident had lost more helium (such as larger cells would have released) than was available for replacement in an operational situation, the outcome might have been different. The airship

[152] Bud Cowart, survivor of the three taken aloft, could not actually tie himself up, as he also held his own weight. Careful study of the motion pictures shows his two feet firmly planted on the spider line's wooden toggle handle. The brave young man kept his wits and secured himself to the "spider" lines.
[153] O. Gault, "Goodyear's Airships, Their Tragedies and Triumphs," *Air Classics*, April 1973, pg. 19
[154] *Gasbag Journal* No. 28, June 1996
[155] W.G. Swistak, BLIMPS, BALLOONS AND BOMBS, Dutch Island Press, 1993. Courtesy Walter Swistak.

operators were not unrealistic about the problem even if it meant defying Congress. Like Lakehurst, the airship mast tender ship USS *Patoka* carried a hydrogen generator that could have rescued a troubled rigid even if the time period's helium allotment had been used up.[156]

Gas cells, valves, and other "gastight" components aged, they tended to leak at a greater rate. Helium, an atom, is smaller than hydrogen; it escapes more readily. When leakage exceeded the replacement allotment, there was no flying, period. Smith observed, "In her two years of existence the *Shenandoah* made 57 flights only totaling 740 hours. It is likely these hours would have been doubled had it not been for the helium shortage." Sadly, the ZR-1's overall leakage rate reached the point she was going to miss commitments as her irreplaceable helium escaped. Desperate, officers ordered the securing of one of the major causes of the leaks —the cells' over-pressure safety valves. These homemade 'jam-pot' covers were designed to be cut in an emergency. Unable to react to the rapid change in atmospheric pressure while riding upward and downward currents during a violent storm, it is thought the ZR-1's cells expanded to break some girders. This resulted in a spectacular in-flight breakup of the vehicle that was similar to the ZR-2 accident, only in this case the *bow* was the minority fragment. Engine cars wrenched free and fell clear. Since there were no broken fuel lines in the proximity of still-sparking electrical cables fed from a running generator, there was mercifully no fire. Only 14 of the 43 men aboard were lost, most of those killed being unlucky enough to have been in subassemblies of the ship that broke loose and fell. The remainder of the crew simply hung on and rode the various semi-buoyant pieces to the ground, coming to rest miles apart. "Shouts disclosed that there were eight men in the rising section,"[157] which was skillfully free-ballooned to safety, supposedly with only one man soaked in gasoline. The in-flight breakup of an air vehicle is rarely survivable even when not caused by an explosion. A Lockheed *Electra* came apart on September 29, 1959, killing the twenty-six passengers and its crew. Before the cause was determined, another *Electra* flying for Northwest Orient over Indiana broke up in March of 1960, slaughtering sixty-three souls. Even after it was believed the fluttering problem was solved, on the 3rd of May, 1968, a Braniff *Electra* disintegrated over Texas, killing eighty-eight persons.

The lack of fire in the ZR-1 accident was seen as a triumph of using the nonflammable gas. Often touted as evidence of helium's safety, the accident was as much a testimonial to the rigid airship concept allowing some survivors in case of an air vehicle's complete structural failure. The accident might also be seen as the false economy of placing extrapolated fire safety over proven operational safety. Even in the uproar that followed Army General "Billy" Mitchell's insubordinate and political struggle, at least one man stood up to imply the lost airship and fourteen dead men *were actually victims of the Government's helium policy*. Anton Heinen, a German Zeppelin Captain who came over with the LZ-126, had been a flight instructor, and was largely responsible for saving the ZR-1 in the mast breakaway incident. Heinen made himself unpopular by pointing out the ship's structure had broken right where valves had been tied shut. He was quoted as saying "Those valves were put there for a purpose!" Not aware existing valves were covered to stop helium leakage, authors Geiling and Sauter can today write, "Absence of high-pressure relief valves is said to have been the cause." However, at the time, the wisdom of the necessities of helium operations was not supposed to be questioned publicly. Heinen told the New York *Times*, "I could have taken the *Los Angeles* filled with hydrogen on that same morning and followed in the wake of the *Shenandoah* and brought her through without the slightest danger… [*Shenandoah*] alterations were designed as a measure of economy to save helium. It was poor economy for it wasted the most precious material aboard - 14 human lives."

---

[156] Robinson and Keller, UP SHIP, US Naval Institute Press, 1982

[157] C. E. Rosendahl "Flying with an Airship Captain," *Popular Science Monthly*, March 1930

The Navy came out with a technical report citing "large, unbalanced forces" in the ship's destruction, with some discussion of operational mistakes. Alcoa feared corrosion, bought all the pieces not carted off, and set about making a superior alloy.

There would unfortunately be more deadly examples to follow even though no rigid airship flew in America for the next seven months, since ZR-1's helium was irreplaceable at the time. It was more than five years before there were actually two rigid airships helium-inflated at the same time, and they actually shared the sky exactly one day. Sadly the magnificent carrier USS *Akron* (ZRS-4) was lost late on the stormy night of April 3rd, 1933. Extremely powerful downdrafts first nearly forced her into the sea, then possibly did the second time. Her inability to take a higher altitude under such violent conditions certainly contributed to her loss. While many insisted there was no worthy reason to be out in that storm that night, all would agree it was suicide to fly at such a low altitude from which recovery from wind shear was impossible. Here again helium gave its airship a very narrow operation band, the difference between pressure height and the ocean surface not being sufficient to operate a 785-foot long air vehicle under such barometric and vertical current conditions. While the storm only allowed spotty gasoline fires, and there were probably no ZR-2 like death by gasoline fire on the water, many men injested gasoline.

Later, following the LZ-129 fire, there was much talk of rescinding the law and offering the Germans cheaper helium on humanitarian grounds. Why had there been no similar discussion following the horror of R.101's digging into the French countryside? "For a moment there was only the noise of falling rain, then came the roar of 5 million cubic feet of flaming hydrogen."[158] Britain had been an ally in the Great War, and no one in America had heard of Adolph Hitler. As the Airship Heritage Trust explains, "The American Government had offered cheap or even free helium to inflate the [R-100] in return for the British technical experience and data."

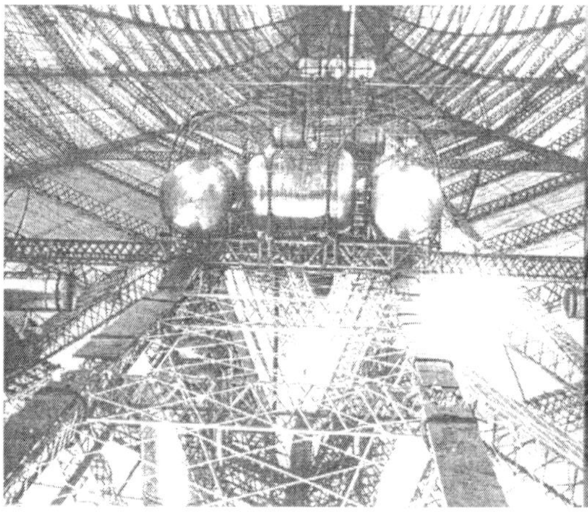

In LZ-126, wind-driven pumps pushed fuel from keel reserve tanks up to 110 gal. (672 lb.) engine service tanks on special athwartship frames 115 and 85, ensuring positive gravity feed out to the engine cars even at intense pitch angles. The Maybach type V. L.-2 was offered to American yatch buyers as "the 550-600 hp directly reversible motor powering the *Graf Zeppelin*" V. L.-2's most common problems were broken valves and snapped springs. From the Hepburn Walker, Jr. collection.

In a word, the British *did not want* helium. P. L. Teed's paper concluded helium's adoption would in itself reduce useful lift 13.85%, decrease altitude 2,200 feet and decrease full speed range by 40.6%. Indeed, a brochure advertising R.100 explains, "The cost of helium is very high in comparison with hydrogen, and its "lift" is less, so that a six-and-one-half million cubic foot

[158] L. Deighton, AIRSHIPWRECK, Holt, Rinehart & Winston, NY 1978. One wonders by what mechanism the hydrogen was retained and mixed with air so as to be ignited and thereby making a "roar."

ship inflated with helium is necessary to do the work of a five-million [cu ft] ship inflated with hydrogen. For these and other reasons it is doubtful whether we should make use of helium until we increase the size of the ship to about 9,000,000 cubic feet."

The "other reasons" meant adding water recovery tonnage, which would worsen those figures even more. The British had a further requirement than just the commercial rigid airship and hydrogen being like hand in glove. "[T]he 1922 Report of the Airship Stressing Panel began with a broad problem, focused upon narrow aspects of the problem, and came to some very questionable conclusions. Worthy of being filed and forgotten, its was brought back by Vincent Richmond in 1923 for reasons political, not technical. Finally it became the basis for the Report of the Airworthiness of Airships Panel in 1924. These criteria, combined with other factors, resulted in an absence of hydrogen capacity in both R-100 and R-101. The comparison of R-101D to R-102 demonstrates that the Air Ministry recognized and intended to correct these problems in the next airship. Discomforting details disappear after a disaster."[159] The R.100 was thereafter sold for scrap; work began to have her dismantled on 16 November 1931.

Revisionist historians also try to retro-fit wisdom after a disaster; some would have readers believe the Germans were anxiously awaiting a helium price drop. Such nonsense includes quotes like "Even before the *Hindenburg*, the Count's successors knew they should be filling their airships with helium rather than flammable hydrogen."[160] A contemporary account reads "Hydrogen, of course, is highly inflammable. But for a third of a century, German air-ships have been operating on hydrogen without a single casualty directly attributable to hydrogen, tho [sic] on several occasions they have been struck by lighting." Extrapolated safety predictions could not override aerostatic law. "If ultimately the much-coveted helium were to inflate the *Graf Zeppelin* the loss of payload would virtually eliminate the ability to carry even twenty passengers for more than a short distance. The options therefore were simple; either you carried on flying with twenty passengers surrounded [sic] by hydrogen or you grounded the ship permanently."[161] Von Schiller was more to the point. "If the 'pay load' of a hydrogen filled ship today amounts to 10%, it would approach the vanishing point in the same ship using helium... Thirty-seven years of experience with this gas may be said to have proved its safety and reliability." Years later, for the ZRCV, Barkley and Buckley state flatly, "The development of the rigid airship will require the use of hydrogen, for several reasons." One was the expected maxium purity of 96%.

With all the extra *gasoline* tonnage required for helium ships, was the point of using the rare gas *safety*? Contrary to popular literature, fire safety was still a great concern aboard the helium filled ZRS ships. On the first flight of the USS *Akron*, there was a brief panic when something smelled like smoke in an engine room. The engineman had not fully released the propeller brake. The band style brake, designed to quickly stop the 16-foot prop, was overheating and it gave the impression something was on fire. On the 48-hour endurance flight, a water-cooling line broke; to save the engine, Adolph Schoepe pinched the line closed, stopping the leak.[162]

As to the single helium advantage, fire safety, Teed emphasized the unlikely combination of a hydrogen leak in the presence of an ignition source. For warships, if hydrogen was really such a problem, Teed suggested an outer ring of nitrogen cells to prevent the lifting gas mixing with

---

[159] C. P. Hall, "Reconsideration of the Causes of 'Overweight' in Airships R.100 and R.101." Following R-101's lengthening, she was designated R-101C. A plan to add another bay would have resulted in a R-101D configuration.
[160] R. Kunzig, "Dirigibles on the Rise'" *Discover*, Nov 2000, pg.95
[161] G. Chamberlin, AIRSHIPS-CARDINGTON, Terrance Dalton Ltd., Lavenham, Suffolk, 1984
[162] Recollections of Adolph Schoepe, Goodyear ZRS engine installation and checkout technician.

oxygen. Indeed, Frenchman Boerner proposed building a multi-cell keel airship with all the hydrogen cells enclosed in a nitrogen envelope.[163]

The master, Dr. Karl Arnstein, gave a talk in December of 1930 in which calmly suggested "...adding small percentages of hydrogen to the helium. A mixture of 10% hydrogen and 90% helium would be absolutely fire-proof, and yet would have 3,000 pounds more lift than helium alone in a 6.5 million cubic foot ship... [if] we store the hydrogen in a ballonet... it would be possible to valve this hydrogen at will, thus disposing of the surplus lift which arises when liquid fuel is consumed... The hydrogen forms a sort of negative ballast and permits removal of a large portion of the water recovery apparatus."[164] Therein we see some thinking along the lines of a rigid airship dual-gas design, which has strong implications today. Unhappily, the heavy loss of R.101's passengers had a profound effect on the only other passenger-carrying rigid airship operators. Construction of the LZ-128 was scrapped in favor of a much larger ship that would allow the same performance with partial inflation using the less efficient helium. In another manifestation of how the heavier lifting gas would necessitate enlarging the size, and correspondingly the difficulty in handling and overall cost, the natural choice was a size somewhat larger than the previously "high water" mark, the American 6.5 million cubic foot ships. Therefore, theirs would be 7 million. With little money, the design and then the construction plodded along through to 1934. In a memo found by Professor Meyer, German government officials urged nationalizing the Zeppelin Service as an obvious example of unique German expertise – which could be exported. The memo admitted the Americans still had the (helium inflated) *Macon*, but predicted the inexperienced Americans would not be able to preserve her.[165] A short time after the memo was written, *Macon* settled into the Pacific.

The concept of starting one-third empty - the enclosure of the rigid not completely displacing air - was seen as unacceptably inefficient. So, the Germans adopted a dual-gas design. The idea was still called "hydrogen anti-ballast" when the Americans discussed building the next flying carrier rigid airship following the loss of the USS *Macon*.[166] The *Macon's* 1935 fire predated the LZ-129 accident by two years, but the sailors rejected the hydrogen anti-ballast idea as simply being too weighty and complex. After all, if the shooting started, they could just switch back to hydrogen and get their performance back. Indeed, when the War Department wrote a manual on hydrogen in 1929, at least one paragraph did not need revision in its 1940 update. "At present helium has entirely superseded hydrogen for use in airships operated by this Government. Its production, however is limited and in case of emergency, when large quantities of lifting gas would be necessary, hydrogen would in *all probability* be used."[167]

Still, the Germans persisted with doughnut-shaped helium cells to encompass inner hydrogen cells, and then suspended them in the center. This was the origin of the LZ-129's axial corridor, allowing crews maintenance access to the inner cell's overpressure and maneuvering valves. These original plans, copied by Addison Bain at the Friedrichshafen Museum Archives, were

---

[163] A. Kapteyn, "Principle of the Boerner Airship," *Premier Congrès International de la Navigation Aérienne*, Vol. 2, Nov. 1921, translated by NACA, courtesy Eric Brothers.
[164] Dr. K. Arnstien, "Engineering Developments in Commercial Airships," 12/12/30, courtesy Eric Brothers
[165] Prof. Meyer made these facts public during a speech given at the first gathering of what would become the Naval Airship Association, at Pensacola, Florida, in 1980, the author's first meeting with the historian.
[166] "U.S.S. *Macon*, Doctrines and Policies for Operation and Upkeep," April, 1935, prepared in accordance with Chief of the Bureau of Aeronautics' letter Aer-La-MW, 5/1, ZRS 4&5/A4 of 28 FEB 35, to Commanding Officer, U.S.S. *Macon*. Courtesy George Wright.
[167] Emphasis added. HYDROGEN TECHNICAL MANUAL, No. 1-315, War Department, Washington 1 Oct 40, supersedes TR 1170-30, 29 JUN 29. Courtesy James Shock.

recreated by airship historian/artist Herman Van Dyk. A 1936 photo labeled "Inside A Zeppelin" released by Bruno & Associates[168] has the caption that read "...There are sixteen gas cells in the hull ... The gas cells have a hydrogen center enclosed by helium. More than a million and a quarter cubic feet of hydrogen will be supplied the airship at Lakehurst by special contract with the Standard Oil Company of New Jersey."[169] The *U.S. Air Services* magazine echoed that April, "The gas cells of the *Hindenburg* are made to take hydrogen, but can be changed to use both hydrogen and helium. As the ship goes into service hydrogen filled cells will be enclosed in larger helium-filled bags, additional hydrogen to be supplied at Lakehurst terminal from the Bayway plant of the Standard Oil Company of New Jersey under special arrangement with the Zeppelin operators." Hydrogen professionals were not shocked to hear a member of the Navy Lakehurst Historical Society point out that hydrogen trains were pulled along by coal-fired steam locomotives, dropping flaming ashes from their ashpans, and shooting sparks and cinders from their stacks.

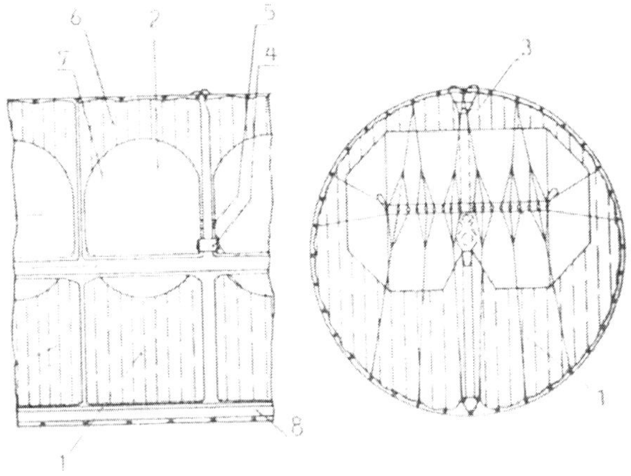

Gasbag Journal #44 says, "In 1934, with the framework virtually complete, an actual trial was made in the ship with a pair of main cells inside which were made from the old cells from the LZ-127 (nearly six years old by then). Hydrogen was used in both cells for the trials, which although they were considered to be largely satisfactory, did not convince those who would be actually flying the ship. Their view was that with the long and successful experience built up with hydrogen, they should stick with it [only], and that view eventually prevailed." Had it been adopted as shown in this publication, the dual cell design would have not only resulted in a less efficient airship, but probably a less safe one as well. Had chafing from cell-to-cell contact, or their elaborate suspension system rubbed some holes in the inner cells, a hydrogen leak would have been surrounded by cell material. Admittedly, with so much dual gasbag experience from the LZ-127, perhaps the design would have made allowances for chafing leaks.

Professor Henry Meyer states, "Still, there is a remarkable ambiguity about [Eckener's] forcefulness of his search for the essential nonflammable gas... To date there has appeared no document recording a formal application for helium by Eckener, or by any other German authorities prior to May 6, 1937."[170] So, contrary to all the authors who give the players 20-20 hindsight and have interpreted early discussions as serious requests to have the law overturned, the Germans weren't anxiously awaiting some drop in the price of helium so they could inflate two-thirds of their ships with it. Litchfield and Allen would recall: "The Zeppelin Captains wanted to go ahead with [only] hydrogen. All their flying for 25 years, in peace and war, had been with hydrogen. Their Zeppelins were expertly bonded against electrical disturbances, and hydrogen airships had flow safely though lightning storms many times. The crews wore felt shoes and buttonless jackets to avoid creating static electricity. Operated carefully, hydrogen was just as safe as helium, had better lift, was cheaper, [and] could be made anywhere. They did not

[168] Address given then was 220 W. 42nd St., NYC.

[169] Another advertisement brags, "The four diesel engines of the *Hindenburg*, capable of developing 4200 hp, are *Tide Water* Lubricated!"

[170] H. C. Meyer, AIRSHIPMEN, BUSINESSMEN AND POLITICS, Smithsonian Institute Press, Wash. DC, 1991

need to ask the favor of helium from America." Indeed, given its unaffordable cost and the rather bleak German foreign exchange imbalance, there was little point in even asking.

Had the Germans actually wanted it, even their so-called allies were not volunteering to give it up. An Associated Press article of October 10, 1930, quoted P. W. Litchfield warning unrestricted exportation of helium would weaken our national defense. Litchfield obviously realized hydrogen ships were at a distinct competitive advantage, but favored an economic explanation. "The lower labor costs in Europe have demonstrated that ships for international trade can be built and operated more cheaply than those built and operated on the American wage scale and standard of living. Some of these advantages can be offset by proper and fair restrictions in the use of helium outside of this country to enable American ships to operate on an even basis with foreign commerce." One can hardly blame Litchfield for being envious. According to *Aeroplane* magazine, in 1935 alone the hydrogen/blau gas *Graf Zeppelin* flew some 3,520 hours in 82 trips covering 222,013 miles while carrying 14 tons of freight and 5,219 persons aloft.

Goodyear's design team had actually paralleled the German development with another dual gas design, a million cubic feet larger than the ZRS and with lessons learned applied therein. Its prospectus refers to the hydrogen bags oddly, as simply "ballast," under the heading 'Maintenance of Equilibrium.' "In this proposed design the lift efficiency and performance of the conventional helium inflated airship have been greatly improved by the addition of a nominal amount of 'Hydrogen Ballast.' The hydrogen ballonets are carried within the lifting gas cells, where they will be surrounded with inert helium and the installation rendered fireproof. The use of this 'Hydrogen Ballast' will make the present elaborate water recovery installations unnecessary..." The flight loading chart shows a fill of 79% helium and 18% hydrogen, and the drawings show the hydrogen cell in the top third of the bay, vs. the center as in the LZ-129.[171]

Only 4 rigid airships ever completely inflated with helium. ZR-1, ZR-3, ZRS-4 and ZRS-5 had a sum total of less than 5000 hours aloft, and no ocean crossings. No direct comparison could be made with the more than 150 hydrogen filled rigids, their tens of thousands of hours aloft and the nearly two hundred ocean crossings. Although it has never been published, an interesting parallel is seen in the history of wartime non-rigids.

Americans flew French airships like the *Capitaine Caussin* (left, at Guipavas, France) in the Great War. 26 years later Americans returned (right, French Morroco), squadron ZP-14 representing the greatest fleet of helium-filled airships in history. Since it took about 1,600 k-bottles each, only two envelope erections seem to have taken place at the former Zeppelin hangars at Cuers. Courtesy Phillipe Nicodeme and Gordon Vaeth.

[171] *Aerostation* 23/2, in *Gasbag Journal* #44, June, 2000

The Great War found the UK unprepared to use airships against U-boats, and an entire fleet had to be created and built in a hurry. In his paper "An Enigma Of Airship History," R.H. Hillsdon wrote, "Up to early 1915, non-rigid development had continued to be ad-hoc and purely experimental, with minor military interest. The advent of the U-Boat in the war, and the early depredations they made, caused much more immediate concern. On the 28th of February 1915 the British Navy`s First Sea Lord, the redoubtable "Jackie" Fisher, called an emergency meeting to find a solution: that solution was patrol airships and a preliminary specification was evolved. It was seen that only non-rigid types could meet the timescale and within three weeks the first prototype flew and the Sea Scout class of airship was born. If consisted of a redundant fixed wing scout aircraft and engine suspended below a redundant airship gas bag."

As the Airship Heritage Trust explains, "There were eventually 3 types of "SS" or submarine scout class ships after the initial prototype was built. Each was quickly and cheaply assembled by attaching the wingless fuselage of a B.E.2c aeroplane beneath a simple [60,000 cu ft] envelope. Each ship was provided with a [spark gap transmitter] wireless for communication back to base and with ships, a small bomb load was also carried along with extra fuel. The demand was so great for these Scout ships that various versions were constructed. The following designations were given: SS / SSP / SSZ (Zero's)/ SSE / SST(Twin). The SS Zero's were fitted with machine guns. 77 of this class were built and were very popular with the crews. The last class to be designed were the SS Twin's which could carry a crew of 5, with a top speed of 57 mph and stay airborne for up to 2 days." The SS types actually consisted of three sub-types, depending on which of the three aeroplane fuselages were slung underneath the hydrogen envelope. Of the Sea Scout BE type, one was lost to fire, the prototype SS-1. Sea Scout AW and Sea Scout MF types suffered no losses. No SSP ships and no SSE ships were lost due to fire. So of the 49 Sea Scouts, only *one* was lost due to fire. Of 77 Sea Scout Zeros, six could be said to have been lost to fire: SSZ-5; SSZ-7; SSZ-10; SSZ-17; SSZ-38; SSZ-54. Closer examination shows SSZ-7 and SSZ-10 collided at Polegate on December 20th 1917; one man was killed as the ships were destroyed in the resulting fire. (Fiery collisions are still a hard fact of aircraft life today, without lifting gas.) SSZ-38 and SSZ-54 were destroyed in a hangar fire, caused by the cleaning gasoline fumes being ignited when the American wireless operator tested his spark transmitter. (A single casualty was a rigger who fell from the catwalk in the roof of the shed.) Of 12 larger, twin-engine Sea Scout Twins, one met fire. SST-6 crashed on her maiden flight August 30 1918, killing her crew of five, and was destroyed by the fire that ensued. (Sadly, aircraft today also crash and burn on their test flights.) In total, 158 'SS' Class ships were built and operated during WWI. Smaller and with less range than WWII ships, they were subjected to a greater number of landings and takeoffs, those most dangerous times for air vehicles. Yet the record shows that, during the Great War, only those eight SS type airships were lost with 'fire' having some role, even after the fact. By popular wisdom, these ships would have not been lost if helium-inflated.

Likewise, World War II also found the US unprepared to use airships against U-boats. At least there was a modern patrol airship design first discussed in 1935, roughed out the next year, contracted in 1937 and first flown late in 1938. A 1940 contract for six production copies allowed the manufacturing line to be in place before the war started, and in fact K-3, K-4 and K-5 were delivered before the Pearl Harbor attack. Even with helium, the design was far from ideal. Particularly dangerous was the necessity to carry tons of gasoline in a loft, in thin tanks directly above the crew's heads, a practice no longer permitted by the FAA. Nonetheless there were 134 helium-inflated K-type airships built, and their extensive service in WWII has been greatly under-rated. Of these, K-25, K-42, K-51, K-57, K-58, K-62, K-64, K-94, K-102, K-109, K-113 and K-131 were *all lost to fire*. Examination of details does not help at all. K-64 caught fire after colliding with K-7. All aboard the K-64 were killed. K-94 might have been lost to *enemy* fire, but the Navy records it as a fire accident, likely having caught fire while fuel was transferred between onboard tanks during a long ferry flight. More than 400,000 cubic feet of helium in the envelopes made no difference in slowing their destruction In contrast to the seven men lost in hydrogen SS ship and hangar fires, *twenty-five* men suffered most horribly painful deaths in the helium-filled K-ship wartime gasoline fire accidents.

Obviously there were great differences between both the design and the operation of the British 'SS' and American 'K' type airships. Their circumstances were only roughly similar; the US Navy claims only one K-type was hit by enemy action, starting an engine fire, whereas many British airships were fired upon, but only one, not an SS (it was C.27) caught fire in combat. Materials and techniques of 1914-1918 were obviously less advanced than 1942-1945, so the advantage should be with the newer, less numerous, less fired upon, fewer ground cycle helium ships. In the only such comparison possible, that of production airships operated for several years in wartime conditions, we see:

Hydrogen - filled SS ships	Helium - filled K ships
158 built	134 built
8 lost to fire	12 lost to fire

In other words, the same type of fiery accidents actually *increased* after helium was adopted. Just as P. L. Teed predicted when he wrote in 1925, "... it is considered that the danger of fire on a helium filled petrol propelled airship is actually greater than on a hydrogen-filled and hydrogen-kerosene propelled vessel..." Was Teed clairvoyant? Hardly; the helium ship, necessarily larger for the same performance, required larger fuel tanks and more petroleum fuel. And after consumption, bomb-like ullage would also be greater. It's still a big problem today even with less volatile JP fuel. "The FAA is considering airlines to pump inert gas such as nitrogen into fuel tanks to reduce potential for catastrophic explosions similar to the one that caused the loss of TWA flight 800 in July 1996."[172] If the record is expanded to include all Royal Air and US Navy airships even after the shooting stopped, the safety record still goes in favor of the hydrogen filled airship. In total some 214 various types of hydrogen filled non-rigids were built by the British in WWI. After WWII, the Cook Electric-Goodyear magnesium-car but standard helium envelope (not Dacron) ZSG-4 Bu No. 131920 burned. The totally new ZS2G-1 helium design lost *two* examples to fire, Bu No 137490 and Bu No 137487. Someone even managed to record a ZP-2 fire with an 8mm movie camera.[173] Home video recording equipment was on hand to catch the vibration-driven destruction of the *Helistat*, a hybrid craft featuring a spare helium-filled ZPG-2 envelope that fell victim to an injurious and fatal gasoline fire.

The makers of R.100 were frank about the danger of her gasoline system. Under the sidebar "Petrol, the chief danger, will soon be eliminated," they quoted Burney, who wrote: "The motorist of to-day has become so accustomed to the use of this explosive vapor that it is rarely that the thought of danger from it enters his head. His petrol tank is usually in a well-ventilated position, and the designer of his car has taken a great deal of unobtrusive care that there shall be no danger of fire. Equal care must be taken by the designer of the airship, but it is unfortunately a fact that the problem set by petrol fuel is more difficult to solve for the airship than for either the aeroplane of the motor car. R.100 carries 30 tons of petrol in the tanks disposed over a great part of the length of the ship. These tanks are necessarily inside the outer cover of the ship; that is to

[172] World News Roundup, *Aviation Week and Space Technology*, June 12, 2000
[173] Part of the Charles E. Rosendahl collection at the University of Texas at Dallas.

say, the mass of petrol with its tanks and pipe-lines is contained in a confined air space… Under tropical conditions, however, petrol becomes very much more volatile. It is by no means certain that in the tropics the passage-ways of the ship can always be kept clear of inflammable vapor, especially if the ship can be imagined to be re-fueling at the mooring mast… In these circumstances it must be frankly admitted that the presence of petrol in the ship constitutes a potential danger, and it is for this, amongst other reasons, that the R.100 will not fly to India until various alterations can be made." In the original, THE WORLD, THE AIR AND THE FUTURE, Burney included: "…adequate ventilation has to be provided to disperse the small amount of vapour that may be generated from the inevitable small oozing and weeping of petrol at some one of the many joints in the system." Unlike quickly disappearing hydrogen, the heavy gasoline fumes would permeate and linger in combustible concentrations.

Buckley and Barkley predicted, "The problem of transporting helium to and the return of empty bottles from outlying airship bases would require a veritable fleet of cargo ships with nothing else to do." Since one K bottle (even at 2640 PSI) only yields 285 ft^3 of helium gas, the Army and Navy each developed rail transport cars; it took four later Navy tank cars to fill one M-ship. A great number of Navy tanker cars were constructed, still serving decades into the shuttle program. Air Products, Inc. has mastered the art of transporting helium as a liquid; it only weighs about a pound per gallon, but is very close to absolute zero in temperature. Since a tube trailer can only hold about 180,000 ft^3 the more practical method is a semi-trailer load of liquid helium. Pricey, but it would expand to over a million ft^3.

Loss of speed, altitude, payload, and endurance, all at a stiff increase in price: the heavier-than-air lobby that hated airships could have hardly asked for a better ally than helium. While we're conditioned to think of airship fire in association with hydrogen, by the time you have finished this chapter, it should be obvious the major problem with airships is their petroleum fuel. K-113, actually just car C-192 at the time, burned on a railroad car in transit and was replaced to maintain sequence. These examples should also remind us it does not matter what is *inside* the bag if the *bag itself* is flammable. Replacement envelope D-528, designed for a ZSG-4, caught fire *without gasoline* at Akron, just testing it with air inflation on 21 JAN 1955! Likewise, the outer covering of a rigid airship was very flammable as well. Few, however, appreciated just *how much more* flammable the rigid's covering material really was. "In a later chapter will be told the strange results which came out of our helium conservation policy, how we drove Germany out of the airship field by denying helium to its ships, but have up to now failed to make any use of it ourselves." Like Allen and Litchfield had said in 1945, so shall we.

Chapter Four: Fireproof vs. Fireprone Airships

As civilization evolved from hunter-gatherers toward permanent settlements, the requirement for moving materials from where they were found, to where they were needed, provided endless inspiration for invention. An immeasurably large percentage of all human action has been directed at overcoming what physicists measure with their "Big G," the gravitational constant. The mass of rock that is our spaceship Earth tends to draw everything toward its center with an attraction that is probably the most unwavering force in the universe. Ancient records have references to planetary movements becoming irregular. The Bible tells of a day the Sun stopped in the sky. Yet even through the worst cataclysms envisioned by Emanuel Velikovsky, the Earth's gravitational pull has never faltered. Nor is it going to.

Man quickly adapted to the observation of which hard ground would, and which soft bog would not support his weight. Very early on he fashioned water-displacement vessels that supported him against the gravity that reliably wanted to pull him below the surface of the waves. It is easy to believe early men reasoned it would also be possible to "float" above the ground as well, just like particles caught up in the smoke of cooking fires. Although historians can not agree just how many thousands of years passed before someone fashioned a machine that implemented that idea, the notion of actually building a container that would preserve a sanctuary of simply nothing probably dates back to the discovery of the atmosphere.

A balloon of air tends to rise to the surface of a water ocean. Likewise, air pressure tends to push an enclosure of nothing to the top of our atmospheric ocean. An ideal air-displacement vessel would be based on a perfectly empty hull, but atmospheric pressure tends to crush any evacuated vessel light enough to overcome gravity. Therefore, lighter-than-air material is employed to *displace* air. Until we can perfect a solid, LTA materials exist at shirtsleeve temperatures and pressures only in gaseous form.

Of the available gases, there is a common misconception that is repeatedly published. "Flying machines that are propelled by engines but lifted by gasbags are termed dirigible balloons or airships," Patrick Abbott wrote. "Such aircraft have nearly always used hydrogen, the lightest of all gasses, as their lifting medium, but this possesses the disadvantages of being inflammable when pure and explosive when contaminated with air." As we shall see, these given parameters are completely inaccurate. This modern misconception stands in contrast to the actual practice of the day of the hydrogen airship, which was forecast in a 1909 issue of *McClure's Magazine* to be 'The Aerial Battleship.' It reasoned, "From the popular standpoint it seems a highly dangerous thing to fire such weapons in the vicinity of so great a body of hydrogen as is contained in these ships. The same popular belief existed for years concerning the explosive gasoline motor; yet this has been used exclusively for ten years to propel dirigible balloons…"

In describing Santos-Dumont's No. 16 of 1907, *Scientific American* reported, "The balloon envelope is of varnished silk… Good protection is afforded to the motor by a wire gauze covering which surrounds the carburetter [sic], thus avoiding the risk of fire to the gas escaping from the balloon." Yet one can see this description is somewhat upside down. Since the ignition temperature of oxy-hydrogen is hundreds of degrees higher than varnished silk, a safeguard against the gasoline motor belching out a shot of flame was more effective in keeping the lacquered silk bag in one piece, regardless what was in it. However, for the sake of argument, let us assume for the moment the lifting gas is non-flammable.

Airship efficiency decreases with every element added to a design. Additional weight aside, if it adds complication inside the envelope, there is less space to displace air. In other words, the great bulk of the finest airship built to date is an enclosed space with very close to nothing actually inside it. Every piece that is allowed inside is vital; there are few reserve parts; there are almost no superfluous elements.

The state of the airship art necessarily employs construction materials of the lightest practical weight. The best available materials in the 1930s had evolved to longer lasting formulations than the short-lived varnished silks of old. With their longevity came the downside of also being from mildly to explosively *flammable*. So, while the best airship design is one that employs the fewest components, one can easily see that making a large percentage of the parts out of flammable materials would greatly reduce the survivability of said airship.

The only successful non-fabric airship, the ZMC-2 at Lakehurst. Note the ZR-3's hook-on plane near the door. Tupper photo from the Hepburn Walker, Jr. collection.

In contrast to Zeppelin's concept, designers including Schwarz, Tsiolkovsky[174] and Slate, all sought to make their entire ship a fire-resistant container. Their so-called "metalclad" constructions suffered from an inherent efficiency penalty, but, like the adoption of expensive helium, offered that safety advantage. It paid off in practice. All the early metalclads suffered massive hydrogen releases, and not one burned. Schwarz was plagued with impure gas and rupturing batiste bladders, suffering a fireless failure in Russia in 1894. Building a nearly identical ship in Germany three years later, Schwarz died just before it was hydrogen filled and walked out for a tethered flight. Its drive belts slipped off, the mooring line broke and after an uncontrolled flight it was landed, where it collapsed without fire. Thirty three years later, the

[174] H. Van Dyk, "Tsiolkovskiy's Metalclads," *Aerostation*, September 1996.

Slate Aircraft metalclad *City of Glendale* suffered structural failure on its second rollout. The metalclad was leaking lifting gas profusely, but did not burn.

Ralph Upson had left the company famous for "rubber bag" pressure airships to pursue metalclads a year *before* rare and expensive helium was adopted for American airships. Upson said "Fabric covered ships... always present a fire hazard." A newspaper reporter quoted him saying "The all-metal ship can use hydrogen rather than the nonexplosive helium and be as safe as humanly possible."

Upson sought to provide a near-ideal fire resistant container that isolated its hydrogen from air, insuring gas flame safety, as well as more fire-resistant components to increase his airship's survivability. Upson never got to prove his design. When the *Shenandoah* crashed, Alcoa's aluminum was suspect for corrosion. Exonerated, Alcoa had none the less developed their new 'Alclad' brand. While an improvement, 'Alcad' was heavier. "In connection with the paperwork done in 1925, it is noteworthy that the completed craft's empty weight, 8,800 lb., exceeds the detailed estimates made at that time by only 100 lb., despite the substitution of Alclad for duralumin about fifteen percent thinner."[175] The final blow came as helium had long been adopted by the time his prototype, the ZMC-2, first flew in 1929. Criticized as an under-performer, and unwanted by the Navy, it was the only airship Upson and Detroit Aircraft Corp. would ever build. Since the prototype's available load was reduced by some 50% from the design concept, it was easy for a Navy that did not want the airship to just blow it off. Still, Upson would have the last laugh. His metalclad airship design was tested at the University of Michigan following the LZ-129 fire. The results were published in the *Wayne Engineer* in an article entitled "Fireproof Gas vs. Fireproof Airships." The report concluded that the metalclad, even if hydrogen-filled, *would not have caught fire* even given the LZ-129's fatal conditions. It is no surprise Goodyear-Zeppelin had obtained a patent for an all-metal rigid airship by 1940.[176]

Upson may have been among the few of his day that realized a major problem with Zeppelins. All vital elements would not only easily ignite, but enthusiastically *spread* fire to other vital components. Lacking Upson's wisdom, hydrogen has been blamed as the agent that starts *and* spreads fire. "Dragging the ground crew behind, L57 was driving away over the field. In an attempt to stop the ship from drifting away, Bockholt valved off gas, and upon leaving his ship, finally had riflemen shoot holes in the gas-cells. Besides being absolutely useless when it come to get rid of the gas – a bullet hole punches open no more than the forty millionth part of a gas cell- this measure was igniting the highly flammable hydroxygen mixture and set the ship on fire." One must remember that L57 was at that time loaded with 311,000 rounds of ammunition, 230 machine gun belts with 57,000 cartridges, 54 machine gun boxes containing 13,500 cartridges, rifles with 5,000 rounds, 30 machine guns and about three and a quarter tons of other materials ranging from bandages to sewing supplies. She was returning from a test flight so her gasoline tanks contained fumes. Not surprisingly, this same accident was given a completely different interpretation by another author. "The derelict drove across the field into the fence. Stars flickered across the surface of the gas cells, and there was a burst of flame."[177] That a

---

[175] J. T. Nevill, "First Flights of the Metal-Clad Airship," *Aviation*, August 31, 1929, pg. 474

[176] U S Patent 2,197,568 filed by Lloyd H. Donnell of Akron, Ohio

[177] Doctoral thesis of Wolfgang Meighorner, later published in book form and translated into English by P. A. Schmidt, as PIONEER OF INTERCONTINENTAL AVIATION BY CHANCE, from the Zeppelin Museum, 1992. Both quotes are traced back to their origins. Various theories as to the mystery fate of the "Africa Ship" on its final mission three months later are also discussed, including scenarios of gasoline or motor ignition that still manage to blame the lifting gas for an "explosion."

fueled and loaded war-supply ship accident could *still* be blamed on the lifting gas shows the general thought pattern in explaining Zeppelin fires. We must set this habit aside, and it is to the construction materials themselves to which we must look. Putting your lifting gas in a flammable container would seem to be a questionable design in any case. Even if cells could be made so sheer as to be little potential fire hazard themselves, they would thereby be more vulnerable to compromise. With the cells went any structural integrity not already burnt up. A few compromised cells – torn, melted or burnt – would likely ruin a ship as surely as the USS *Macon*, lost after only two of her smallest cells were ripped.

Perhaps what seems so foolhardy today – flying under a bag of hydrogen – could be more easily understood if one keeps those facts in mind. If your gasoline caught fire, you were doomed; if the bag caught fire, you were doomed. *It made little difference what was in the bag.* The Army airship C-2, wind-slammed against the hangar, its gasoline tanks ruptured and consumed by fire, would not have lasted many seconds longer if a big bubble of helium had escaped when the bag burned through.

In a tragic accident caused by atmospheric electricity, a ZP5K (ZS2G-1) helium-filled airship is lost to fire. The newspaper story read in part: "...the airship was inflated with helium, which does not burn, but the flames from the gasoline quickly consumed the fabric of the bag, allowing the gas to escape." NASA tests showed blimp envelope fabric burned at about one foot per minute. Earlier K-type airships were lost to fire when dumped fuel found ignition sources below. US Navy Photo courtesy Naval Airship Association.

The Goodyear *Pilgrim* is often touted as the first example of the great progress made with helium, allowing the flush mounting of the car up against the envelope. Rarely mentioned is the fact its first flights were hydrogen inflated, showing the importance of a more reliable, non-flame-producing motor being operated in close proximity to hundreds of yards of flammable fabric. Likewise, sold surplus, the ex-L-19 was hydrogen filled and enjoyed a ten year long, fire-free life not terminated until she was finally helium-inflated.

Just like the dozen helium K-ship fire victims mentioned, the L-19 would not have been able to survive a gasoline fire radiating heat and catching the envelope on fire regardless of what was displacing air. The lesson is more easily illustrated on single-celled airships. A Baltimore *Sun* article headlines "LIGHTNING HITS AND DESTROYS NAVY BLIMP." [178] It reports "The airship is called a 5-K type...the airship was inflated with helium, which does not burn, but the flames from the gasoline quickly consumed the fabric of the bag, allowing the gas to escape... The Navy several years ago lost a blimp at the Santa Ana (Cal.) station while several hundred feet in the air and the cause of the loss unofficially has been laid to lightning." Made of earlier

[178] The person who donated this reference made no note of the date, but historian James Shock has suggested this was the ZP5K Bu. No. 137490 which was lost to fire at NAS Glynco, GA, in 1957.

materials, the British NS-11, long thought lost to lightning igniting vented gas, could very well have been a victim of a gasoline fire that spread to the envelope or gasoline tank explosion.

In a dramatic demonstration of World War One type combat, air show crowds would thrill to the sight of a biplane fighter attack and destroy a kite-type observation balloon. In contrast to various incidents where escaped barrage balloons actually could not be brought down by aircraft, one setup demonstration captured on motion picture film is dramatic. The special bullets pass through the envelope, and the bag begins to burn as expected at the exit point. Smoke is seen rising from the wound, then a bright flame of the fabric. Since little gas has escaped, the vehicle remains aloft and attitude is unchanged. The flames advance in all directions for seemingly endless seconds, bright and smoking. Finally the balloon's structural integrity is destroyed and the gas escapes; the remaining fabric folds in on the fire like a limp rag. The payload plummets.

"[Zeppelins] proved remarkably resistant to aircraft bombs and machine guns and even direct hits from high-angle artillery. After the introduction of Brock-Pomeroy-Buckingham incendiary bullets in 1916, however, they suffered disastrous losses to British planes."[179] Rogers had predicted balloons would be immune to such attacks when helium filled. It is fortunate no trusting observer was sent aloft under helium with the commanding officer's assurance he was invulnerable to airplane's flaming bullets! Likewise, had any of our helium-filled rigid airships fallen victim to an enemy armed with ammo capable of setting the fabric on fire, their fate would have been little different than the Zeppelin bombers. Sowrey's shooting up the L32 was described as "...using drums loaded with the mixture of new Brock, Pomeroy, and tracer... first two drums apparently had no effect, but the third one caused *the envelope to catch fire* in several places in the center and bow." [180]

Years earlier, the fabric developed by Hodgeman for Baldwin consisted of two layers of Japanese silk separated by a layer of rubber. The Signal Corps was to supply the fabric for the first military airship of 1908, described as rubberized silk, covered with an aluminum preparation, without varnish, weighing 5.842 ounces per square yard and having a minimum breaking strength of sixty-two and a half pounds per unit measure, likely per linear inch.[181] (When the Army purchased the SC-1, they also bought the hydrogen gas generator Curtiss brought to the airship's trials.) The first Navy rigid, the ZR-1, is described as, "The silver coating of the big tube is stretched over a skeleton of girders, and comprises 16,000 square yards of heavy cotton cloth. It is painted black inside to mellow the rays of the sun. On the outside, a coat of waterproof aluminum paint shrinks it tight on the frame."[182] As we shall see, the outer covering suffered from rather formidable problems all its own.

Next inside were the rigid airship gas cells. By contrast to pressure ships, cells do not have to maintain an aerodynamic shape, survive the weather, or keep control cables in tension. The great advantage of the rigid type is the ability of the gas container to expand and contract in reaction to the ever-changing conditions required for operations in our atmospheric ocean. Protected by the structure and outer cover, the more lightweight cells can "breathe" inside their bays. The design allows use of much lower pressures, typically one inch of water or less, which is directly related to less gas leakage and lower cell seepage rates. In other words, the rigid airship cell is as close to a trouble-free simulation of a vacuum as man is likely to build.

---

[179] R. Largess, "Airships at Sea," *Warship* 44, Conway Maritime Press Ltd., 1987, courtesy R. Largess.
[180] Emphasis added. W. Cross, ZEPPELINS OF WWI, Barnes & Noble, New York 1991
[181] "The American Aeronaut" (I), January, 1908, courtesy the Curtiss Museum.
[182] J. B. Wood, "Seeing America From The *Shenandoah*," The National Geographic, Vol. XLVII, No. 1, Jan. 1925

Perfecting a balloon material as impermeable and lightweight-sheer as possible was a key rigid airship element. The earliest cells were rubberized cotton, but since rubbing it could generate sparks, an improvement was developed using bovine intestine glued on cotton. Accompanying an excellent photo of a ZR-1 cell on page 81 of Robinson and Keller's UP SHIP is the caption "Note the transparency and fragility of the material, made up of separate longitudinal gores of light cotton fabric, with goldbeaters' skins shingled on." Of the S-L airships, Karl-Heinz Hoffmann wrote, "15 gas cells provided buoyancy. The cells were made of double rubberized fabric or of goldbeater skin. Gold brocade or goldbeater skin originated from the upper skin layers of the appendices of sheep or cattle. After appropriate treatment, seven staggered layers were glued together to form one surface from which the gas cell was made. This extremely gas-proof cell only weighed 150 g/m². However, it must be mentioned here that this material is very costly and susceptible to bacteria with a decomposing effect."

Gastight material has been the most basic building block of all air displacement vessels and had long been a problem. When Umberto Nobile went to Russia to build airships, he noted the crude quality of the materials. "In fact, the production of rubberized material in the Soviet Union at that time gave results that were anything but good; this was mainly due to the technological process adopted for the rubberizing procedure. It often happened that the new envelopes after a few months' wear – sometimes only a few weeks – were fit to be thrown away, as in fact happened in the case of the V-7 (II). The envelope that gave the best results was that of the V-6."[183] (Indeed, in spite of many challenges, V-6 surpassed 11,000 feet altitude and in 1937 established a long standing endurance record of 130 hours, 27 minutes.[184])

Von Schiller wrote, "All the cells consist of an artificial filmy material not unlike cellophane covered on the outside by a thin cotton fabric, to prevent damage from friction against the girders and wires." A major break came as the Americans, desperate to contain precious helium, sought to improve the permeability of gas cell fabric. As Dr. Dale Topping wrote, "To replace the goldbeater's skin gas bags used on earlier Zeppelins, Goodyear had developed a rubberized cotton fabric, and the Bureau Of Standards had developed for the Navy a gelatin-latex compound and impregnated a cotton cloth with it. Both appeared promising, and half the gas cells in the [USS] *Akron* were made of one fabric and half of the other, as a service test. The gelatin-latex proved to be best." It's said the Germans grudgingly accepted the patent trade deals with Goodyear specifically to get the excellent gelatin-latex cell material, which is listed as loosing less than 100 cc of gas per square meter per 24 hours. This allowed the Germans to abandon the cattle-gut shingled cotton cells, whose greater permeability lost hydrogen at the rate of *one liter* per square meter every 24 hours.

Another giant leap in efficiency was effected by borrowing a lesson from non-rigid airships to use the gas for some measure of structural support. The rigid airship was from its inception merely an enclosure of hydrogen balloons. The structure, light enough to fly, could not also be built strong enough to actually be moved without support, in flight by its gas cells. "My first experience with the so called 'rigid' airships occurred when I happened to be at Roosevelt Field, L.I., on 6 July 1919 to see the arrival of the giant [R.34] after its crossing of the Atlantic... In photos and when first seen it seemed a massively strong structure justifying the name 'rigid.' But

[183] Gen. Umberto Nobile, MY FIVE YEARS WITH SOVIET AIRSHIPS, translated by F. Fleetwood, The Lighter-Than-Air Society, Akron, Ohio, 1987

[184] The record of the 671,000 cubic foot hydrogen V-6 was broken by a US Navy 1 million cu.ft. helium-filled ZPG-2, which stayed aloft 200 hours about 20 years later. Harassed by a storm, an inexperienced pilot crashed V-6 into a hillside in February 1938.

close up one was astounded to see how the frame squeaked, bent and shivered with the cloth covering almost flapping in the wind gusts...Next day a storm came up... I was shocked at its flimsiness and the wicked wrenching of the gusts..."[185] Nevil Shute echoed concerns with R.100's power tests, "All my life I shall remember the sight of those engine cars leaping and straining at their cable drag wires with terrific force, suspended from a hull that was completely full of hydrogen, each car with smiling men gesticulating with thumbs up out the window..."

When gas cell replacement was necessary, the rigid airship would have to be suspended from the hangar above, and shored up from below. Zeppelin hangar window glass was tinted yellow to cut down on the sun's degrading effects on the airship materials. During construction the lightweight structures utilized a counterweight system of block and tackle against the solid hangar walls to eliminate sag or distortion before wiring.[186] Evolution of the rigid airship was in part measured by builder's ideas in structural design that balanced integrity with lift - and loss of cells to accident or combat. "It is protected by a cover of tough rubber cloth, stretched over aluminum rings and ribs, each strong enough to support a man's weight; and the whole is greatly strengthened by the upward pressure of the hydrogen inside its balloons."[187] Variations on the structural theme were many, even given the mere four decades of the rigid airship's existence in our world. By contrast, outer covering itself had not changed very much since the first sheets of Pegamoid [188] were lashed to the girders of the LZ-1. (To save weight, LZ-1 underside sections in between bow and stern, equator and below, were only silk.[189]) In a book explaining the new ZRS-4, the reader is told, "By the use of an outer cover, it protects those cells from rain, hail, snow, and other external injuries during flight."[190]

The 1924 Rigid Airship Manual spells out the required qualities of the outer cover. "The fabric must have 1) Sufficient strength to withstand the air loads imposed on it in flight... 2) Good elasticity so that satisfactory tautness is maintained over a long period of service life. 3) Water soakage resistance so that the inside of the ship stays dry during rains. Also, the cover should absorb and hold as little moisture as possible... 4) Opacity, that is, to be impervious to light." The Goodyear rigid airship promotional book says "Rain is hardly even an inconvenience, since the outer cover is water-proof and today's fabric refuses to absorb moisture from the air."

Eventually only the finest tightly woven fabrics were used, imported from the Middle East. This turned out to be a godsend when times got especially bad in post-WWI Germany during "... the inflation which peaked in 1923. During those years of hardship and scarcity, *Luftschiff-bau-Zeppelin* would allow employees to buy small quantities (five yards or less) of the fine Egyptian cotton cloth that was used for the outer covers of Zeppelins."[191] The lashing work was glued over with a fabric strip. Major strips' edges were reinforced with wooden slats. The manual explains

---

[185] G. Loening, TAKE OFF INTO GREATNESS, Puntnam, 1968.

[186] Author's research for the script of "The Flying Carriers" video included an interview of ZRS riveter Edward Fielding, then 94 years old, in Hemet, California. The illustration of this seldom mentioned necessity brought praise from sponsor Adolph Schoepe, then 93, who had worked ZRS engine integration.

[187] C. Dienstach & T. R. MacMechen, "The Aerial Battleship," *McClure's Magazine*, August 1909

[188] Don Woodward: "...Pegamoid, a proprietary waterproof fabric used for the folding tops of carriages etc." *Gasbag Journal* #44, June 2000. LZ-1 covering is also described as a "waterproof cotton cloth."

[189] This is visible in the most clearly reproduced photos of the first flight, July 2 1900, identifiable by the presence of the sliding trim weight. Useless, the weight was removed before the flight on October 17th.

[190] AMERICA IN THE SKYWAYS, hardcover book issued for ZRS-4 ring laying ceremony

[191] WHEN GIANTS ROAMED THE SKIES, 'Karl Arnstein and the Rise of Airships from Zeppelin to Goodyear.' Dr. Dale Topping had completed a substantial portion of the text when he passed away in 1993. Eric Brothers finished and prepared the book for publication by the University of Akron Press in 2000.

"Sealing strips are secured only by means of dope and consequently can be easily removed for the adjustment of the lacing."

117 Zeppelins later, *Baumwolle* material grades as heavy as 130 grams per square meter were employed. The weight decreased to 90 grams per square meter as location in the pressure of the air stream allowed. The lacing edge of each panel was made of a slightly heavier grade, doubled over around a strong herringbone cotton tape sewn together with four parallel lines of stitching. "The outer cover of the ship is ... made of cotton cloth...covering an area of more than seven acres. To facilitate the lacing, Duralumin eyelets,[192] averaging more than 800 to the panel, are placed along the edges of the panel. The panels are laced as tightly as possible, and allowed to stand for a time, after which the lacing is again "pulled" or tightened. This process is repeated before the first coat of clear acetate dope is applied by hand... Application of the first coat of dope by brush allows every part of the fabric to become permeated. The effect is to shrink the cover and make the panel taut. To diminish the effect of ultra-violet radiation, the top side of the outer cover is treated with a special red paint. The next coat[s] of dope... contain aluminum powder, which gives the ship her silvery appearance."

The "special red paint" was actually iron oxide, applied as the first coat from the equator to the top. (Various sources have misstated the red iron oxide was applied after, or even on the inside.) It had been another item added to the improvements list, obviously influenced by the last British rigids' practice Its purpose was to help minimize cell degradation, also making the ship less vulnerable to sun-then-cloud variations in superheat. Lacquered on top of that was a chemical mix of dopant that shrunk the fabric tight and made it water-resistant. In the late 1920s book AIRSHIP DESIGN, Cox and Collins published this recipe for about a hundred gallons of aluminum dope:

Nitro-cellulose syrup.....260 lb.
Butyl or amyl acetate ....20 gals.
Alcohol .... 15 gals.
Benzol.................... 15 gals.
Acetone .... 16 gals.
Castor oil .... 25 gals.
Aluminum powder .... 100 lb.

Cellulose acetate dope was certainly a less flammable, and had better water repellent characteristics. However, it was hard to produce, and it required expensive, rare chemicals. As a result, the readily available and lower-priced nitrate dope was used whenever practical. The 1924 manual was hopeful stating "Cellulose acetate dope is nearly always used on airship outer covers although cellulose nitrate dope has sometimes been used. Cellulose acetate dope gives a longer life to a cover than cellulose nitrate dope and the former does not burn as readily as the later. Cellulose nitrate dope gives a slightly better initial shrinking of the cover ..." Applied by hand using buckets and mop-like brushes, the constructors stroked toward the stern to lay the *Aussenhautstoff* nap down in the slipstream direction. They strived for about four and a half grams per square meter for the first coat, probably thinned for better penetration, increasing to slightly over six grams per square meter in the second coat. Another twelve grams per square meter of aluminum powder dopant was mopped on top of the initial *zellon* coats. Not that there was any

[192] The LZ-129 drawing specifies *brass* eyelets.

choice for the classic Zeppelin design at the time, but the component's assemblage yielded a veritable powder keg of combustibles.

The R.101 was another milestone in airship history; much has been written about her troubled creation and her tragic end. While her outer cover loss is said to be the cause of her fatal plunge, the resulting details bear a re-examination. The necessity to carry gasoline for the big locomotive powerplant's starting engines was obviously not meeting the requirement for gasoline-free tropical operation. As P. R. Stokes wrote, "The first reversing 'Tornado' arrived on the 12th September, the second on the 16th. They were installed in the forward power cars, together with the one diesel starting engine in the front starboard car, leaving the port unit with its petrol starting arrangements in common with the other three cars. All these arrangements coinciding with the fitment of R.101's additional gas bag lift capacity. On push out on the 22nd September proceedings were delayed by the inability to start the new diesel starting engine... It was proposed that the high speed trial be carried out near home on the flight to Egypt, whilst written instruction given to the ship made no reference to undertaking the full power test after departure." It might seem incredible that such a flight was undertaken with so many major airship components untested, but off they went, next stop Egypt.

Fabric panels were laced to the aluminum framework using twine-like ramie cord. Run between eyelets squeezed in the fabric and corresponding holes in the girders, the fabric had to be cinched up a couple times before it was doped. LZ-127 photos courtesy NAA and Zeppelin Museum.

P. R. Stokes continued, "...the surviving voyagers simply spoke of steady cruise conditions at 825 rpm on all five engines, which with their fixed pitch propellers equated with an air speed at 52 knots or 60 mph. When at 2.05, after change of watch at 2.00, the two stage dive resulted in fall to earth and subsequent destruction by fire, the evidence spoke of a sequence of engine room telegraph signals being heard or observed in sequence fore to aft, initiated supposedly just before or in the diving phase. In the aft car the order to 'slow' at entry to the second diving phase, with the order carried out. At port midships signal was received, 'slow' and reacted too as the diving attitude commenced, the engineer succeeding in shutting down 'engines' as the ship first struck. At starboard midships no signal was observed and the engine continued operating at cruising condition, the engineer advising that signals could not be heard, only observed. We have, of course, no evidence reported from the forward cars with the forward port being at the probable primary impact point and described in the French report as exploding, and the fore starboard reported as having its propeller still rotating with gouge marks in the earth and the car then twisted through 180 degrees under the wreckage. Ground observers forward and to starboard spoke of sparks emanating from an engine car prior to impact. The engineers forward were among those killed."

The R.101 fuel tanks had a cutting device to rip open the tanks in an emergency, perhaps a throwback to the time of ballooning. (Instead of throwing out an entire bag of sand, ballast would be discharged a handful at a time least someone below would be injured.[193]) If any of these devices were activated, fuel oil would have sprayed in the 40+ knot windstream against the ship. In any case, R.101 came down in a pool of her own fuel. P. R. Stokes wrote in closing, "In civilian operation the practice only appears debated in terms of the fuel dispersing as mist ensuring that those below were not affected, fire risk factor appears generally set aside in the classic accident investigations." Insult was added to tragedy in the aftermath, as each school of thought reached its own conclusion. Jerome Hunsaker wrote Dr. Eckener, "I have just had a copy of your testimony on R-101 which agrees with what we think here except we suspect a lack of skill on the part of the elevator coxswain combined with suspiciously small horizontal tail surfaces which may be contributing causes of the trouble." With so few diesel powered aircraft operated to that time, there had been no way to compare accident details. While time would prove diesel-powered crashes were no less frequent nor more survivable, the L-Z team would somehow, in the face of the seemingly obvious fact the R.101 victims died in blazing diesel fuel, reach the fateful conclusion they wanted diesel fuel for their new ship.

The choice of gasoline starting engines added a most dangerous combination. According to the FBI, diesel fuel with a gasoline topping is the current instrument of destruction preferred by terrorists of the ELF, whose goal is the maximum property damage with the fewest clues remaining. No contemporary air vehicle, with a full fuel load and impacting the ground that way, could possibly have survived. Second only to the LZ-129 itself in misunderstanding, the accident of the R.101 – torn outer cover, deflated forward gasbags, engine and tank smashing impact, with tons of diesel and some gasoline released on materials that were highly flammable themselves - still somehow worked out to be a *hydrogen fire*.

Shute had pointed out earlier that "...there was no great danger of fire in the air, for any gas escaping would go upwards and out of the top of the ship, remote from any likely source of ignition." Yet he'd eventually go with the more mainstream thought by saying, "The chief danger, I thought, was of fire after hitting the ground, when broken electrical cables could make sparks in the presence of large masses of escaping hydrogen or petrol; in that case the fire would spread instantly." Indeed, his prediction had not even mentioned the airship would likely fall into a pool of its own petroleum fuel, being jettisoned in an attempt to save the ship. Few remembered the findings of the R.38 accident investigation, and blamed hydrogen as well, not realizing the great difference between both suspects' ability to *spread* fire.

Images of the elaborate funerals and bandaged burn victims were devastating. Yet it did not seem relevant that crash photos show that the cover, which was linen coated with iron oxide and painted with nitrate dope impregnated with aluminum, had almost *completely burned away*. A popularly published view shows the nearly intact horizontal fins as all but naked skeletons. The fins neither contained hydrogen, nor were they particularly close to the diesel fuel tanks. It is more difficult to understand how this went unappreciated when images of downed Zeppelins must have been fresh in the English mind. L33 was practically fuel and gas-less when its crew set it ablaze, but its thin outer cover burned away to the last stitch. The intense flame was withering, but over so quickly there was enough structure to allow an exact copy to be reverse-

[193] A. Smith, THROW OUT TWO HANDS, Companion Book Club, London 1963, courtesy Kent O'Grady.

engineered. The motors were not only available to be reverse engineered, they were refurbished and used to propel British airships!

If we are to understand the famous airship accidents we must first come to grips with the fact the great majority of materials used in their construction were highly flammable. Even if emptied of lifting gas, the Zeppelin did not need any accelerant to assure its destruction once set afire. The LZ-129's cloth, adhesives (such as that used to glue reinforcing strips in panels), pigment, wood (strips and dowels at seams), the cells, and ramie cord was enough to make a terrific bonfire. Add interior comforts like bed and table linens, cushions, and even silk tapestry and you have more fire hazard. Even the alkyd/acrylic phthalocyanine pigment added to deter corrosion of the aluminum structure was flammable, exacerbated since "it is dissolved by the cellon-lacquer of the covering material and adheres to the material."[194]

Unfortunately, starting where nature would be first looking for any weakness, we come to the most vulnerable part - the outer cover. Cotton and flax used in making the fabric is the nature's purest form of cellulose, containing more than 90 percent cellulose by weight. Cellulose has approximate 60 percent of the energy content of bituminous coal. In other words, without anything else, the LZ-129 linen itself had the Btu equivalent of 9,720 pounds of coal. Coal is listed as producing 94.1 Kilocalories (thousand calories) per mole. Before we conjure up images of a soothing bed of coals in the home fireplace, we must remember this was in thin, sheet form, all hung at angles.

Chemicals used to increase the fabric's life and water repellence add more flammables to the cotton. In his 1921 book *Airship Fabrics*, J. Dyer states "The use of cellulose nitrate dope instead of acetate is because it is less affected by humidity. It is actually being employed now for some parts of outer covers, and is not used throughout only because of its greater flammability." Not so distantly related to gunpowder, more docile nitrate film used to offer cheap thrills. Naughty boys were known to roll up nitrate film and cram it into soda bottles. Lighting a fuse and taking cover, they would compete for local shrapnel yield records.[195]

The Industrial Chemist, March 1927, says the "inflammability[196] of doped fabrics is one of great importance," observing "It is when the airplanes are stationary that the nitro-dopes may exhibit their greatest inflammability," as if predicting the airship's mooring as its most vulnerable time. It lists the ignition temperature of undoped linen as 353° Centigrade. The various finishing chemical additives reduce that 353° flash point until nitro-doped "fireproofed" fabric lights off at only 195°C. Nitrate dope soaked linen fabric caused many an airplane to burn down to small piles of metal parts. A part of Naval Aviation history not generally emphasized was the very real problem of trying to place airplanes on fighting ships. A few unlucky glowing cinders floating down from the coal-burning stack, and another flaming airplane had to be hastily dumped over the side. Ralph Upson was ahead of his time, obviously. Their danger was not lost on the constructors, of course. Read a caption entitled "Hangar for the *Graf Zeppelin*, Friedrichshafen, Germany: The huge, specially built hangar for construction and storage of the German airship *Graf Zeppelin*. A sign in large print near the entrance (*Rauchen Verboten*!) forbids smoking at

---

[194] "Analytical Research of Covering Material of Zeppelin Airships LZ 127 *Graf Zeppelin* & LZ 129 *Hindenburg*" (1937)

[195] Recollection of engineer Roy Gibbens.

[196] Webster's Dictionary says "usage: *Flammable* and *inflammable* are identical in meaning and are equally acceptable in standard English at all levels."

the site, due to the *highly flammable fabrics* and gases used in the airship."[197] There was little else they could do about the dangers of what was the best material of the day.

Of course spreading all that cloth between frames in the wind, one would expect there would be some flutter. Flutter is undesirable because the fabric quickly wears out and fails. Fluttering fabric is as old as the rigid airship itself. T. S. Ede wrote of a 1910 vision of a Zeppelin in *Savage Skies:* "The propellers, which murmur like innumerable bees, made the canvas shiver along its rigid structure..." The 1942 Bluejackets' Airship Manual (lifted from the 1924 Rigid Airship Manual) asks "Q. What prevents the outer cover from flapping? A. The longitudinal wires to which the cover is sewn and the anti-flutter ties. The longitudinal wire also provides some support for the air loads on the cover...A reinforcing strip of balloon cloth about 4 inches wide is sewn to the inside of the cover panel and centered over the longitudinal wire... Later a 1 ½ inch wide strip of balloon cloth is doped on the outside of the cover over the sewing after the first coat of clear dope has been applied to the cover." It is not known if this exact system was still in use on the LZ-129 but close-up photos do show reinforcing cloth strips in place on her cover. Also, the weight schedule shows 5,500 pounds of cover support wiring.

***Schmeiröl* belches forth from the *Auspuffsammeln* at startup as *werft* number 95, one the Great War's 'stealth bombers,' displays its night camouflage. LTAS photo.**

LZ-129's first flight showed the cover nonetheless had some fluttering problems, and according to Harold Dick, more anti-fluttering wires and/or additional doping was still being worked toward the last of her existence. During the last landing attempt, LT R.W. Antrim saw a flutter aft of the port engine where the cover seemed to him to be very loose. This is even more surprising when the 1924 manual states "Interior outer cover reinforcements are provided in the wake of propellers, in order that any ice or other material thrown from the propellers will not penetrate through the gas cells above. These reinforcements are made of canvas."

In motion pictures taken at landings and launches, the cover of the ZRS ships can be seen flapping in the vectored thrust of their huge props. The massive LZ-129 twin two bladed props were certainly capable of fluttering the cover. Perhaps the wind gust from port exacerbated the cover flutter as Antrim was looking. If seen elsewhere, the fluttering is actually an indication winds aloft were considerably greater than the reported six to eight knots measured at the

[197]Author's italics. www.corbis.com/E.O. Hoppé Image ID: CA007229

weather tower. The weather tower was 96 feet high, while the airship was never lower than 150 feet before the fire started, probably closer to the 200 feet reported in the station log.

Even the Germans tore LZ-127's horizontal fin fabric. "The fabric covering still presents the most serious problem in airship construction and operation," Fritsche told the Royal Aeronautical Society just two months prior to the British airship R.101 cutting a shallow trench into a hillside near Beauvais, France, during a thunderstorm in October 1930. The R.100 suffered considerable outer cover failure on its otherwise successful trip to Canada, necessitating some heroic repairs as men climbed out inside the fins. Nevil Shute told of a full power trial whose value "… lies in the effect it was disclosing weaknesses in the outer cover of the ship. In the case of the R.100 the outer cover took up a curious wave-like formation when the ship was at full speed, which was not evident at cruise speed. These harmonic deflections ran longitudinally down the ship from bow to stern… The outer cover was a weakness in both R.100 and R.101..." The latter ship was put on the high mast after modifications, and, "in a very light wind, the outer cover split longitudinally, making a tear one hundred and forty feet long...A second, shorter split occurred the next day...They reinforced the cover with tapes stuck on the inside with dope, and made a short test flight. They did no full power trial, probably because they were afraid of the outer cover… in both ships the outer cover was the main weakness." Indeed, even with the more successful the cover was a huge disappointment. R-100 was deflated on 11th December 1930, hung and supported in the Cardington hangar. The outer cover was under inspection, and was seen to be deteriorating in places.

R.H. Hillsdon also wrote, "For both ships, the outer cover consisted of pre-prepared fabric panels finished with current "dope" and stretched in place by lacings; the same practice as used in R34 and all British rigids. In both cases, extra fabric tension was applied by "reefing" the panels in place… In the R100 this reefing tension was applied via patches which were laced to wiring and pulled the panels inwards, i.e. towards the centre of the ship. It so happens that we have a very good calibration on the fabric stretching of R100. Early in her flight trials it was noted that the fabric distorted in a static "standing wave" made at speeds of over 70 miles per hour. This was photographed independently by Farnbourough experts. It was deemed not to be of sufficient importance to prevent R100 flying to Canada; that journey being the final phase of acceptance. While approaching her destination, R100 suffered a fin fabric failure; probably the result of aerodynamic overload in extreme turbulance. While stopped in the air for repairs, in a prevalent wind known to be of the order of 25mph, the ship was again photographed and the standing wave was clearly visible…" In the case of the R.101, it was a fatal weakness. R.H. Hillsdon continued, "In the R.101 the reefing tension was applied by supplementary beams pushing outwards. The deterioration of her fabric occasioned some repair work, which was ill performed and throws some doubt on the final event. That event was her crash en-route to India. The formal enquiry was not conclusive but surmised a major failure of the outer cover causing secondary rupture of the forward gas bags." Shute agreed, "As to what caused the gasbag failure, the Court suggested that a large outer cover failure had occurred on the top of the ship in the forward region, thus exposing the gasbags to the violence of the air flow at fifty-four knots."

Here is the description of a newsreel of yet another downed Zeppelin: "Wrecked German Zeppelin outside Salonika (Thessaloniki) in a field; with the crew that shot it down present; guys are climbing around it; looks like a giant Erector Set. Slow Pan to left; in FG men bending under wires and around metal supports of the wrecked craft which lie topsy-turvy all over the field CU of Soldier finding the propeller. Group of Greeks who brought down the Zeppelin pose with the large German flag..." Or L45, "…a roar of flame, and L45's duraluminum bones were quickly bared of their fabric integument." (Niderost:) "Beside 'Pomeroy' and 'Brock' bullets, there was also "Buckingham" ammunition, which was a phosphorus incendiary bullet...The RFC pilot decided to concentrate his fire on one section of the hull. Soon a tiny glow flickered tentatively inside the airship's hull, growing brighter and brighter with each passing moment. Then long tongues of flame appeared, greedily consuming the ship's outer envelope." The wreck, first to hit British soil, burned for a few hours after impact. Common element: *the fabric rapidly burned away, gas or no gas,* just like the helium filled USS *Macon*.

Of course these seemingly lopsided combats must be taken in context of the price the defenders paid for victory. One of many examples the first victor, Flight Lt. R. A. J. Warneford whom Kenneth Poolman quoted in ZEPPELINS AGAINST LONDON as noticing "...there were pieces of something burning in the air all the way down." Following his bombing of Army airship LZ-37 over Belgium, which fell from 5000 feet with only one survivor, Warnerford climbed into another airplane that killed him.

History has given us many clues to what was actually burning in these airships, had we been perceptive enough to decode the messages. The last Zeppelin to be attacked in the Great War, the L53, was still a challenge to its airplane adversary even armed with the three-part ammunition. Rimell: "There was only one way he was going to get his guns to bear; pulling the control column into the pit of his stomach, the Camel almost at the brink of a stall, he let fly. All the while Culley was firing he could see the incendiary bullets striking the back flanks of the airship, which showed no sign of igniting." When it finally did light off and burn, "...the remains of the doomed Zeppelin plunged down, blazing furiously. The gaunt, scorched framework had been stripped of all its covering well before it struck the water and sank..." (The scenario would be almost replayed in the Pacific in the helium-filled USS *Macon's* final sinking minutes.)

The clues include the conduct of the combustion and the appearance of the fire. British defender Gunner Leckie used the three part ammo on L70. Rimell: "The APT [ammo] was seen to blow a great hole in the fabric and a fire started which quickly ran along the entire length of the Zeppelin." *Hydrogen doesn't burn sideways.*

Rimell's witness described L32 thus: "The light from the blazing Zeppelin was colossal; it was beyond imagination or description." Another said, "As it began to fall the countryside for miles around was lit up by the flames..." *When burning, hydrogen does not produce significant visible light.*[198] SL-11, a plywood-framed Shuttle-Lanz rigid airship, first to be set afire above England, was photographed burning in the sky over London. The SL-11 "...expanded and exploded with a terrific flash, bathing the whole of the capital with its brilliance." Lt. W. L. Robinson had attacked the SL-11, firing upwards from below. Before the terrific flash, Robinson was first to describe a seemingly strange phenomenon: "A dull pink-stained glow deep inside the airship was the prelude to fiery disaster which was about to engulf Hauptmann Wilhem Schramm and the 15 members of his crew..." Since hydrogen doesn't glow, what was he witnessing?

Remembering the Zeppelin is an enclosure of basically empty space covered with thin materials, we can realize it would be rather translucent. Also remembering the exploding and fizzling ammunition always ignited the wounded vessel only at the exit point, we realize Lt. W. L. Robinson was the first man to witness the effect of the outer cover burning brightly on the side *opposite* his view. As he looked up from below, the fire was showing through at least a few

[198] Most of hydrogen firelight is near ultraviolet, around 310 nanometers, outside human eyesight limits.

dozen feet of airship. Its original brilliance was of course diminished ("dull") and its original color was filtered. When the witnesses were at the same altitude, like fellow Zeppeliners, they never reported a rosy interior glow because they were looking at the actual flame directly, not through a depth of airship. The fact that Robinson and others had time to perceive the glow is another clue. Others, looking from below, made similar observations. "A dying Zeppelin was a hypnotic spectacle. A British reporter saw one fall to French coastal artillery: 'The Zeppelin's bow heaved suddenly upward... the long cylindrical envelope went all red inside, glowing like an immense Chinese lantern. I saw three crewmen fall or leap from the aft nacelle (gondola). There was a vast blossom of liquid flame and a deafening concussion as, with much the force of a volcano, the great aerial cruiser erupted amidships.'"[199]

When burning, hydrogen cannot provide lift. This is an easy concept to grasp; oxy-hydrogen fire would instantly incinerate or melt the top of the cell and the remaining gas would escape. In the book ZEPPELIN, Heinrich Mathy reports witnessing the L32's final moments: "First the bow burned, and then the flames tongued over the whole envelope. The aft gondola broke off, and the wing cars followed. For eighteen terrible seconds the blazing ball hung like a fateful planet in the sky; then it burst asunder." In another action from his LZ-98, Lehmann saw "The flaming mass hung in the sky for more than a minute; then single parts detached themselves from it and preceded it to the earth." L34 was attacked by a BE2c and took 8 minutes to fall. This would have been possible only because its lifting cells were still somewhat intact. Meanwhile, the covering burned off so quickly that even *paper* inside the control car survived.

Most incredible was the L48, witnessed by their sister crew in the L42: "'The airship suddenly begins to glow bright yellow-red. A darting flame shoots out the hull. Then a fine smoke creeps over the ship, the stern drops, and the ship plunges down into the depths. It is a column of fire, like a meteor hissing out of the heavens. They are finished over there.' In fact, the L48 fell slowly, *kept buoyant by gas cells that remained intact*." [200] If there was any doubt about what was burning, author Chamberlin pointed out in a caption, "Looking down from an aeroplane on the blazing L48 showering the sky with a fiery confetti of fabric." Two of the survivors, Mieth and Ellerkamm, indicated it took 7 minutes for the still largely buoyant ship to reach the ground.

In spite of the formidable three-part ammunition, exactly five Zeppelins were brought down on British soil: SL-11, L31, L32, L33, and L48. Of course SL-11 was a wooden framed S-L airship, so its great bulk was totally flammable. L33 did not burn until its crew crash-landed it, stepped out and set it on fire. Two others took 7 and 8 minutes respectively to fall, meaning some cells remained intact. These actions must be kept in mind to properly appreciate the visual record of the LZ-129 accident.

"How the accident happened is not known, but the fact that the airship had been cruising in a heavy thunderstorm leads to the suggestion that she may have been electrically charged, but it is unwise to accept this as the true explanation because special precautions against such a contingency were embodied in her design and it may be assumed that there was some other cause."[201] Indeed. The most well-recorded accident in history offers many more clues to what is really going on in the infamous fire.

---

[199] B. Cox, interviewing Bernard Peake, "Zeppelins, Instruments of Divine Wrath," *Florida Today*, (newspaper) 3 April 1987, pg. D-1
[200] Emphasis added. P. Fritzsche, A NATION OF FLIERS, quoting Marben, *Ritter der Luft*, pg. 103
[201] M. J. B. Davy, F.R.AeS., AERONAUTICS: LIGHTER-THAN-AIR CRAFT, Her Majesty's Stationery Office, London, 1950

Chapter Five: Fuels and Oxidizers vs. Smoke and Mirrors

Sir Arthur Conan Doyle had his character Sherlock Holmes say, "I can discover facts, I cannot change them." Today, if one looks up "hydrogen" in a typical encyclopedia one will probably see a photo of the LZ-129 burning. In this case the facts are not subject to change, but interpretation of evidence is surely subject to preconceived notions. Certainly disturbing are the many photographs taken during the accident. The hull shrinkage visible suggests a partial vacuum attributable to the fire convection column, and the plume reaches a peak bursting from the bow and the stern hits the ground. Yet, there was nothing present that would provoke or allow hydrogen to explode, and burning hydrogen *doesn't make a fireball*!

With airplanes of the Great War, there never has been much mystery about what caused combat losses. (Fritzsche:) "The most effective way to destroy an opponent was to aim for the fuel tanks, which in two-seaters carried up to 300 liters of highly flammable fuel. A fuel leak released a cloud of white vapor, which tracers or incendiary bullets easily ignited, or splashed fuel directly on the hot engine, which caught fire in front of the pilot's face. Headwinds fanned the flames toward the pilot and quickly engulfed the entire plane. Even when a pilot managed to shut the engine off, a fire, once started, was almost impossible to contain and usually ended in the pilot's gruesome death; many preferred to jump to their deaths rather than be burned alive." Not necessarily second to someone shooting at them, early airplanes failed in *accidents* at a greater rate than combat losses. Two of Germany's most famous aces, Max Immelman and Oswald Boelcke, both survived many combats, only to die in airplane accidents.

By contrast we have the somewhat contrived newsreels shouting "*HINDENBURG* Explodes! ...The inferno of blazing hydrogen cremates all living beings in its path!"[202] Instead of only just such a jaded description - "blazing," indeed - we have a visual record of the last part of the infamous event. The newsreel cameras were at Lakehurst to cash in as they did with the many 1936 interviews of the rich and famous traveling on the world's fastest transoceanic vehicle. Said to have been hastily activated by a single apprentice,[203] their films recorded only the final seconds of a conflagration that *does not look anything like* a hydrogen fire. There was no synchronous sound possible in the hasty startup, so sound was added later. The tracks - some quite laughable when heard today, some complete with horrific screams adding to the crashing and burning sound effects - made each newsreel unique even though they had shared each other's footage.

Already after 7 p.m. on a rainy night, it had not been practical to get lots of landing footage for "establishing" shots to be run with the interviews. The backlighting at twilight was awful, as the cameras were facing into the clearing sky. Like taking a facial picture in a house with a sunlit window in the background, the subject would be dark, as if in silhouette. In the 1994 book *HINDENBURG:* AN ILLUSTRATED HISTORY with text by Rick Archbold, Toronto's Madison Press ran a series of landing photos across pages 180-181. The airship is little more than a dark shape against the clearing sky. Without telephoto lenses the true altitude of this unique landing is also appreciated in this series. Only the last photo in this series has been

---

[202] Hot Shots/Cool Cuts item # CC 2439. Even impure hydrogen so loosely contained can't actually explode.

[203] The ABC TV show GOOD MORNING AMERICA, on both the 45th and 50th anniversaries of the crash, repeated the story: the cameramen, unable to get useable footage in the stormy dusk and chilled by the drizzle, left an apprentice by the cameras and went off in search of hot coffee. Courtesy Alan Gross.

"dodged," a darkroom technique used to correct underexposed portions of a photograph. Faced with the need to lead into the crash film, any amount of stock footage from various 1936 landings was spliced in ahead for "establishment" shots. The ship thereon is often brightly lit and usually so low the '36 Olympic rings are easily visible.[204] Others would be later convinced of the ship's "unexplainable" tail heaviness as some scenes show water ballast being dumped aft, without realizing this common occurrence was filmed in 1936.[205] Scriptwriters would conclude the dropping of water ballast was "a sign of trouble," even *impending doom*, and many TV programs would splice several occurrences of this typical procedure together to make it appear urgent. Actually, as the airship slowed and lost dynamic lift, dropping ballast was necessary most every landing, hence the large amount of film recorded of this action.

If all that sensationalist media were not enough to convince mankind airships were dangerous, then Herb Morrison's legendary record was the final nail in the coffin. One newspaper account reads: "The National Broadcasting Company last night at 11:35 broke its ten-year ban against the use of electrical transcriptions to broadcast a recording of the *Hindenburg* disaster...The completed disk was flown to Chicago where network officials heard it and decided to make it available to stations on both networks." (More than just bad press for airships, some saw this as a giant ethical step backwards, when for the first time news stooped to becoming entertainment.) Modern video programs on the disaster discard the corny 1937 sound-added tracks and pipe in Morrison's record, of course without any timing clue, as though it were synchronized sound.

As recently as 1998 the producers of a TV show hawking "the dangers of hydrogen" contacted the author about royalty-free footage of the LZ-129 fire so they could make some money like everyone else. (Such people are always disappointed to hear that the film does not show hydrogen burning, but it will make money for them anyway.)

Well respected airship historian Dr. Douglas Robinson would share his discovery of eyewitness Mark Held with Lighter-Than-Air Society readers. " 'Oh heavens, the thing is afire...up along the top ridge' – before there was a big burst of flaming hydrogen from a point ahead of the upper fin." [206] With all due respect to Professor Held, and Dr. Robinson quoting him, they did *not* see a burst of flaming gas. Hydrogen just does not work that way.

In recent decades, people who manufacture, move and consume hydrogen look at the crash photos and wonder how the bright, smoky forest-fire like flame canopy, direct evidence of cellulose combustion, could ever have been mistaken for a hydrogen fire.[207] LCDR Frederick C. Sachse testified he noticed "a burst of red-orange flame and black smoke which first met his vision; first color appeared to be a highly intense yellowish red, which seemed to deepen into an orange red with a gradual increasing amount of black smoke." Another author concluded hydrogen must burn with an orange flame. "However, during World War II, over the Thames

---

[204] The ring decoration was removed during the 1937 off-season refit. To see a true, not dodged or otherwise enhanced photo of the airship as she approached the mast on May 6th, 1937, see THE *HINDENBURG*, Madison Press, 1996. The two photos on page 180 and first on 181 are believed to be unretouched. One is identical to the Hoeling book. Note the contrast problem and the ships' height. The second, well-published photo on 181 has been "dodged" to enhance the ship's details.

[205] Rosendahl chided ignorant people who bemoaned tail heaviness at landing as being some forbearance of doom, saying the condition "was old as the rigid airship itself."

[206] Wingfoot LTAS Bulletin, Vol. 10 #4, Feb. 1963

[207] Bain performed in depth study at the Fire Sciences Laboratory in Missoula, Montana, working with the accident photos and witness descriptions.

Estuary, I witnessed barrage balloons set on fire by lightning or Messershmitts on at least 50 occasions – but never a blue flame, even for a second – only orange."[208]

Hydrogen does not make red, yellow or orange flames. In daylight hydrogen flame cannot be seen with the naked eye. Also, hydrogen's clear combustion, if added to another burning material, is not known to change the color of its flame. During the Moon Program, hand-carried brooms were once waved out ahead of line walking troubleshooters. Invisible, and radiating little perceptible heat, one can unwittingly *walk right into* a hydrogen fire. Surprisingly, NASA only recently developed a reliable hydrogen fire detector. "The sensors we currently use at the launch pads can only detect the ultraviolet emissions from a flame. When they work properly, they alert pad crews to invisible hydrogen fires…"[209]

It was the solid materials, not the gas, which made visible fire. No one had any problem understanding what caused fatal fires in airplanes competing for the Orteig New York-to-Paris Prize. In TRANSATLANTIC FLIGHT, author Joshua Stoff shows one picture of the Sikorsky S-35 and its fifty drums of gasoline, then a picture of the melted slag that remained after its fatal takeoff accident. "Early on the morning of April 26, 1927, Davis and Wooster took off on one last test, with the plane carrying a full load, including 17,000 pounds of fuel. The *American Legion* got off the ground but in veering away from a stand of trees went nose-down and crashed in a nearby marsh. Both Davis and Wooster lost their lives."[210] Yet in airships, we skip over the obvious and try to blame nature's closest approximation of a vacuum – even when it really isn't there. "The crew attempted to destroy their airship [L33] after landing but there was little hydrogen left to burn, and it was only the doped fabric that flared up when they fired signal flares into the ship."[211] There certainly was nothing invisible or undetectable about LZ-129's solid materials. The weight schedule published for the ship can be amended as follows:

Envelope	18,000 pounds (cotton, aluminum powder, dope)
Gas cells	24,500 pounds (cotton, coating, and film)
Gas cell netting	1,000 pounds (doped ramie cord)
Diesel fuel	7.5 metric tons (77 hours consumption of 50.5 ton fill)
Lube oil	1,000 kilograms (consumed from 4000 kg at liftoff)

Visible or not, it is easy to blame hydrogen for the fire's severity based on energy by weight. One example given speculates that if there was 28,000 lb of hydrogen left, each pound containing a whopping 51,552 Btu, the gas would contain about 1.443 billion Btu. If 65,000 lb of diesel was left and it was the grade containing 18,113 Btu/lb, the liquid fuel would contain a seemingly lesser 1.177 billion Btu. Yet this is a game of densities and reactions, with the solids and liquids winning the misleading Btu picture in detailed analysis of how energy is released, or transferred from, energy carriers.

Starting with the outermost line of defense, the outer cover faces the static-filled atmosphere. Of rigids' outer covers, Norman Peake wrote, "These needed to be as light as possible, due to the

---

[208] N. Peake, "What Happened to the Hindenburg?" *Dirigible*, [AHT] Vol XI, No. 2, Summer 2000, pg. 14

[209] *Spaceport News*, June 30, 1995, in an report detailing a new hydrogen flame detector that was to pioneer the use of digital signal processing to analyze input from both ultraviolet and infrared sensors yielding a much more reliable alarm system.

[210] J. Stoff, TRANSATLANTIC FLIGHT, A PICTRE HISTORY, 1873 – 1939, Dover Publications, Mineola, NY, 2000.

[211] G. Kinsey, PULHAM PIGS, Terrance Dalton, Ltd. Suffolk, 1988

vast area, yet adequately strong and weatherproof though gas-proofness was *not* a requirement. In fact, World War I Zeppelins had their upper panels of a weave calculated to pass hydrogen rapidly, so dangerous mixtures with air could never accumulate beneath their covers. Later, hooded vents were provided instead." As mentioned previously, the LZ-129 linen itself had the Btu equivalent of 9,720 pounds of coal. Coal is listed as producing 94.1 Kilocalories (thousand calories) per mole. There has been no specific calculation on how much extra energy was added by saturating this cellulose with nitrate or 'cellon' dopes.

Yet in airships, intermittently at least, since the Hodgeman fabric made for Baldwin in 1907, outer coat(s) of the dope were impregnated with *aluminum powder*. The metal helped reflect the sun's heat and decrease control problems associated with temperature swings entering and leaving cloud banks. Also used where there were no gas cells to worry about, such as the fins, the metal extended the life of the lightweight fabric in the harsh atmospheric environment. The 1924 manual stated: "Light and heat are enemies of the material used in the gas cells... Heat rays from the sun passing through the outer cover heat the gas and the air in the ship and effect its state of equilibrium. To counteract this, aluminum powder is used in the last coat of dope, which gives the dope a reflecting surface and also makes it largely impervious to light. The pigment (aluminum powder) thus increases the life of the fabric many times... about ¼ pound of 140 mesh aluminum powder is usually mixed with each gallon of thinned dope." As such, this desirable innovation overcame its weight penalty and was used on all Zeppelins from LZ-126 onward. Von Schiller explained, "Over the framework of the outer cover, of flax or cotton fabric, is stretched, covered with a coat of varnish in order to secure as smooth as possible a surface. This varnish is mixed with aluminum pigment in order to make it reflect the sun's rays and thus prevent the absorption of heat, which would *reduce* the lifting power of the gas."[212]

Unhappily, such metals are also used in high-energy *propellants* to increase the total heat released during combustion. The most commonly used *propellant metal* just happens to be aluminum, because of its high heat of combustion and combustion efficiency. A study performed at the Los Alamos National Laboratory had shown that, in oxygen, a test aluminum rod burned at the rate of 25.4 centimeters *per second*. Water makes things worse. The Material Data Safety Sheets (MSDS) for aluminum powder all emphasize that, when burning, *not* to fight its fire with water least a violent reaction ensues.

Even worse is aluminum powder's explosivity. A jar of polishing powder exhibits the warning that the contents are explosive for good reason. The list of materials with which the powder has bad reactions is a page long, including "Oxidizers, fire and explosion hazard, " and "Oxygen, spontaneous ignition and possible explosion," to name a few. Following several spectacular fatal accidents, industrial safety practices forbid ordinary oxygen to contact it; oxygen tanks can no longer be made of aluminum. The MSDS warns "Dispersed particles are explosive; hazard increases as fineness of particles increases."[213] The MSDS also warns of an explosive mixture

---

[212] Emphasis added. This may seem like an error, since he continues, "Thus it can carry considerably heavier loads in winter or in temperate zones than in summer or in the tropics." But we must remember von Schiller, like all hydrogen airshipmen, immediately felt the effect of superheat as loss of lift because the cells were normally full. He had written, "With an increase of 1 degree C. in the thermometer, gases expand by 1/273 of their volume, which means, in the case of the *Graf Zeppelin*, a decrease of some 880 pounds of lifting power." Hotter gas of course gives greater lift; but a full-cell ship would lose lift right away through overpressure valves as the gas expanded.

[213] The author received an e-mail insisting the outer cover fabric, "no matter what it was doped with," could not be set alight by static electricity. That same night, watching the motion picture THE SIXTH DAY, the author noted the main character criticizing his clone as they made a bomb to blow up the bad guy, saying "you are not putting enough *aluminum* in there!"

when ammonium nitrate is exposed to it. (A TNT/ammonium nitrate mixture called Amatol had aluminum powder added to enhance effectiveness.) Ironically, it was powdered aluminum mixed with phosphorous (in a hollow Buckingham bullet) that brought down the Zeppelin bombers!

Worse yet, aluminum has the ability to react explosively and powerfully with water under suitable high temperature initiation. The reaction generates 2.5 times the heat as the same weight of natural gas. In other words, soaking from the rainwater actually made the LZ-129 aluminized cellulose fabric burn more furiously – even explosively. Even if the aluminum had burned alone in the air, the metallic fire would have contributed immensely to the radiant heat available and the corresponding ability to spread thc fire. Firefighters knew not to fight such a fire with water; in those days, one could only isolate the burning material, ringing it with sand.

As an element in the cover matrix, aluminum likely acted as an accelerant for the cellulose component. Mother Nature provided more fuel to the fire, ironically in the form of rain. However it was not only exacerbating the metal burning: "...cellulosic material when wet becomes more flammable..." Hawley's Condensed Chemical Dictionary states: "Hazard: Cellulosic materials when wet with water are a dangerous fire hazard. Yellow label is required for all rail freight; not accepted by express." One could argue this effect was minimal due to the selection of cellon specifically for its waterproofing properties, but some moisture was still involved.

While it might be assumed the nitrate dope would be similar to lacquer paints of today, the label would have to read something like "Avoid strong oxidizing agents, strong acids or bases, and strong alkalines." Between the fabric, dope and aluminum, the matrix was a formidable hazard when wetted by the rain that fateful night. Once ignited, the metal-impregnated covering assembly would produce a flame temperature of more than 3,000° F. The propellant-explosive aluminum would lead the fire and would be self-sustaining to the limit of the available oxygen. Emmons shows with the famous fire triangle that combustion is kept in check by the amount of oxygen available. But the LZ-129 covering material fire was *not* limited to how much oxygen it could suck up out of the air. The doped fabric provided some of its own oxygen to support the imbedded aluminum's combustion.

Ramie cord netting spanned the interior structure, shown here on LZ-126, which transferred the cells' ascentive force to the framework with minimal chafing. It had also been nitrate-doped. LZ-129 contained about half a ton's worth of the cellulose-based netting. Photo courtesy the Naval Airship Association.

Tens of miles of twine-like ramie cord was keeping the cover lashed to the structure. Just inside, acres of cord netting were employed to prevent cell chafing on the girders as they lifted the airship's weight. Some 72 miles of ramie cord had been consumed making the much smaller LZ-127. Doped as well, ramie cord added more weight of cellulose to the available combustibles. One of the most difficult concepts to grasp is the lifting gases' inability to make highly flammable materials burn even faster than they otherwise would without hydrogen escaping around them. To burn, hydrogen must have oxygen; oxidizers do not themselves burn, but adding more will increase a fire's intensity. Oxy-hydrogen is used in welding and cutting, largely in brazing operations, underwater cutting, and working on glass. Anyone who has worked a cutting torch will tell you, adding more fuel to the torch only makes an improper mixture. Mashing the toggle that pours in pure oxygen really gets the cutting going. And, *even more* oxygen than the usual 20% or so was readily available that night, owing to the thunderstorm.

Electrical storms like the one that passed over Lakehurst that fateful day produce huge quantities of *ozone*, an allotropic form of oxygen. This much more unstable triatomic oxygen gives the air a "clean" smell after a storm passes, but it is no good around tons of aluminum. If industrial safety no longer allows aluminum to contact ordinary volatile oxygen, one can understand why the triatomic form is even more dangerous around both the metals and the nitrate. Of greater importance to a fire sucking up the oxygen in the surrounding air, ozone is a very powerful *oxidizer*. Unlike adding more fuel to a fuel-rich fire, a radical oxidizer would certainly make a bad fire much worse. Since electrical arcing produces ozone, it is only logical that the area of the highest voltage and greatest charge would also be the richest in ozone.

Gas cells were cellulose films sandwiched between two cotton cloth piles. Then several layers of mixed compositions of latex and gelatin were applied. About the color of khaki,[214] the cell material was so thin as to be translucent when backlit. "There has eventually been produced a gelatine-latex fabric in which many coats of latex and gelatine are spread over fabric; this matches the goldbeaters' skin in weight and shows considerably higher resistance to gas diffusion. More than twelve acres of this fabric are required for an airship's gasbags." Its major component was *cellulose* again, and there was more than a dozen *acres* of it hovering that night.

Since the accident, much speculation concerning the gases' conduct revolved around it necessarily escaping the cells, then being retained by the outer cover. Hydrogen's ability to diffuse through seemingly solid materials was not emphasized, but recognition of this exposes an important fact. This diffusive nature permitted the construction of early gas leak detectors. As the 1940 manual explains, "The action of the instrument is based on the different rates of diffusion of hydrogen and air." Running the detector over the cell or balloon, "...when it is exposed to an atmosphere containing hydrogen, the hydrogen diffuses through the porous wall of the closed chamber at a rapid rate and a pressure is built up within, which is indicated by the pressure gauge." In other words, the fabric outer cover offered no serious barrier to hydrogen heading for the stratosphere.

The heaviest energy carrier aboard any airship is its petroleum fuel. Danger aside, its ponderous weight posed a big problem for designers. Karl-Heinz Hoffmann, wrote of the S-L airships, "The weight of 18 voluminous petrol tanks in the interior of the airship was spread over 18 ribs. In each of the two nacelles a small petrol container served as an intermediate depot from which petrol was taken directly when the need arose... Another major function of the frame was to hold

---

[214] Author asked *Macon* rigger William Clarke the cells' color when there was an effort to make a very accurate model kit of the ZRS ships. It was an example of the type of question usually not conceived in time to ask surviving veterans. Sadly, there will be many more after it is too late. Powerful odors make unforgettable memories, many of those involved with the LZ-129 accident recall the odor of singed hair staying with them.

the main gas tanks which had the form of spherical balloons. The spaces between the balloons were filled with balloons which had the form of toroids with a triangular cross section. They were arranged concentrically about the main longitudinal axis of the airship and were encaged by steel wire rings." Hugo Eckener agreed with British publications and had said he considered gasoline to be the airship's greatest safety hazard. Barkley and Buckley wrote an editor's note, "The following paragraph contains the French suspicion, which was accepted as fact in Germany, that dangers of gasoline-vapor fires were far greater that any hydrogen fire hazard." Thierry followed, "The use of high grade gasoline as fuel is a dark point, perhaps the worst one. On board all air vessels the most frequent fires are caused by this fluid." A concern justified, as gasoline produces a whopping 1,302 Kcal/m. Its lower flammable limit is 1.4% and "flowing gasoline can be ignited by self-generated static electricity; containers should be bonded and grounded." Frenchmen examining German hangars after the Great War noted, "Gasoline is never handled in the open air. It is stored in subterranean reservoirs of 30 to 60 cu m capacity. The upper part of these cisterns is filled with some inert gas, generally carbon dioxide..."

Generally perceived as safer owing to its higher ignition temperature, diesel fuel actually contains more energy than gasoline, thereby helping justify its weight. The Zeppelin tanks designed to carry it are wafer-thin, motion pictures showing a man lifting a huge tank with one arm. Their mountings are also fragile, and not just to save weight. In an emergency, airship tanks were designed to be cut loose so their weight would punch through the outer cover to clear the ship. On the west coast flight of the USS *Akron*, a gasoline tank collapsed and filled the passageway with dangerous fumes that lingered for a considerable nervous time in spite of efforts to vent them. When sister *Macon* settled into the Pacific, gasoline went everywhere, floating on the surface. The heavier diesel fuel would do the same when its tank mounts were smacked. A caption reads "The engines of the German airship burned a light oil that was less flammable than gasoline."[215] The exact grade of diesel used in the LZ-129's DB 608 engines is not known, though the enginemen remembered it as having a very low flash point. So those fragile, delicately mounted tanks, could have carried the least dense diesel grade. Lowest on the cetane number chart is naphthalene, containing about 1,462 Kcal per mole.

Hydrogen is a fuel as well, and like those mentioned, is a secondary energy source, or energy carrier. While it burns with much great efficiency – some say 50% greater than gasoline – when consumed in an engine it yields only 57.8 Kcal/m. This means hydrogen carries about the same energy as the basic building block of the covering, the cloth itself, without the explosive-propellant metal added. From an energy standpoint, hydrogen was about the least of the evils even if *all* of it burned, for good reasons.

In contrast to the luxury promenade, LZ-129's heaviest energy carrier rode along in fragile *Behälter* suspended along the seldom photographed *Laufgang Keil*. The 66 tons of 'light' diesel oil could have pushed the monster airship from San Fransisco to Manila non-stop and, as suggested, used a hook-on airplane for limited contact with Honolulu along the way. Ludwig Dürr supposedly found a few torn-out pieces of the ship's structure some ¾ mile from the mooring circle. Today we think nothing odd about airplanes blown to bits from fuel tank explosions from far fewer enclosures, but in LZ-129's case it somehow worked out to be a 'hydrogen explosion.' Photo courtesy Hepburn Walker, Jr.

215 G. DeAngelis, THE *HINDENBURG*, Chelsea House Publishers, Philadelphia, 2001

Before energy is yielded in combustion, an ignition source is necessary. The self-ignition temperature of aluminum powder alone is 959° F, but ozone *reduces* aluminum's ignition temperature. The fabric covering lights off at 195° C and the doped ramie cord would no doubt be similar. It is not known what temperature would get the featherweight gas cell fabric to burn, and this should have to be measured in the real world condition of gas containment's cooling effect. We can assume, like the covering, it is somewhat lower than its components taken one at a time. By contrast, oxy-hydrogen needs a temperature of 457° C to begin combustion. In other words, most every part that made up a Zeppelin would light off *before* its lifting gas would reach burning temperature.

Airship lovers shudder to think of what they fear was a gargantuan energy source in the lifting gas. They tend to forget the gas was used because it is the closest thing to a vacuum available. Again: *hydrogen is the closest thing to nothing as nature has to offer.* People who are pushing for the use of hydrogen as fuel are ridiculed because it takes so much volume of nature's simplest element to yield any realistically useable energy. Called three orders of magnitude too small for acceptable power, there is another problem with getting enough energy out of the gas.

The key to understanding hydrogen's role in the LZ-129 fire is directly related to how it is used as a fuel today, and is related to the real secret of the LZ-127 *Graf Zeppelin*. While it is desirable to burn renewable and clean hydrogen in our automotive engines, we find we can mix no more than about 30% hydrogen with another gaseous fuel before a normally aspirated engine delivers unacceptably low power. To use a higher percentage, right up to 100% hydrogen, we must add more air, typically through super- or turbo- charging. The unmatched efficiency of the *Graf* was directly related to its mixture of hydrogen and propane in its fuel, which not only weighed the same as air, but put in as much hydrogen as a normally aspirated engine could use.

If the lifting gas got away from its containment, and an ignition source was present, we must realize nature has no blower to provide pressurized air. Lacking a turbo, nature simply cannot supply enough air for all the gas to burn before it dissolves into harmless percentages. If allowed to, a fire will draw as much oxygen it needs from the surrounding air. When one of the 1000-foot long timber hangars of the WWII Tillamook NAS burned down in August of 1992, the suction drew air into the firestorm at 80 mph speeds. Since hydrogen in the cells was the innermost of two levels of flammables, it is not clear how much oxygen would be available *within a fire* to mix and possibly support another fire.

Airship lovers may not want to believe their eyes, but what was seen and photographed burning in the LZ-129 fire *really was* the greatest source of energy in that accident. What you see is what they got, and the witness statements are telling. Some passengers looking out of the airship's windows were first aware of a problem when the landing field's grasses were illuminated below. "In the control car Heinrich Bauer…from his position on the port side, monitoring both water ballast and hydrogen, noticed Hangar One suddenly lighting up brightly."[216] "Mrs. William Craig, wife of the Standard Oil magazine writer, who was now standing against the spectator's fence, thought the sun had finally burst through the storm clouds…When Mrs. Spah first saw the flash of light she thought it was fireworks. Chief Emil "Babe" Klaassen, a veteran of the *Macon* who had been assigned to the tail end of the ship, noticed the sky light up... Monty Rowe…saw the fins light up strangely." [217] Others recalled "flashlight snapping on" and "like a flashbulb

[216] R. Archbold, THE *HINDENBURG*, Madison Press, 1996, pg. 182

[217] J. Toland, THE GREAT DIRIGIBLES, THEIR TRIUMPHS & DISASTERS, General Publishing Co., Canada, 1957; also Dover Publishing, New York, 1972

suddenly going off. "Vincent Sheridan, the line handler, noted 'The first and only inkling we needed that the ship was gone was a sudden lighting of an otherwise gloomy afternoon.' ...Henry Roberts, the radio editor from *Aero Digest*, observed a 'yellow flash', followed by a bright explosion'... Another said 'brilliant as a thousand, ten thousand, magnesium flares...' Picture-snapping passenger "Otto Clemens did not even know he had just photographed the reflection of the fire on the wet earth..."[218] Ground crew member Leroy Comstock relates to the "flash of light."

Hydrogen flame wavelengths are mostly in the ultraviolet region, 4340 angstroms and beyond, with some energy in the infrared region. It cannot produce serious illumination or even light visible to the naked eye, or expose ordinary photographic film. Yet crewman Helmut Lau said "It was extremely bright in the lower fin." Since hydrogen could not light up the grasses, brighten the normally dark interior or reflect off the cloudy sky, what was making all that raw light just as the fire began?

Hollywood special effects men use *handfuls* of aluminum dust to create a brilliant fire, simulating flashes of lightning off-camera. Likewise, aluminum dust - *several tons* of aluminum dust - was lighting up Lakehurst the night of May 6th, 1937. A testimonial at the time, the excited voice of radio announcer Herb Morrison, immortalized via the record he was cutting, vividly demonstrates the astonishment caused by the brilliant light. "It's broken into flames! It's flashing – flashing! It's flashing – terrible!"[219]

Just as we have no photos of a hydrogen fire, neither have we heard about one: participants and witnesses, who were promptly interviewed, *did not feel* a hydrogen fire. *When burning, hydrogen does not radiate heat perceptible at a distance.* Toland's interviews yielded "Chief David 'Doc' Safford, another *Macon* survivor, who had his back to the ship's bow." Safford was waiting for the nose mooring cable to touch ground so it could be safely grabbed without becoming a discharge path, a lesson illustrated in the *Macon* ground handling film produced at Moffett Field years earlier. "But before the cable could reach its destination, Safford felt a terrible blast of heat." Philip Lenz, listed as chief electrician on the ship, told the inquiry of using a metal shield to ward off the heat. Rosendahl called it "The intense white heat of the hottest flame man knows." Others tell how rescuers couldn't advance because of the blistering heat. We've just never realized this furnace-like blast *was not* provided by hydrogen.

Like other thrill ride operators, Universal Studios uses petroleum fuel mixes to flash their fires' blistering heat into the faces of happily terrorized patrons on the "Jaws" and "Earthquake" attractions. Emissivity, or heat transfer, from a gasoline fire is anywhere from 30% to 42% of its impressive 17,780 Btu per pound. Its face-tightening radiant flame energy is a memorable experience. The bright light and searing heat horrified the onlookers.

Terrorized, many LZ-129 victims died not from burns, but from impact with the ground, having jumped from too great a height trying to escape the horrific inferno. We see photos of burn victims published all the time; at least the Hoehling book shows a fallen victim, and says ground crewman W. J. Barnard chose to escape toward the mast because "... falling debris and bodies seemed to be on the sides." Rosy wrote: "George Grant, a British shipping man, escaped without a single scratch or burn by jumping through a window, and then as he was picking himself up to run away from the fire, suffered the great misfortune of having another person leap from the

[218] A. Hoehling, WHO DESTROYED THE *HINDENBURG*? Little, Brown & Company, 1962 pg. 108
[219] This is John Toland's transcription. Author hears it as "It's *burst* into flames!"

same window and land squarely in the middle of his back injuring him severely." Amid the Doehner family, almost half were lost; Mrs. Doehner saved both her sons by throwing them out the windows, and jumped herself. [220] Of course one would not know this from reading the coroner's reports, since in most cases they simply list cause of death as "zeppelin explosion" and contributing factor as burns suffered. A moment's thought would reason a badly injured or unconscious person unable to run clear would be badly burned as well.

Devoid of its supporting cells, the fragile Zeppelin brought down in the Vardar marshes had to have its framework propped up for a victor's display. While the flammable materials, ignited by the defender's special ammunition, burned completely and ferociously, it was over so quickly even the undropped bombs (foreground) survived the relatively gentle, hydrogen-slowed impact. Even paper survived in some cases. At Lakehurst, Robert Rutan found a book that appears to be a LZ-129 gas cell log book; the front page is burned away, but interior pages are still readable. By contrast today, NTSB investigators will tell you a very common element in airplane crash witness reports is a loud explosion *before* the plane impact. The ullage in large petroleum fuel tanks are not unlike bombs themselves. Author's collection.

The heaviest loss of life among the crew was in the nose, where the bright geyser of the outer cover's aluminum fire and confetti of fabric (widely captioned as pillar of hydrogen fire) supposedly incinerated everyone sent forward for landing. Some writers have suggested the axial corridor acted as a hydrogen stove pipe baking all the occupants inside. Again, no death by hydrogen inside the ship: electrician Leibrecht had a good grip and survived to tell how he endured the sight of eleven other men, waiting in the nose, *loose their footing* as the steeply sloped nose keel walkway, already at a steep upward angle, approached and then passed the vertical angle. A little further aft on the centerline passageway stair, Albert Grotzinger was able to hang on and lived.

Inside engine car IV, Eugen Bentele would hang on to the roof spars only to be thrown over the engine and out the open end of the gondola. [221] Sadly, with the heavy diesel and its associated drive train elements spinning while impacting the ground, several other enginemen would be killed, most heavily on the starboard side. It is impossible to say how many persons were victims of falling, or secondary impact, rather than jumping, amid the crew or the guests. Passenger Joseph Spah would later tell of two men that were hanging on to him and how they jumped through the windows from more than a survivable height. Spah himself, a trained acrobat, waited a bit longer but still broke an ankle when he jumped.

One might ask, so what if they couldn't see it or feel it? That was *hydrogen* in there! Almost thirteen *tons* of it remaining at that point in the trip! We are reluctant to give up the long held belief that some five and a half million cubic feet of flammable gas – imperceptible to our senses

[220] Interview of Nurse Mary Shannon, who treated the surviving Doehners.
[221] EUGEN BENTELE, THE STORY OF A ZEPPELIN MECHANIC, MY FLIGHTS 1931-1938, Zeppelin Museum, Friedrichshafen, translated by Simon Dixon, 1992

or not - greatly hastened the airship's destruction *and* fatally injured most of the victims. This belief is based on the common misconception that the pure gas could burn from *inside* the cells, outward. The characteristics of the airship and the true nature of hydrogen make such wanton destruction by gas less much likely than imagined.

Unlike the outer cover and cells, completely surrounded by oxygen in the air, the lifting gas was isolated from oxygen. *To be flammable, hydrogen must be uniformly mixed with more than 25% air by volume.*[222] The often-quoted 4.1 lower limit is the gas-in-air figure; introducing air into a massive gas cell is just the opposite situation.[223] A cell of nearly pure hydrogen gas is not extremely easy to ignite, contrary to sabotage theorist A. A. Hoehling who suggested a flash bulb provided both ignition *and* the required air. While the 1975 fifteen million dollar movie dramatizing the "lone nut" theory was seen by millions, one only needs to consider the lessons of WWI - including how the crew of the L49 was unable to set their cells afire even by firing flare pistols into the cells. The power for the light by which you are reading this might very well come from a generator bathing in pure hydrogen. Electrical generators are run in a hydrogen bath to reduce windage losses; the gas also acts as an insulating coolant.

By some chance, was it possible the LZ-129 gas had suddenly become impure? Least we believe some air-exchange contamination was possible, we must remember, airships are air-displacement vessels. The masters of the rigid airship were not fanatical about gas purity for just flame safety reasons; *gas purity was their livelihood.* Every ten cubic feet of air allowed to leak in meant more than half a pound of payload could not be charged a fare. A rule of thumb sometimes quoted says for 1% contamination the cell looses 1% lift. Small wonder when their frequent purity checks detected air contamination of just two percent in any given cell, it was dumped and refilled with fresh gas. Impurity of even three percent in just one 300,000 cubic foot cell – let alone all sixteen - would mean two people - stout people, with their luggage, or that equivalent weight - would have to be thrown over the side. Simply stated: any hydrogen airship whose lifting gas was so contaminated as to be flammable simply wouldn't fly!

Had the Zeppelin's materials been less volatile, pure gas busting from melted or ruptured cells would tend to *smother* the fire. The national hydrogen balloon race begun on May 31, 1928, provided ghastly proof. Helplessly entering a lightning-rich thunderstorm, two balloons were struck and one occupant was electrocuted in each. Ward Van Orman's balloon envelope caught fire from the bolt that killed his basket mate Morton - yet it appears the bubble of pure gas actually snuffed the flames leaving enough fabric to act as a streamer parachute, thus saving Van Orman's life.[224] Dr. Bain recalls the early days of the moon program when hydrogen fuel trailers would occasionally vent. At night, specs of light caused by contaminants in the air would suggest the presence of a fire. "We simply opened up the vent valve and snuffed the fire out with [pure] hydrogen." [225]

---

[222] A common reference figure is 25.4%. C. deF. Chandler and W.S. Diehl: *Balloon and Airship Gases*, Ronald Press, New York,1926. (Courtesy F. Marc de Piolenc) quote a lower limit of about 15% air in gas.

[223] It is also harmless when lean: TRW employs a lean hydrogen-air mix in its automotive safety airbags. Yet there is no cause for concern. In the real world, 8% gas in air is the lower limit. The people who wrote the book on H2 safety - one of whom was Addison Bain - bumped it down to 4.1% for sensor calibration purposes, i.e. set the sensor for lower than flammable so inaccuracy will err on the side of safety. Chandler and Diehl (above) quote a more real world 9% to 64% gas-in-air, verified by real world experience.

[224] THE WIZARD OF THE WINDS, Ward Van Orman as told to Robert Hull, North Star Press, St. Cloud MN, 1978, courtesy Eric Brothers.

[225] *Do not* try this at home! Of course NASA safety did not condone such practices and trailers have long since been helium purge-equipped.

The insides of a Zeppelin provided no exception to the rule of pure gas snuffing fire. (Cross:) "A shell burst squarely inside one of L-33's hydrogen cells but for some inexplicable reason did not set the gas on fire." Not unexplainable at all; pure hydrogen is such a good fire extinguisher, "atomic hydrogen" welding has been perfected. In this method, the welding is carried out in an atmosphere of high-temperature hydrogen. The hydrogen bath is very effective in reducing oxides or in preventing them from forming, thereby eliminating porosity in the weld.

On the ZRS ships the deep main rings supported the engines' cooling water tanks (left). The adjacent heavy lube oil tank (center) fed a smaller service tank and separator in the engine room below. Design of the reinforced keels had to accommodate smaller 110 gal "reserve" tanks, six of which were equipped with remote dumping valves. Electric pumps fed main 400 gal. (2400 lb) service tanks (far right) that gravity-fed the engines. *Macon* pitched up enough to starve them anyway. G-Z photos courtesy LTAS.

However, one thing is of even greater importance in the fire's analysis than the total calorific value of the combustibles, or their ignition temperature, or natural extinguishing abilities. That is *the process by which their heat energies are transferred to neighboring materials.* With hydrogen, it's a story of diminishing returns. As one e-mail correspondent put it, "aside from the extremely low calorific effect of hydrogen, the gas would have burnt off immediately and expended most of its energy heating air and surroundings. Thus the heating value of the diesel [listed as 42.5 MJ per KG] was tens of thousands of times greater than that of the value of the hydrogen. It is one of the reasons hydrogen is such a safe energy transfer carrier... but rather inefficient as a heating medium." Furthermore, hydrogen burns six times faster than gasoline; such short duration also figures into its lack of ability to raise the temperature of secondary materials until they emit flammable vapors. Energy can be transferred to or between combustibles by radiation, sparks and flames. *Heat radiation is considered the most important process in fire spread.*[226] Fire aboard steel ships would not seem to be such a problem, yet heat radiation makes fire even more dangerous in that environment. "Given the right conditions, a small fire quickly can ignite the paint or wall covering of an adjacent bulkhead, and within a few more minutes, the burning paint or covering can add sufficient heat to cause a fire to erupt suddenly – to flashover – and engulf an entire adjacent space. Searing heat levels can contribute to additional flashover in nearby compartments and passageways, creating a chain reaction and rapid fire spread that, if unchecked, can be catastrophic." [227] Unlike the doped canvas interior

---

[226] Schultz, FIRE AND FLAMMABILITY HANDBOOK, Van Nostrand Reinhold, NY, 1985, pg. 46

[227] Dr. S. C. Truver, "Fire! Fire! Fire! All Hands Man Your Battle Stations!" USNI *Proceedings*, Jan. 2001, pg. 74

walls in the airship, whose fire is intensely radiative, hydrogen fire radiates only 17% to 25% of its flame energy. This actually poses a safety hazard of a different kind; firemen *can't find* the fire. The Material Safety Data Sheet for hydrogen discusses fire fighting in a modern pipe and tube storage facility. "Hydrogen can burn with almost invisible flame of low thermal radiation. Most hydrogen fires will have the characteristic of a torch or jet and originate at the point where the hydrogen is discharging. If a leak is suspected in any part of a hydrogen system, hold paper before you and approach the suspected leak so that an invisible flame will strike the paper before striking you." Moreover, hydrogen fire benefits from the absorption of atmospheric moisture: 45% of the fractional remainder is absorbed by water in the air within a short distance of 8 meters.[228] Rainy nights were not tested but would logically reduce this further; the NASL log records the relative humidity at 98% that fateful night.

Let's go over that again, so we get it straight. The gas was stored in loose bags at less than .03613 pounds per square inch, not in much hurry to escape. If it did, an ignition source would have to be present within the time it mixed with O_2 in the air in a flammable ratio. Properly oxygenated hydrogen burns in a straight line from the source. It radiates less than a quarter of its energy in radiant heat; almost half of that quarter is absorbed even in normal humidity air very close to the flame; and it's all done in one-sixth the time of a gasoline explosion.

As if that were not enough, two other factors reduced hydrogen's role in a fire to an even greater degree. In NASA studies where hydrogen was released to typical atmosphere, in presence of an ignition source, it was found *less than half* of the rapidly dispersing molecules actually burn! This would come as no surprise to experimenters who have found they can't use more than about 30% hydrogen in their fuel mix in an engine without a blower or turbo. Make no mistake, the *Hindenburg's* lifting gas was *not* released into calm air like some NASA study. It was haphazardly released into a firestorm that would have already consumed much of the oxygen around the cover. Just as adding more fuel to a flooded engine won't get it started, the proper air/hydrogen mix allowing combustion would have been more difficult to find during the scant seconds before the liberated hydrogen ceased to be there at all.

Familiar with the mess created by a smashed gasoline tank given an ignition source, we wrongly conclude the airship's lifting gas would soak into components as well as slowly promote a flame front across the cells. As stated, most of the suddenly released bubbles of hydrogen just got away largely unharmed to dissolve in the air, especially since the area was oxygen-depleted. The small percentage that burned did not radiate more than one quarter of its flame energy. Of what little heat was thereby radiated, more than half of its energy was dispersed within a few yards, and all that would have been done six times faster than gasoline fire.

This paucite addition to the heat flux could be compared to the proverbial flatulence in what we will see was a formidable hurricane. It could not have caused serious secondary ignitions in anything that was not already burned nor would it have ignited anything that might have otherwise survived the fire. Of the greatest importance is the fact that mostly *all things that constituted a Zeppelin were flammable*; once they themselves were burning, the airship was finished even if more fuel actually could have made the fire more intense. One would be hard pressed to name, from the short list of materials not already aflame, those topside airship components specifically damaged by this additional energy, which was certainly *no where near the passengers.*

[228] "Selected Properties of Hydrogen" National Bureau of Standards monograph #168, issued 1981.

There was certainly no sense that those in high society were traveling about on a conveyance that itself was highly flammable. To the contrary, when asked what he thought of the 1975 movie, Harold Dick replied, "Where did they get that sense of impending doom? The atmosphere was light and gay on board."[229] And why not? With a perfect safety record, these privileged few were traveling on the equivalent of today's supersonic transport – only the speed came with comfort and luxury.

"The New Queen of the Skies," proclaimed a newsreel announcing the launch of Germany's new pride and joy in 1936. Indeed the ship was a marvel, "the pinnacle of man's dreams to conquer the air." The successful, record-breaking season of 34 Atlantic crossings closed in the US with "The Millionaire's Flight," on which notable and moneyed Americans were given a taste of luxury in the air. "*Hindenburg* Demonstrates Qualities to U.S. Industrialists," proclaims an article showing the beautiful Lady Ashley and her Scottie dog on the promenade. The big-business guests of Standard Oil of New Jersey's Chairman of the Board W. S. Farish cruised over six states. WWI ace and airline mogul Eddie Rickenbacker spent a little too much time in the bar, becoming too inebriated to negotiate the exit ladder. As that happy crowd left (or was carried from) the world's most advanced transport vehicle, the gay atmosphere certainly contained no sense of impending disaster.

***Kegel* secured to Lakehurst's extendable *Ankermast*, LZ-129 takes on some tonnage of *Gasöl* through the bow connections while the ship rode the *rundlauf kreis*. Tupper photo from the collection of Hepburn Walker, Jr.**

Likewise for the crew that took him home and docked him for the post season overhaul, it seemed surely their team had built a successful design. Each engine was said to consume only $300 worth of diesel fuel per trip, while each passenger was forking over $400 per trip. Just as LCDR Scott E. Peck had been aboard for the first crossing, LCDR Hugo Fischer was the USN

---

[229] Author's interview with Dick at the NAA Reunion, Jacksonville, Florida, 1989.

liaison on that last eastbound north Atlantic flight. Like his predecessors, Fischer made a report. He noted the airship consumed about 20.9 pounds of diesel fuel per nautical mile flown. Extrapolated, that would mean the LZ-127 would have had about the same range as the much larger 129! But surely diesel seemed much safer; there would be no going back to dangerous gasoline. Fischer also detailed the German preference to land the vehicle on its taxi wheels and approach the mast on the ground, in contrast to the American practice of the "flying moor," utilizing a smaller crew and heavy ground handling equipment.

The road to hell, as the parable says, is paved with good intentions. So also it seems the path to airship destruction had only the most seemingly wholesome improvements insisted upon by past experience. Guarding against another R.38, R.101's cutting edge diesel propulsion inadvertently made the post-crash aftermath more deadly. "Mr. Spanner communicated… 'Have you entirely eliminated the risk of gaseous explosion in the engine cars of R.101 by the adoption of heavy oil fuel?' The answer is 'No' – because in the same cars as accommodate the heavy oil engines there are small petrol engines, and a small petrol tank storage for driving these auxiliary engines. The gas from the petrol tank and engine in so confined a space, may easily provide the much dreaded 'seed of fire' and a fire, once started, would feed ravenously upon the heavy oil being used for the main engines."[230] So, little gasoline would be carried for fear of duplicating the R.101 disaster in LZ-129; an advertisement for the 'Vedol-Lubricated *Hindenburg*' proclaimed "It not only gives more economical power to the *Hindenburg*, but permits higher speed and greater safety. In fact, safety is promoted to such a degree that smoking is permitted."[231]

For the L-Z team that designed and built him, LZ-129 was a work in progress, incorporating as many improvements as experience had shown were desirable. The wish list had been long, and had been revised during the long period of design and construction. One of the items on the list, helium outer cells to encompass the hydrogen, had not been practical. Every other item had been checked off. Other airships' structures had failed due to corrosion; contrary to popular lore, R.100 was not retired solely due to the depression-era economy, but because her structure was corroded, and attempts to arrest it failed. As if in answer, LZ-129 would have a coat of new blue anti-corrosion primer on its framework. Other airships' gas cells had withered in the sunlight, LZ-129 would also benefit from British experience and have a base coat of red iron oxide to slow down the sun's destructive UV rays.

The *Graf* 's skipper pointed out, "The surface of the ship is something over seven acres in extent. If only a sixteenth of an inch of rain adheres, we have thus immediately about six tons of extra weight to carry." LZ-129 responded with a tough new water-repellent paint, which would make such sudden weight increases impossible.

Sadly, it was *those very improvements,* as well as the last-minute change in landing technique, that would have far reaching effects on our civilization. If we now have a better understanding of the materials' true behavior, we should be able to form a clear picture of what actually did happen when we listen to the testimony and examine the evidence from this most famous of aircraft accidents.

[230] P. Stokes, "Power Plant and the 1924 Airship Research Programme," *Dirigible*, Vol. XII No, 1, Spring 2001
[231] 'Tidewater Oil Company, 17 Battery Place, New York City'

Chapter Six: The Fire Felt 'Round the World

The *Hindenburg* had set out on its first 1937 transatlantic trip in March, depositing passengers and cargo at Rio and returning more to Germany on the 27th, much in the same way he (using the German male reference) had done seven times in 1936. *Graf Zeppelin*, preferred on the less stressful South Atlantic run owing to more favorable winds and weather, picked up the next two trips down to Rio. LZ-129 was put back in the hangar for installation of a U-shaped bar for airplane hook-on experiments. Frau Bentele, wife of the engineman who would later write of his experiences, and her friend Frau Fiederer, were told "they did not weigh much and could come aboard" for a flight.[232] Ernst Udet, who had finished the Great War as Germany's top surviving ace, tested "landing" a skyhook-equipped Focke-Wulf 44 *Stieglitz* aboard as the ship flew over Darmastadt (Hessen) on flights 61 and 62 on 27 April. The printed schedule called for LZ-129 to next make a passenger round trip to Lakehurst and be back to Frankfurt on May 9th following the removal of the "trapeze."

Just 148 kilograms of cargo was loaded for the May 3rd flight; only 36 passengers boarded, with their 879 kilograms of luggage and two dog baskets. They were accompanied by 61 crewmembers needed to attend the booked-up return flight.[233] Just past 3,000 flight hours in her log and just passed 3,000 passengers in her register, LZ-129 left Frankfurt at 16 minutes past 8pm. He then dropped some mail as he passed over Cologne. (Had the hook-on flights been followed to incorporation, the plan was to bring up last-minute mail, a major source of revenue, before making the crossing.) Saying goodbye to Holland over Flushing (Vlissingen) on the Dutch island of Walcheren, *Hindenburg* had to turn 40 miles north to avoid some thunderstorms in the Channel. This 63rd flight was further hampered by headwinds in the cloudly, rainy weather. 41 hours into the flight, a severe thunderstorm was encountered and the ship tried to climb over it, encountering the highest elevation of the trip, 1400 meters (4600 feet), logging a major valving of gas near Nova Scotia. Ice formation on all exposed surfaces accompanied the 19°F temperature drop.[234] "On the third day we sighted Newfoundland. Binoculars and cameras appeared, and my wife's delight grew when the white dots along the coast turned out to be icebergs. The captain ordered the ship to fly low and to steer onwards them. Very slowly we passed over the most beautiful, which looked like a magic marble statue. The sun came out and laid a double rainbow around the airship. The giant iceberg turned into a monument of sparkling brilliants."[235] Joseph Spah took some home movies.[236]

Later resuming his cruising altitude of 200 to 300 meters, radio call sign D-EKKA[237] was ready to report in, having reached New York about 2 PM. 31 year old WLS Radio broadcaster Herbert Morrison and his engineer Charles Nielsen had flown in from Chicago via Newark on an American DC-3 for the morning landing. "It was to be a simple story, another safe crossing, an

---

[232] Bentele's response to the author's question concerning the hook-on flights.

[233] Many of these well-heeled people were paying $400 for what was the last chance to get to London for the May 12th coronation of King George VI.

[234] H. Van Dyk, "From Coast to Coast, *Hindenburg's* Last Voyage," *Gasbag Journal* #39, March 1999

[235] L. Adelt, "The Last Trip of the Hindenburg," *Reader's Digest*, November, 1937. Both journalists Leonhard and Gertrud Adelt survived the fire.

[236] These icebergs are briefly seen in an episode of the TV series "In Search Of..." Seven minutes of home movies taken the previous year have been committed to DVD by the National Film Preservation Foundation's "treasures from American Film Archives." (www.filmpreservation.org) Info courtesy Art Paulson.

[237] P. W. Brooks, ZEPPELIN: RIGID AIRSHIPS 1893-1940, Smithsonian Institution Press, Washington,1992

interview or two." [238] They waited...and waited. Ground crewmen were employed with the stipulation they would be needed either in the morning or the evening, the times of the calmest winds, but that they would be paid for only the actual landing. Coming in after 77 hours at about 3PM and witnessing storms that might have jostled the passengers during a landing, LZ-129 went back out to the coast and loafed around for the next couple hours. A thunderstorm blew through Lakehurst about 3:30 and the clear was finally given about 5:15. LT George Watson, who had made five trips the previous season, stood by answering questions.

Operating from the Frankfurt airport base gave proximity to road and rail terminus as well. For the airship, the lower elevation there allowed greater payloads to be lifted off than in the thinner air on the Bodensee. Several new hangars, including a rotating *Halle*, were planned. Courtesy LTAS.

Herb and Charlie moved out of the hangar, recording what was to be a lead in for the interviews to come. They resumed cutting the record saying "...last minute preparations are being completed for the moment we have been waiting for *so* long." (Author's italics, Herb having emphasizing the word. The author's own Aunt Florence and Uncle Gerard had already given up, returning to their homes so as not to violate their sunset curfew.) LZ-129 became visible below the 2000-foot level clouds at about 6PM and approached the mooring circle from WSW. The winds changed again, and the ship had to go around and make a second approach from the north.

As the ship approached the circle the final time, gas was released from the forward cells, first for fifteen seconds, and then again for five seconds. This action, along with dropping of water ballast at frame 77, brought the airship into trim while reducing the altitude to 30 meters, where it was trying to stay. Second officer H. Bauer later provided notes showing angle of attack of 3/4° positive, or up inclination, of the bow. The elevator man, Eduard Boetius, kept her there, and the altitude never varied more than fifty feet during the attempted mooring.

According to later recollection, Movietone, Metronone, and The March of Time had film crews at Lakehurst, but the experienced cameramen had left an apprentice in charge of the cameras and

[238] Joel Siegel's lead-in description prior to the appearance of Herb Morrison on *Good Morning America* on the 45th anniversary show. Courtesy Alan Gross.

gone off in search of hot coffee. It was already too dark for good exposure, and it was obviously going to be a while before they could set up to do interviews. (Oddly, one of the still photos taken later would only show two cameras mounted on their trucks, and one appears to be manned. Even stranger, one version of this photo has appeared with a third camera noticeably *inserted later.*[239])

Morrison records: "The mighty diesel engines just roared, their propellers biting into the air and throwing it back into a gale-like whirlpool." [240] The heavy (98%) humidity and drizzle no doubt made the wake swirl more visible even in the fading light of cloudy dusk, and as stated, ground observers noticed the covering flutter in its wake. The passing storm was clearing, so people looking at the *Hindenburg's* port side viewed the ship silhouetted against the clearing sky. No one with a view of the starboard side appears to have taken photos, but their view looking in the direction of Hangar #1 saw the ship against the dark backdrop of the stormy sky. "The ship is riding majestically toward us, like some great feather, riding as though it was ...uh…mighty... mighty proud…" ad-libbed Morrison. After stumbling through a few more sentences, they stopped recording as he was obviously stretching for words while little was worth describing and someone was talking loudly enough in the background to be heard. "We had it so we could stop the recording…had a gear arrangement…"[241]

"Like the other phases of ground handling, mooring also became mechanized at Lakehurst in the later 1920s. Prior to that time, except for operations at the high mast, which was discontinued in 1928, the airship landed in the open space on the ground in the hands of a ground handling crew of about 200 men and was walked to and docked in the hangar using only manual ground handling methods." [242] Shipmates remembered Captain Pruss usually approached the landing with a slight nose up attitude.

A writer who watched the 1936 landings verifies what is seen in newsreel film, "...her motors idling until she drifts down and her big, rubber-tired wheel bumps lightly upon the ground… once the bumper wheel touches the ground (tho [sic] it need not touch), a cable drops from the ship's nose to be attached to the cup atop the mooring mast."

This May landing, however, was something new. In order to reduce the number of ground crewmen needed to fly the airship down directly into their hands, the Americans for some time had been trying to get the Germans to put more trust in the expensive mechanical ground handling equipment gathering rust around the station.[243] The Americans had money tied up in ground handling equipment and the sailors were getting little training operating it. The LZ-129 had only been docked twice – once bow east, once bow west – and even then the Germans refused to drag the stern beam in and out of the hangar. The Germans were finally talked into putting more trust into less labor-intensive methods. The Navy had hired only 139 civilians to

---

[239] The third camera was there, since there are three distinct viewing frames of the accident. Oddly what appears to be a fourth port view, unhappily also very late in the fire, was found by the author in NARA RG 428 (NPC) footage, different mainly in its hand-held appearance.

[240] Morrison's recording is available for purchase on the Internet, at sites such as www.realaudio.com

[241] Herb Morrison, interviewed on ABC's *Good Morning America*, on the 45th anniversary

[242] CAPT C. W. Roland, USN, "HANDLING RIGID AIRSHIPS ON THE GROUND, With Emphasis on the Use of Mechanical Equipment," unpublished commentary put down for NLHS, courtesy ZRS engineman John Lust.

[243] Remarks of Norman Mayer to the author via telephone. While they had smaller mobile mooring masts, the Germans viewed the stern beam built for the ZRS ships as overkill. They allowed it to be used to pull the tail around to align with the hangar on two occasions, but did not believe the spreader gear could be trusted to carry the beam's 90 ton weight. Indeed the USS *Akron* had been wrenched on one occasion.

supplement the workforce that day, paying $1 an hour for the landing work.[244] For the first time they would perform a "flying moor," or high landing, in which the airship would hover at EQ and be winched down to the mast. Sixty-three years later Eugen Bentele would still relate how peculiar this seemed to them, saying through the translator, "we did not land," just stopped and dropped ropes.

Albert Grotzinger, one of the cooks, had just finished cleaning the kitchen. Proper landing configuration was aided by the routine action of sending some unoccupied crewmen up to the bow. Among those called was Grotzinger, one of the usual four without a landing station, tapped for the purpose. This time he was accompanied by two young helpers, Mueller and Flakus, who had enthusiastically volunteered to go forward during landing in order to see parts of the ship previously unavailable to them.[245] Harold Dick would tell how such easily moveable "ballast" was referred to as "galloping kilos."[246] The action was more flexible than pumping water forward or aft, which upon dumping, would be lost. The available men could be relocated more quickly than pumping ballast water fore or aft.

In the bow, Grotzinger was the last man in the line climbing forward, and the advancing men happened to stop so he was on a step close to a hatch. Not having had much of a view from his duty station in the galley, he laid down in the passage, trying to watch the procedure through the hatch.[247]

About 7:20 PM the flying moor began with the airship all but stopped and the yaw ropes being dropped from their bow hatches, and Morrison noted them "dropping out of the nose of the ship..." and the crew picked the soppy hawsers off the soaked ground to connect them to the winch lines. Experienced hands knew better than to grab the steel mooring cable before it touched the ground. As W. S. Diehl wrote, "Balloons and airships that descend rapidly from considerable height are charged electrostaticaly at the potential of the upper atmosphere, which may not have time to dissipate. Especially if damp, the hemp drag-rope of a free balloon begins to discharge the acquired potential as soon as it touches the ground. The wire cable dropped from an airship approaching the earth is effective in immediately equalizing the potential unless the ground is rocky or exceptionally dry." Historical literature has mentioned the wise precaution of not grabbing the descending wire rope, least the man become part of the circuit that discharged the ship's accumulated static charge to ground.[248]

The airship was typically kept on such even keel that legends grew; one had wagers being placed on the number of *hours* a pen could stand up right on a promenade table. A popular misconception that the stern was low or "heavy" was given credence by people interpreting the viewer's angle as if there was a noticeable downward slope to the ship's keel. As noted, officers had sought a half a degree up by the bow. Rosendahl testified "The wind velocity on May 6 was between six and eight knots and hour..." which would have easily allowed attitude correction via

---

[244] Associated Press report May 3, 1987.

[245] Albert Grotzinger told Dr. Bain about the two young men. Sadly, their enthusiasm cost them their lives as they were certainly among the 11 who were seen loosing their footing and falling back into the keel.

[246] Author first met Dick during the latter's 1977 lecture circuit. Dick told of crewmen having to run aft along the *Graf's* keel after Brazilian Army ground crewmen had released the stern car to pose for a photo. Author (and no doubt many others) urged Dick to chronicle his experiences in a book, which Dick promised to – and, happily, did so, long before passing away in 1999.

[247] Author's short conversation with Albert Grotzinger via translator on 3 JUL 00.

[248] J. Leasor, THE MILLIONTH CHANCE: THE STORY OF THE R-101. New York: Reynal & Co., 1957. P 52. Courtesy Kent O'Grady

Boetius working the elevator wheel, as he likely did; the gust might have caused a slight over control. Rosendahl noted "The stern, however, did not drop but actually rose somewhat while the ship hung on the landing ropes" for the next couple of minutes. Nonetheless the notion that the tail was "sagging" would become a fixation with people later trying to extrapolate a gas leak. Some would actually testify it was *very* tail heavy for the seemingly obvious reason: there was a fire, there must have been a leak; since there was a leak, that portion of the ship must have been heavy. First officer Sammt would later state fuel consumption had caused more weight to be aft, this was apparently unnoticed until crewmen had taken landing stations aft; hence the order to send some unoccupied men forward.

R. H. "Ducky" Ward, in charge of the port forward landing party, saw a fluttering of the outer cover. Hoehling: "...correspondent Alice Hager saw much the same manifestation. She thought it as a "rippling." Rosendahl would explain it was "...caused by the passage along the hull of the ship of the light gust of wind which struck the ship on the port side...It is a matter of common knowledge that airship fabric which becomes slack in places in damp weather will flutter." Indeed, fluttering had been a problem, and had even been captured on motion picture film.

Fluttering covers had been a great problem for the British. Captain G. F. Meager of the R.100 wrote his memoir LEAVES FROM MY LOGBOOK (published by WLTAS in 1961). "I spent the time during the full speed trials inspecting the cover at the bottom of the ship from the corridor. There were the same undulations in the cover, which in places, became enormous bulges and must have put great strain on the fabric. In addition there was much more flapping and chattering of the cover on the transverse frames, apparently caused by gauze ventilating panels in the bottom of the cover... After full speed trials... the cover seemed to be full of small holes... some caused by petrol tanks coming into contact with the cover..."

This gust of wind from starboard not only had enough force to ripple the cover, it also was threatening to blow the airship back away from the mooring circle. The gust was at first handled by the experienced rudder and elevator men at their wheels in the control car (although one aft observer would later tell of how he thought the upper and lower rudders were out of sync). The bridge responded by ringing up the engine enunciators for thrust to counter its push. Hoehling: "While Francis Hyland, the line handler...was momentarily startled by what he took to be a backfire: flame spurting out of the after port engine as it was reversed to aid the ground crew holding the weight and tide like surge of the airship." Under a port engine, young Robert E. Rutan, son of ZRS-4 CPO Lucius Rutan, also remembers hearing a backfire. (Enginemen had noted the seemingly impossible phenomena did actually occur, even in an engine without an ignition system, if the cam lever was thrown over and the engine restarted very quickly.) Frank Wainwright would later tell of also witnessing the backfire.[249] Other witnesses testified they thought the engines were running at full speed for about 30 seconds following their reversal.

Ficher and Deutschle survived in the port aft car; some enginemen would later dispute that hastily reversing caused anything notable, so this odd diesel backfire would seem questionable. Bahnholzer and Schreibmueller, in the starboard aft car so close to the fire's origin, did not survive to testify. Bain would later interview Robert Buchanan, then 17 and under engine one, who had written he not only heard the backfire but also saw the exhaust plume which seemed to include glowing chars. He agreed the winds would have carried the plume along the side heading aft toward the fins.

---

[249] In a 1977 *Asbury Park Press* interview, courtesy Robert Rutan

During this time, something else was happening that even the most experienced crew could do nothing about. Once dropped from their hatches, the 413 foot long mooring ropes had immediately began to absorb water via capillary action from the soppy ground. These hawsers were hygroscopic, meaning they would absorb water. The manila lines also took on water from the breeze-driven drizzle; Herb Morrison notes, "It's started to rain again..." Later National Bureau of Standards tests verified the ropes became conductive in no more than 170 seconds. The *Hindenburg* structure then became a lightning rod, a "piece of Earth" aloft, like Ben Franklin's kite.

Again we must remember what nature saw was not an airship, but a capacitor. As a 1936 account read, "They are metal condensers; lightning cannot enter. And here is why: The 803-foot hull – 93 feet longer than the largest battleship afloat – is composed of huge rings made of aluminum-alloy girders with steel wires for bracing. Every section of these rings and the longitudinal girders are connected so that the whole outer structure of the hull make a single electrical conductor." The 'condenser' analogy here is particularly appropriate when one remembers condensers as made with thin sheets of aluminum foil.

"There was no proof either way at the time, but in later years two reliable witnesses were found who claimed they had seen a blue flame of St. Elmo's fire flicker along the backbone girder of the *Hindenburg* a minute before the ship caught fire."[250] Douglas Robinson would later publish an interview with Professor Heald, of Princeton, in which he says "Heald... undoubtedly saw St. Elmo's fire flickering along the airship's back a good minute before the fire broke out." [251] No one else seems to have reported electrical activity from port; but against the darker sky from starboard, more was visible. Not part of the live testimony, *starboard* witnesses dictated seeing a visible electrical activity ranging from what was described as "a real light blueish" indication from starboard aft to Kirkman's description of a "neon light type of activity" from his starboard view some 400 yards off. To be visible at those distances and for such a duration as a full minute, as one NASA electrical expert, Dr. Ray Gomph put it, "That's a lot of coulombs!" Professor Alex Dressler stated to be visible, the charge must have reached a minimum value of 10,000 volts per square centimeter.

This electrical display was even visible underneath the ship. Hoehling: "And W. W. Groves, the Yale & Towne engineer, was noticing something he thought interesting; a small spark 'like static electricity' dancing somewhere overhead and underneath the hovering airship, not far from the tail. He opened his mouth to comment to a companion standing beside him. But Groves never formed his words..." because about four minutes had elapsed from the point the bow ropes had hit the ground.

Mrs. Elizabeth Tobin, wife of famed USS *Shenandoah* Chief Petty Officer "Bull" Tobin (himself then on landing station) had crossed the street from the Officer's housing area.[252] Standing on the field with a clear view of the airship's starboard aft section, she testified everything seemed perfectly normal until she saw a "...small flame immediately in back of the top fin, in back of the fin, in back of the whole surface and the rudders..."

Luckily, Herb Morrison had elected to start recording again, the change in tone and clarity being evident. "It's starting to rain again, the rain had, uh, slacked up a bit..." He continued "The

[250] THE GIANT AIRSHIPS, Time-Life Books, Alexandria, Virginia, 1980

[251] Robinson, Douglas H., LZ-129 *HINDENBURG*, Aero Publishers, Fallbrook, California 1964.

[252] Retired NASL fire chief Henry J. Applegate assisted Bain in finding Lizzy's position from the record.

back motors of the ship are just holding it, uh…" and then seems distracted, continuing "…just enough to keep it from…" when a scream is heard in the background. "A burst of flame, its origin unknown, shot from the stern of the ship."[253]

The aluminum-rich fire's light began to reflect off the clouds and illuminate the area; some people thought the sun had burst through the gloomy cloud cover aft of the tail. "A burst of flame from the stern, and the entire sky was alight."[254] Small wonder Mrs. Spah thought the Germans had set off fireworks heralding the ship's arrival; ironically, fireworks makers use a "german black aluminum." Henry W. Roberts, editor of the *Aero Digest*, said "it was as bright as 'a thousand, ten thousand, magnesium flares.'" In stark contrast to transparent hydrogen combustion, such a retina-burn producing flare-like fire has no direct comparison today outside of rockets accompanied by solid boosters. The aluminum-fueled SRBs of the Space Shuttle so enlighten the night sky their launch is routinely visible in Miami, and sometimes visible as far north as South Carolina. Burning aluminum makes a brighter, whiter light than the sun itself.

The fire was of course immediately visible to men *inside* the fin, since the "thousand, ten thousand magnesium flare" brilliance was close to them and filtering through the thin materials. Crewman Helmut Lau [255] was on landing station in the lower fin. He testified "The first burning out of the fire was on the starboard side, above and up." Here was a witness who seemed to have seen the actual origin, although from his station he would have to have been looking aft to see it. He explained after hearing a "whump" sound, he looked *up* and "...saw from the starboard side down inside the gas cell a bright reflection on the front bulkhead of cell number IV."

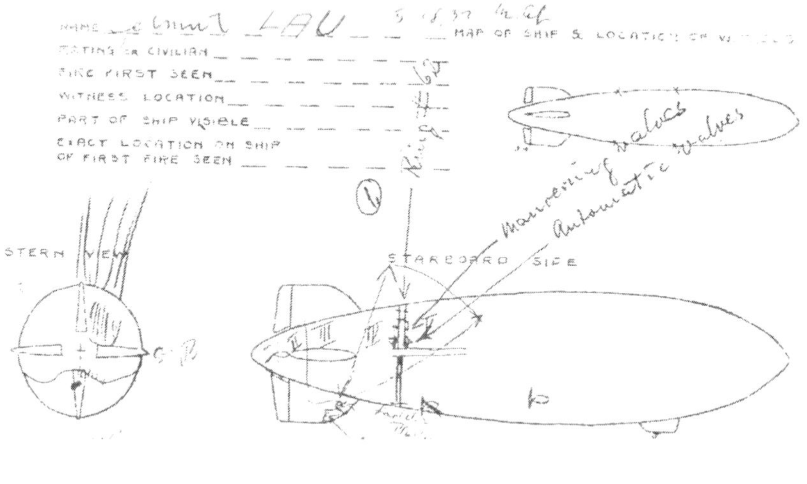

Like other witnesses, Lau filled out a diagram, now in the National Archives. The original, with color, diagram's stern view shows red and yellow flames on starboard aft, between the upper fin and the starboard horizontal stabilizer, not "forward of the top fin," and how Lau saw the light. NARA courtesy Addison Bain.

Diagrammed just like he said, Lau was questioned on this point again, and responded "I did not see the absolute center of origin of the fire, because it must have been further up, since I saw the *reflection* of the fire *through* the cell wall." (Emphasis added.) Produced by the aluminum-rich fire, its flare-like brilliance was bouncing off the forward bulkhead of cell IV, which was perpendicular to the starboard aft fire. Lau used lines to illustrate how his viewpoint made it possible to see the light traveling through the airship to bounce off cell IV's bulkhead. The

253 *New York American*, Saturday, May 8th, 1937
254 E. Horton, THE AGE OF THE AIRSHIP, Henry Legnery Co., Chicago 1973
255 Lau's testimony in its entirety was copied from the National Archives by Addison Bain. NARA also supplied DIGEST OF TESTIMONY OF ACCIDENT TO AIRSHIP *HINDENBURG*, the basic witness reference document, which contains an abridged Lau testimony.

aluminum fire's light was so bright, what little bounced off the forward cell wall was still visible through the cell's lower material. Hoehling: "In the tail, the four men trapped there agreed, it was 'extremely bright,' but none of them was conscious of heat." Bouncing off the thin cell walls and reflecting off interior surfaces, there was, at first, only bright light. With several million cubic feet of nature's most perfect coolant between them and the fire, all they were getting was the flare-like brilliance. (Hydrogen gas thermal conductivity is 0.001875 w/cmK; in other words, about as good a cooling medium as you can find.)

If we remember the airship is, basically, thin materials enclosing almost empty space, we appreciate that even people on the port saw the fire almost immediately as well. To port (left, "camera" side) and lower observers, the blinding aluminum fire's light from the flaming starboard side was *filtering through* both sides of the translucent cells and thin outer cover. Hoehling states Bill Springfield of Acme saw "...a small light inside the tail assembly..." USS *Los Angeles* crewman John Iannacone would recall seeing "the red glow" from "inside" as he looked up from below.

It would be an eerie replay of Zeppelins attacked from below during the Great War. That earlier fabric, ignited at the exit point by the three part ammunition fired from the underside, would begin to burn brightly, showing through the manufactured void and through the near, or under, side. In an amazing parallel to the testimony of British defenders, Hoehling wrote that Mrs. W.R. van Meter, of Upper Darby, Pennsylvania, said "The stern lighted up sort of like a Japanese lantern...lighted up the covering of the bag so that I could see the framework of the ship through the covering." It's just like the description of L31's exit wounds catching fire opposite the defender's position: "Even as the British airman was firing, he saw the Zeppelin 'go red inside like an enormous Chinese Lantern...'" The early Zeppelins did not even have aluminum in the final coats of nitrate dope.

At least two port side witnesses made it clear they understood what they were seeing was a fire burning on the other side of the stern. M.M. Ream was near Hangar #1 and said he "saw glow on the other side" and R. L. DeBow, a few hundred yards from the hangar, "saw the pink then red glow on the starboard side."

It is difficult to place Herb Morrison's recording precisely in the time line. He heard and reasoned the aft engines had been throttled up to overcome the gust, saying "The back motors of the ship are just holding it, uh..." but he seems distracted by something. There is a confused pause before he seems to just mouth the words "...just enough to keep it from..." as if now completely enthralled by something he is seeing. Was he also seeing the glow seeing to come from the inside? At some point he must have noticed the bright light illuminating the clouds and field grasses, which at first could have been taken, like the other witness' assumption, as the sun breaking through, or fireworks whose sound had not yet reached him. 50 years later, interviewed by the AP in his West Virginia home, Herb would again state "A fire erupted in the tail section."

Regardless of their position underneath or to starboard, all witnesses agreed the fire enlarged very rapidly. Hoehling states Madeline Lipton detected "a faint pink glow in the lower center of the ship...like some thick silvery fish with a rosy glow in the abdomen...it began small and pale and spread redder and larger..." Another says "Mrs. Leif Neandross...also watched in horror the airship 'light up all inside." Willie Von Meister described it as a "florescent tube flickering to light." Conspicuously absent is any memorable sound, verified by Morrison's recording.

Notable at this moment is the complete lack of similar descriptions from the starboard side. Since the covering had first ignited starboard, aft of the rudder, it was advancing on the starboard side first. Starboard side observers were seeing the actual covering fire, not the filtered light of the bright flames showing through "...so that I could see the framework of the ship through the covering." No one would ever be able to establish an exact time line, but had there actually been some solid material burning from inside outboard, port stern observers like Lizzy Tobin would have also reported an interior "glow" before seeing actual flames. They specifically did not.

Again, lest there be any doubt about what these witnesses were actually seeing, that is an aluminum-rich, gas-less fire, we see the parallel to the USS *Macon's* brilliant finale reflecting off the Pacific's surface. "Suddenly a fire bloomed... The massive star insignia painted on the underside of the bow was *reflected by the glare*, even as it was consumed... the last skeletal remnant of what had been the world's greatest airship slipped beneath the surface."[256]

Inside, as the covering fire advanced toward him, Lau testified "...it was coming down on the starboard side...If it had been towards the port side first and up, I would have seen it..." Lau saw it coming *down* starboard, but outside, to Lizzy Tobin, the initial ignition was orange in color and the flame front was a saw tooth shape, advancing *topside* rapidly on starboard. The covering was actually burning in a rough concentric, creating an enlarging hole in the fabric from its ignition line on starboard.

Sadly, the testimony is nearly silent on starboard side witness accounts. The Air Commerce Bulletin stated "The great majority of the ground witnesses who testified as to the first appearance of the fire were at the port side of the ship." Indeed, if one had not read the entire 800 plus page report, one would assume *everyone* interviewed was standing to the port side of the ship. Certainly most, if not all, the functioning cameras were to port, but one person with a camera at least had an aft quarter view. Obviously taken by surprise, this photographer managed to snap a photo showing the progress of the fire as it advanced around the stern and began to encompass the fins.[257] No book publisher had seen fit to print this impossibly-contrasted blurry but interesting image. Sadly, even this image was taken too late to emphasize the covering fire's initial starboard progression.

---

[256] Emphasis added. RADM H. B. Miller, *True* (magazine) August, 1963

[257] The photo opened our *Buoyant Flight* series and was printed on high quality paper in GASBG #31.

From his bow position, CPO "Bull" Tobin would still realize and testify the fire had started on the starboard side at the stern, and stern origin was later recorded in the station log. Masked by the huge upper vertical fin, the bright light heralded something amiss coming toward the port side observers looking topside. Hoehling: "Von Meister observed the reflection of a very strong light on the entering edge of the upper fin." But it would not be long before even the port side observers *not* looking through the ship would realize what was going on, and what was causing all that bright light would be visible on the left.

Rosendahl would be widely quoted reporting: "At 7:25 p.m., or just four minutes after the landing ropes had been dropped, I saw a burst of flame on top of the ship just forward of where the upper vertical fin attaches to the hull. It was a brilliant burst of flame resembling a flower opening rapidly into bloom. I knew at once the ship was doomed, for nothing could prevent the flame from spreading to the entire volume of hydrogen with which she was inflated." [258] Rosendahl of course realized the brilliant burst was not itself hydrogen, but knew that it would not be long before whatever was burning so brightly would reach the gas. It was a worry, even though the pure gas was neither contaminated nor contained enough to explode, nor could the pure gas burn. It was a worry because the cells were not just keeping some 240 tons aloft at 200 feet; they also were responsible for a great measure of the vehicle's *structural integrity*. A key rigid airship design element from way back, this was described in 'Greyhounds of the Air:' "The rings are joined together by strakes and ribs braced inside and bolted together in one long hull. The frame is tremendously strengthened by the outward pressure of the gas." The structure had held to that point because its main support was still largely intact as the outer covering, and even its anti-corrosion primer paint, was burning aft. Morrison had by then reported the aforementioned "flashing – flashing, terrible" as the blinding bright flames raced around the stern. How long could the cells hold?

The radiant fire's ability to initiate secondary ignitions was hard to resist - unless there was a cooling medium at work. There was; it was the gaseous hydrogen in the cells. While its specific heat per unit volume is nearly identical to air, hydrogen's heat conductivity is *eight times* that of air. Also, its viscosity is about half that of air. As Marc de Piolenc put it, "This makes hydrogen the ideal cooling medium for large electrical generators where its good conductivity ensures rapid heat removal and its low viscosity keeps windage losses low."[259]

The translucent-thin cells offered little solidity to absorb the heat given off by the aluminum-rich fire. As the fire advanced across the aft covering, little besides hydrogen's natural cooling ability prevented the rubbery cells from melting as more aft airship elements joined in the blaze. In contrast to gasoline's autoigintion temperature of 563°F, oxy-hydrogen has to be heated to 1,076°F to torch off. The aforementioned wartime Zeppelins whose outer covers were ignited with the three-part ammunition suffered cover burn off, but enough cells survived to lower the structure from altitude so slowly as to permit some structural – and even crew's – survival. Sadly, the "improvement" of adding aluminum to the fabric since 1924 made for a much hotter, more radiant fire.

Since cells II & I were all but full, the gradually heating gas had no expansion room. As the pressure increased with the rapidly enlarging fire, their overpressure safety valves eventually

[258] Ernst Lehmann, ZEPPELIN: THE STORY OF LTA CRAFT, (foreword and last chapter by Rosendahl) Longsmans, Green and Company, NY 1937
[259] He explained, "Hydrogen cooling is used on all large and 1,800 and 3,600 rpm machines and on synchronous and large frequency-changer sets in order to reduce windage and provide better cooling."

opened. At first these 36-inch valves were likely dumping lift, possibly as fast as the gases' heat was creating lift. However miraculous that cells II and I lasted second after second, the heat most certainly exacerbated the aft cells' topside tension where it is naturally at maximum during flight.

What happened next is familiar to all that have read about the accident, but its severity varies widely with the observers' position and experience. "Then came the explosion. It rocked the entire ship, sending the passengers and crew sprawling." [260] Rosendahl said "There was a muffled report and the flames spread rapidly through the aft quarter of the ship." Hoehling quotes Spah: "There was a blinding light...then an explosion." After regaining his voice, Morrison recorded "I've sort of recovered from the terrific explosion...the tail surface broke into flames first, then there was a terrific explosion." Some supposedly heard one or more explosions as far away as Point Pleasant, miles distant.

Difficult to place in the time line is the action of Morrison being so startled he bumped into the record-cutting recorder. It could be this initial blowout that caused his jump and nudging the record platter. Others would fantasize that the force of the explosion had knocked the needle from the groove! Assuming he had knocked the record off kilter, Morrison probably motioned to his engineer, Charles Nelson. This writer's interpretation of his excited words comes out as "Get this started! Get this started!" Others insist he says, "Get this *Charlie*." Either way the needle was still cutting and getting his excited words, but no particularly loud report is heard on the record. Generally, the perception of loudness or concussion increased with the distance, with variables owing to relative position. Of course this was not a yield of an air-gas mix, as the widely misinterpreted "hydrogen explosion" long surmised. Rather, it was the most colossal *balloon popping* in history. Cells II and I, all but full with little expansion room, and whose overwhelmed vents were tied together, had finally lost to the withering heat. Unable to expand owing to their fullness, the heat-bloated cells *burst*.

This blowout had a profound effect on the LZ-129. Sammt as "conscious of a heavy push," and Lau said, "I noticed that the ship received a definite acceleration downward, and it nearly threw me off the little catwalk on which I was standing. It also shook the fin right and left, and I was holding myself with one hand."

Would helium in these cells have made any difference at this point? Helium would not have cooled the membranes quite as well; they may have melted rather than burst. Sadly, there was a single event that suggests what might have happened. In September of 1996, the unique yellow pine hangar at Weeksville, North Carolina, caught fire following welding operations. Trapped inside was the prototype naval helium airship *Sentinel 1000*. Prevented from trying to rescue the airship as the timber fire raged around it, witnesses reported the lost blimp moaned a sickening sound as it ruptured in the crackling fire.

In the noisy and enclosed engine cars, the jolt was the first indication something was wrong. One engineman would later say he thought the mooring line had parted. On duty in engine #4 car with Ritter and Schadler, Eugen Bentele wrote "A powerful shock ran through the ship... when I looked down to see what was going on, I saw a fiery glow lighting up the landing field..." Up forward, filled by drainage from the passenger's lavatories and shower, were several wastewater tanks whose bases protruded slightly through the outer cover. These tanks were mounted on

---

[260] E. J. White, FAMOUS AMERICAN DISASTERS, Creative Television Associates, Boston 1992

quick-release frames to facilitate the disposal of weight in an emergency. The sudden blowout was powerful enough to jerk two of those wastewater tanks free of their mountings. They continued through the cover and fell toward the ground. Gauging by the distance between girders and knowing the acceleration of gravity, we can deduce exactly two seconds later (Toland): "Murray Becker, Associated Press photographer, had come forward to get an interesting angle shot from the tail side [sic]. Seconds after the first explosion he snapped the first [sic] picture. The ship was then still at an even keel."[261] And so it was, the *Queen Mary*-sized vehicle still level at that instant. Judging from the fire's progress in the later motion pictures, this could have been as little as ten, eleven or possibly as much as twelve seconds after the fire started.

The moment, frozen in time by Becker's shutter, was the last moment the LZ-129 was on an even keel. Ejecta from the blowout can be seen jetting out from the starboard aft. A large bubble of hydrogen must have jetted out of the stern, giving the vehicle a push strong enough to dislodge those tanks. The fins, stern and central structure are already withering from the vacuum created by the fire, certainly straining the balloons underneath that were somehow still offering some ascentive force. There would be little time for the massive vehicle to react to the loss of lift from these small cells I and II, just as there had not been enough time for whatever heat energy these smallest cells absorbed to apply extra lift to the withering structure. This even takes in to account their extreme end position and the leverage they commanded vs. the center of buoyancy. As quoted, Rosendahl had time to worry that the fire would spread to the hydrogen, and others had similar time to first come to the realization the interior glow was a fire on the other side, before wondering what would happen to the lifting gas.

Bentele would recall that the "fire began at the back" and that shortly after feeling the lurch he looked outside his car to see the outer covering already burning.[262] Bentele was certainly aware of what had happened during the Great War - much lighter engines in smaller cars had broken off as their ships' covering burned, melting their suspensions. He was at first concerned his heavy #IV car would break loose and fall as well. Then he would write: "I knew that the connecting struts and wires between the gondola and the ship...were made of high-grade steel

[261] Thanks to Patrick Russell calling author's attention to the Becker reference
[262] Author's conversation with E. Bentele on 3 JUL 00.

and could withstand even this heat – but for how long? If only the ship would loose height! It seemed like an eternity before the stern began to drop groundwards, slowly at first, then more quickly…"

A little more than ten seconds elapses in Morrison's recording between him saying "It's burst into flames!" and him saying "It's falling!" but we must remember, discounting his hesitation via confusion over the interior glow or bright light bouncing around, he was only crying fire *after* its emergence on the port side. Hoehling quotes Spah noting concern over the shudder: "The steward tried to calm the passengers, assure us that everything was all right. Before the steward could walk three steps the stern of the ship sank." In other words, there was enough time to take three steps between the time they felt the lurch and the stern lost buoyancy. These three steps might seem like a short time, but to others, the perception of time slowed. Since Morrison would moments later record "The tail caught fire first…" he was possibly already realizing what had distracted him had been an early stage of the fire.

In the next second following Becker's shutter snap, the fire would melt cell IV. Lau testified "The cell suddenly disappeared because of the heat..." With the loss of lift from this cell, the aft structure accelerated its collapse and the stern, missing the lift from three cells, started to sag. A blurry photo of this exact instant is on file at NARA. With more room to expand than mostly full cells II & I, it is anyone's guess why cell IV did not simply bloat and blow like the others. The next cell would not release its gas by just gently melting away.

Almost immediately after IV's melting, Cell III and Cell V disintegrated as well. Hoehling: "The blistering inferno of hydrogen, after devouring Cell 4 in one molten gulp, blasted Cells 3 and 5 and then tore on like an incandescent tornado through the 803 feet of the *Hindenburg*, roaring toward the bow." While Hoehling obviously did not know hydrogen made no visible light, let alone a glow, he had noted the difference between witness accounts of Cell IV melting and other aft cells rupturing. Yet the next two cells forward, possibly owing to their larger expandable area, would hold long enough to duplicate the rupture of Cells II & I. Then, soon after the first rupture and with greater force, more forward cells cooked off. The second, more powerful rupture would overwhelm the first so quickly that some witnesses, including Herb Morrison, would report only a single blowout. The many times larger Cells VI and VII blew with much greater force than the smaller stern cells. This terrific rupture would blast sections of outer cover, cells, and other debris clear of the falling structure. Other witnesses noted the second distinct rupture. Margaret Mather recalled "The lurching of the ship threw them repeatedly against the furniture and the railing…" With Cells I and II already gone, the melting of IV, disintegration of III and V, and finally the bursting of VI and VII, the stern could no longer defy gravity.

Now devoid of the uniform force that had kept it in shape, the topside and equator girders rapidly collapsed aft. No one captured the exact rupture on camera. The next shutter snap, from a photographer closer to the old crawler-tread mast, caught the two wastewater tanks almost half the way down, clearly showing the gasless aft structure crumpling inward and the debris already well clear of the falling stern.

This is the first point where major quantities of hydrogen were being freed to enter the conflagration in such an position where it would not just be streaming out the top. As explained, its haphazard entry into a oxygen-deprived inferno would not have encouraged combustion beyond the demonstrated minority percentage that ordinarily burns when released into free air in the presence of ignition sources. It is even more important to remember that burning hydrocarbons radiate *ten times* more heat than hydrogen. No one could have felt hydrogen heat in that mass of blistering cellulose any more than one would expect to feel the heat of the light bulb that comes on when you open the oven to check on your roast.

Unburned fabric blown outward can be seen raining down; note starboard yaw rope's tightness and colorful cellulose flames contrasting aluminum's brightness furthest aft. NARA photo.

As the keel twisted under the strain, photographers snapped many images that can be studied in series. Ejecta from the second blowout is seen rising from the stern in a spectacular mushroom geyser. Morrison also reacts to this second rupture in astonished tears "...four or five hundred feet into the sky, ladies and gentlemen..." Large pieces of unburned fabric and other debris can be seen raining down from the blowout.

It is a testimonial to the strength of the LZ-129 keel that it did not break at this point, with the stern's now dead weight pulling down against the buoyant bow's intention to stay aloft. Again, there had been historical precedent. A Sopwith *Pup* was launched from HMS Yarnmouth on the 21st of August, 1917, to deal with the L23. Flight sub-Lieutenant B. A. Smart concentrated his 3-part combination ammo on one point and got a fire going. Rimell quoted Smart: "It struck me as the funniest thing... the engines were still turning away merrily with the Zepp enveloped in flames, and pointing up at that ridiculous angle... [Looking back as he retired] the wreck had just hit the sea, only the very tip still being still intact; it was still burning away merrily and continued to burn on the water for 3 or 4 minutes, the smoke having changed to a much blacker colour probably due to an oil tank bursting and mixing with the flame."

In the dramatic LZ-129 photos, a form of *crown* fire called *plume-dominated fire* is visible. Although it is often labeled as Time-Life Books captioned it, "a geyser of burning hydrogen," the fire's pattern is common in ordinary forest fires. Plumes dominated fires, only found in winds of 20 mph or less, are dangerous in two ways. First, the fire produces extreme turbulence at the base of the convection column; its pulsing increases fire intensity and heat transfer to adjacent fuel. Spreading the fire, the process feeds on itself accelerates as the convection column grows. Second, a downburst of cold air, initiated by evaporative cooling of precipitation, descends from the column, blowing outward along the surface at high speeds in all directions. Unexpected and sudden, it's likely the downdraft will encounter the flames, when the fire becomes wind-driven for a short but violent period where even more extreme spread rates may occur.

Amidships, the structure began failing rapidly. The section mounting the engines failed even more rapidly than adjacent structure, not doubt owning to their weight, but to some degree the force of torque from the aft engine's full throttles and huge twin props still spinning. Without the gas cells to carry the load, girders crumpled. Also in these photos, one can follow the progress of the wastewater tanks knocked loose by the rupture of cells II and I. In the very next photo, the tanks have impacted the field, splashing their soap-foamy contents over a wide area. The lack of injury from this tonnage impact and forceful splash is one of the unsung lucky breaks amid the somewhat well populated ground crew area below.

Rosendahl: "As the stern burned, the ship, having lost buoyancy in that portion, assumed a stern-down attitude and as the fire progressed forward, the ship of course settled to the ground at what we would consider a moderate rate of decent..." Spah would recall on the 45th anniversary that "To me, it was like the earth coming up to me."

One photo shows a ballast bag emptying during the decent. It's recorded that in the control car the officers had the presence of mind to not drop any more ballast, hoping the flaming vehicle could be grounded more quickly. ("If only we would lose height!") It would have been hard to convince these men the lifting gas above their heads was burning, as the buoyant bow strained the port bow rope trying to head skyward. The stern's dead weight pulling against the bow, lightened by the loss of the two wastewater tanks, meant something had to give. By then, we had a new recording medium started up.

On the 50th anniversary, Joel Siegel said "A former Eyewitness news film cameraman named Dick Nedalonge[263]...father was one of the newsreel cameramen here told me this story. Waiting for the ship to moor, nothing seemed to be happening, so the cameramen went out for coffee, as cameramen will. They left one man watching the equipment, and that's when the *Hindenburg* exploded. He turned on everybody's camera, saved everybody's job."

The first camera activated manages to catch a single falling tank passing thought the frame, but the combination of falling stern and the restrained buoyant bow meant something had to give. The airship's mid section slides forward, carring its main bulk quickly out of the frame so only the top spine is visible in silhouette for several seconds. This newsreel would be played repeatedly and subjected to freeze-frame in the 1975 motion picture despite its basic lack of visual interest. (Perhaps it was the only one to which they could obtain rights to use.) The view still does show us an important pause in the ship's plummet.

One of the cameras had a telephoto lens setting, probably with the intention of getting a close-up view of the passengers waving from the promenade as they had during a sunny 1936 arrival. Now the collapsing airship fills the frame, patches of aluminum rich fire overexposing portions of the view. Extremely bright light silhouettes sailors that scurry toward the camera. The camera is hastily re-aimed, first too low, then back in line. A most amazing thing is visible: the dead weight of the crushed and gasless mid section is grounded, and the ship *stops falling*.

---

[263] Author's best guess of spelling, transcribing from video tape supplied by Alan Gross

Here again there was a historical precedent, this in 'The Greyhounds of the Air': "The day before the *Titanic* disaster the German Army's airship Schute-Lanz – Zeppelin type – sixteen miles faster than the *Titanic* – as a result of a mishap to its rudder while it was being driven to the earth, struck the ground at full speed. Yet, with the forward engines disabled, the Schute-Lanz rose, drifted across the Rhine, and finally was brought to earth without further damage or loss of life. It withstood the terrific collision because its gas cells remained intact." (Its wooden frame probably made a difference there as well.)

As the Lakehurst ground crewmen scattered, 51 year old (and somewhat heavy) Allen O. Hagaman tripped on the mooring circle rail. A shipmate saw him fall, but could not stop and come back to help him in time.[264] No doubt in part due to the starboard fire's greater intensity and head start, neither of the enginemen in starboard car #I would survive.

The weight of the engines, the thrust of the aft engines, and their prop vibration was then neutralized. Another of the three cameras offers a wider view. The astonishing half-airship is now almost motionless, since its cell-supported structure is still somewhat intact. It's as though a giant airbag has cushioned the structure's impact with the ground, and for a pregnant moment it simply sits on the gasbag cushion.

The fire had started on the starboard side, and the acute angle the hull had assumed made heat transfer up the starboard side much more rapid than on the camera side. The flames reached the pointed bow on the starboard side first. Just as they had on the blunt stern, the flames raced around the bow and proceeded to consume the covering, now heading aft. Losing cells 16, 15, and so on aft as the covering burned by them, the bow lost its buoyancy and finally began to rotate downward. Uneven burning of the cover is very noticeable at this point, especially from the telephoto lens camera. One strip of covering appears to resist the onrush of flames; it pauses; then it disappears in an instant.

One camera is lined up in time to catch the control car area at wheel touchdown, one of the most astonishing moments in the films. *The ship is still buoyant enough to bounce. Its structure remains intact in spite of the impact.*

It seems impossible! Surely the hydrogen in the bow cells *must* be gone or burnt by now, but here the vehicle's forward section settles, bounces, settles a final time, standing erect for a pregnant moment as if waiting to discharge passengers. Buoyancy aside, there was no way the structure could take it. A Frenchman inspecting the Zeppelin hangars after the War observed, "Lastly there are usually under the roof three longitudinal bridges along which run suspension rails, which serve as supports for the frames of airships held in reserve or being inflated." As the Airship Heritage trust explained, "With the end of the war, the German crews were ordered back to their bases on 9th November 1918. The ground crews were segregated from their officers, and the ground crews then ordered to deflate their ships and leave them suspended from the roofs of the airship sheds, as was the normal procedures for large rigid airships. When the lifting gas was valved off the weight of the ships was taken by large slings in the forms of loops which suspended the craft from the roof, and thus preventing the ships control and engine gondolas being crushed under the weight of the framework above. The crews who flew the Zeppelins remained loyal to their country in spite of defeat. At the same time their waterborne comrades were scuttling their vessels in Scapa Flow, the crews stealthily entered the sheds at Nordholz and

[264] During one anniversary, a relative of Hagaman told a reporter "an engine fell on him." Other sources say the man died of a heart attack trying to run clear. He left a wife, Aleeni Hagaman.

Wittmundhaven on 23rd June 1919, released the straps and allowed the ships to crash down on the hanger floors. Irreparable damage was done to them which prevented the craft being handed over to the Allies."

Indeed, even attempting to deflate a rigid airship without proper shoring a suspension will spell disaster. As von Schiller observed, "Likewise the LZ-120, which as a Navy airship had made a 105 hour voyage under the command of Captain Lehmann, was smashed while being deflated in an Italian hangar." Robinson and Keller noted ZR-3 was "...was shored up in the hangar, and overhead suspensions were installed..." Yet here about half of the LZ-129, slowly settling to the ground, remains defiantly erect with no visible support. Too thin to be visible in the flare-like brilliance of the aluminum fed fire, cells are most certainly still there, giving the structure the support necessary to survive a falling bounce. People can be seen taking advantage of the pause, punching through to escape. Grotzinger fell through the hatch into a puddle. In another pregnant moment, the half-airship settles, then stands on its wheel. The command was given on the bridge, "Everybody out!" Finally settling, the last cell is compromised.... and the structure collapses. Tiny patches of burning material highlight the skeleton. The girders are charred black, since their blue primer coating had also burned. The heavy petroleum fuels would continue to burn for hours in spite of the station's fire fighting efforts. There was plenty of time to go and get a roll of the brand-new Kodak color film and snap some pictures of the fire.

Everyone smiles at the story of cabin boy Werner Franz who had jumped and was protected from burns when a water ballast bag opened up above him. Yet it's too horrific to discuss that some crewmen suffered a similar experience, only with diesel fuel tanks rupturing. "We saw one crewman come out of there he didn't have a stitch of clothes on him, he was burned from his head to his feet, only thing he had on was his shoes, and they were smoldering."[265] Gruesome photos of at least two horribly burned men being lead away from the smoke-belching wreck verify John Iannocone's recollection. Eleanor Zarr "...ran too, screaming, feeling like she was in the middle of a nightmare and hoping to wake up any second. People with their hair and clothes on fire ran past her, and the stench of burnt flesh was heavy. 'You could smell it, it was awful,' she said. 'And screaming...'"[266]

Incredibly, Rimell noted a case that turned out to be ghastly similar to LZ-129, that of Naval Zeppelin L48. Builders' number LZ-95 was completed in May, 1917, as the first of the 'W' type, displacing 1,970,300 cu. ft. Engine trouble forced it down to an unusually low 11,000 feet to try to take advantage of a homeward-bound tailwind. L48 fell victim to three defending airplanes, each armed with the three-part ammunition. The planes proceeded to empty half a dozen drums, setting the stern fabric brightly ablaze. L48 crewman Otto Mieth: "our vessel fell perpendicularly, descending like a mighty column of fire through the darkness, striking stern first..." Rimell states: "Scutze, Eichler and 12 of the ship's crew either jumped or were burned to death. That L48's bow survived intact endorses reports the ship fell slowly to earth and the rear took up the impact, absorbing the shock." Incredibly, enough cells survived the non-aluminized outer cover fire to slow the 11,000-foot drop and allow three to survive! Impossibly bizarre photos after the covering burned away show the lightweight bow girders held their skeletal shape after what must have been a very gentle impact. Just like the USS *Macon's* final minutes, when

[265] Etched into Iannocone's memory, he has recited this experience many times for TV audiences.

[266] M. Karen, "Witness Recalls 'Hindenburg' Inferno" *St. Petersburg Times*, 02 Jan 2001. Bain interviewed Mrs. Zarr in Feb. 2001. Also on the starboard side, she none the less verified details of the fire sequence, saying, "a lot was going on before the fire became visible in front of the fin."

certain cells actually survived the covering fire long enough for the crew to witness "…the last skeletal remnant of what had been the world's greatest airship slipped beneath the surface."

Stories of human tragedy and heroism have been told and retold in the six decades since the LZ-129 disaster. The pinnacle of the airshipman's art had been turned into a charred skeleton, and "From spark to crash, the whole incident had lasted just 34 seconds."[267] Griehl and Dressel: "Naval wireless station at Chatam broadcast verbatim: 'An explosion destroyed the German dirigible *Hindenburg* and it went crashing in flames Thursday night - a loss of nearly one hundred lives – period -'"

The misinformation had already begun. Immediately, everyone wanted to know how the great airship came to grief.

FINAL DAILY NEWS NEW YORK'S PICTURE NEWSPAPER — THE LARGEST CIRCULATION IN AMERICA

New York, Friday, May 7, 1937 — 2 Cents

LAKEHURST ZEP BLAST KILLS 80

Wrecking of the Hindenburg Laid To 'Demon Proton' by Scientist

GERMANS SEE 8 CAUSES FOR AIRSHIP DISASTER

Investigators Say None Can Be Proved Responsible for the Hindenburg Fire

Wireless to THE NEW YORK TIMES.

BERLIN, June 10.—Six members

FABRIC IN ZEP FACES TEST AS BLAST CAUSE

Hindenburg Blaze Laid to Sabotage

Clippings from National Archives

[267] M. Flynn, THE GIANT AIRSHIPS, Carlton Books, Dubai, 1999

Chapter Seven: Whodunit

Awakened at his home, Hugo Eckener was told of the accident. *Graf Zeppelin*, then in transit from South America, was radioed the news, but its passengers had to find out after landing. While plane crashes were routinely reported on the radio news in America, Herb Morrison's grooved platter made a record of its own. For reasons known only to the radio people who made the snap decision, the accident was elevated to status of a national crisis by means of a unique signal. Bill Harris[268] wrote: "The book THE FOURTH CHIME stated that it was originally contrived as a confidential alert to signal the members of the NBC news staff, engineers, and other personnel responsible for broadcasting the news to the people. It was first heard on the air with the crash of the dirigible *Hindenburg*, May 6, 1937 at Lakehurst, New Jersey. It was fortunate that Chicago NBC news reporter Herb Morrison decided to take a recording device with him that day to Lakehurst to record the landing of the *Hindenburg*. With the airing of Morrison's recording the next day over both the Red and Blue networks, NBC's strict rule against using recorded programs for broadcast was broken for the first time in it's history so the radio audience could hear this dramatic eye-witness account of this tragic event. Jack Hartley, a reporter and member of one of the NBC Mobile units that was dispatched to the crash sight is quoted in the book I LIVE ON AIR, 'There were about a dozen of us in the original party that left with Mobile Unit No. 1. Shortly before we started, we broadcast an emergency signal to all of those attached to the News and Special Events Department in the New York Office. This was done by means of the fourth chime.' Both WJZ and WEAF, the Blue and Red network stations in New York broadcast the signal. The fourth chime was sounded again during the Munich crisis in 1938, with the news of the bombing of Pearl Harbor, December 7, 1941, and during the early morning hours just preceding the bulletins announcing the D-Day invasion June 6, 1944."

An accident involving a few dozen fatalities would not seem to warrant the attention of a World War, but that milestone of media history (some say downfall) was only the opening round of the overreactions to come. Captain von Schiller's previous instructions to prepare to make ready for the next scheduled run to South America with the *Graf* was countermanded by Hermann Göring himself, according to an account in *The Daily Express* on the 10th of May.[269] This 'get back on the horse before loosing your nerve' flight, which would have been her 579th, [270] would have had interesting implications. The *Graf* was still quite a capable ship in 1937 and was certainly to be employed until the LZ-130 was to come on line. His Captain wrote, "…the *Graf Zeppelin*, which no longer has the lifting power of its younger days. A screw or a lever has been added here, a new coat of varnish there, and some new installation at another point; all these things together add very considerably to the weight. But the lifting gas capacity and the length of the voyages has remained unchanged, while the number of passengers and freight has increased."

Abroad, without the recording being played as if it were live radio, the accident was initially treated as just another sad aircraft loss, not uncommon in those days and, with only 13 passengers killed, lower-half news today. Geiling and Sauter: "On May 12th DZR report reads: 'It must be generally said that after the accident of the LZ-129 became known, only nine out of

---

[268] http://home.flash.net/~billhar/chimes.htm , called to attention by Rick Zitarosa.

[269] H.G. Knausel, ZEPPELIN AND THE UNITED STATES OF AMERICA, An Important Episode In German-American Relations, Luftschiffbau Zeppelin, translated by M. D. McClellan, 1976. It is interesting to note that most other published sources claim Hugo Eckener stopped the flight, something that clearly would make no sense, Eckener's having survived or overseen return to operations after other accidents.

[270] J. Duggan and J. Graue, COMMERCIAL ZEPPELIN FLIGHTS TO SOUTH AMERICA, Diversified, Valleyford, WA 1995.

one hundred passengers wished to cancel their flights; six passengers requested extra bookings, for example the wife of the Brazilian President for herself and her two daughters.'" Clara Adams, who had made the LZ-129's first north Atlantic crossing, placed a $100 deposit[271] on a ticket for the first flight of the LZ-130 even as the LZ-129 was still smoking.

Perhaps there would be another Vitaphone recording of a airship song – there was one with ZR-1 and ZRS-4, and then we could move on. Indeed, *Modern Mechanix* magazine in August of 1937 noted "Since the World War there have been 69 major submarine accidents, in which 771 men were lost. We still build submarines. During the same period there have been more than 100 'notable' marine disasters in which 12,000 lives were lost. Yet try to book a quick passage on even a lowly freighter! ...There have been 100,000 railroad fatalities since the war, yet the railroads advertise (and truthfully) they have the safest means of transportation." A small number of Lakehurst personnel were killed in 2000 when their commuter plane crashed and its fuel burned, not even making national headlines by its irony.

On the 7th, Göring had sent a telegram to Rosendahl thanking him for "The unreserved help of the American airmen in coming to the aid of their German comrades is beautiful proof of the spirit which links airmen of all nations."[272] But that back-stabbing Nazi Minister of Aviation personally saw to it LZ-127 would never carry passengers again. "On July 19th the veteran of 1,060,000 miles and 13,000 hours flying was ferried to Frankfurt, hung up and deflated, to serve as a museum." Still, no one predicted just the one accident would end the service and delete rigid airships from our transportation options. There was nothing to take its place. A LZ-129 replacement would be needed to keep up the ambitious schedule which had called for *Graf II* to replace *Graf*, relegating the older ship to training the multitude of airshipmen needed for the fleet to operate out of the multi-hangar complex coming at Frankfurt. The Americans proposed building such a ship with their helium experience. In a proposal submitted through liaison Harold Dick, metric blueprints with weights in kilograms showed two proposals. Knowing German distain for the partial inflation of the ZRS, Arnstein proposed the option of using 10% hydrogen inflation and burning lift left over after initial altitude in the engines. Something had to be done; there was no alternative technology to replace the fastest oceanic travel known to our civilization.

There had been an accident; accidents are caused; one need only find the cause, and make sure that particular set of circumstances could not happen again. *What* had actually burned did not seem nearly as important as what actually *started* the horrible conflagration. Speculation on its cause began long before the seventh of May's dawning light highlighted the belching smoke still pouring from the wreck. On that day, Ralph Upson told a reporter "We should not focus our attention on the fact that the *Hindenburg* disaster may have been caused by the explosion of hydrogen gas. As long as fabric casing continues to be used, any spark... might result in such a tragedy." It is possible Upson's theory was considered as the Department of Commerce authorized an investigation team. Sitting to hear testimony would be CDR Rosendahl, Major Schroeder, Colonels Hartney and Lincoln, General von Bottticher, The Honorable G.R. Wilson, Mr. Trimble, and Mr. Mulligan. A German team was also hastily readied to ship out. In addition to Eckener and Dürr, Lt. Col. Breithaupt and staff engineer Hoffman would join professors Beck and Dieckmann to leave Cherbourg, France, on the *Europa*.

---

[271] Clara got a receipt, now in the collection of John Provan; *Correio Aereo* #26, Oct-Dec 2000

[272] Copies of this and other telegrams lauding the US Navy men's role in the rescue were included in letters of commendation for their files from station commander C. E. Rosendahl.

Historian Henry Meyer found that, in route to America, Eckener had written his wife that *General der Flieger Staatssekretar* Erhard Milch, Germany's Secretary of State for Air, had taken he and Dürr to see Göring. The minister said "I was never much for airships, but now we *have* to persist with them." [273] The National Socialists could not end one of their well-publicized endeavors in an embarrassing failure. Wary of the Government's commitment or not, for the airshipmen, their entire future depended on finding out what really happened.

Professor Max Dieckmann was the electrical expert on the German investigating team. As he was in route to Lakehurst, he mentioned LZ-129 had been painted with a different aircraft dope than LZ-127.[274] He was heard to say he thought an examination of its dielectric properties was necessary. At first, as they began their hearings, the dual commission seemed open to suggestion. FABRIC IN ZEP FACES TEST AS BLAST CAUSE shouted a headline.

Upson appears to be alone in his suspicion of a connection between electricity and the fabric, even though it was hardly a new idea. Twenty seven years earlier *The Century Magazine* had warned, "Atmosphere filled with electricity may dangerously charge an air-ship's metal parts and set it afire on contact with the ground, but provision has been made automatically to relieve such tension."

Newspaper accounts covering the board of inquiry mentioned other theories, as had happened in earlier disasters. "Did Sabotage Wreck the *Akron*?" had shouted the June 1933 *Modern Mechanics and Inventions*. During the witch hunt following the drowning of all but four (radioman Robert Copeland later died of injures and exposure) of the USS *Akron's* crashed crew, some would insist the mighty symbol of American power could not have failed, but rather fell victim to a determined madman. Dredged up was the overblown scandal of a loudmouth construction worker. As Richard K. Smith wrote, "...the arrest of Paul Kassay... an admitted Communist...[to the public] From coast to coast, the *Akron* had been 'sabotaged.'" Also like the *Macon's* Japanese plot, there would be conspiracy advocates who reasoned the giant swastika-emblazoned vehicle could have been the victim of an anti-Nazi fanatic. Since the final solution offered seemed so unlikely, it was only natural to believe a clever saboteur started the fire that erased all evidence of his crime. Behind the scenes the FBI actually was checking out the possibility of sabotage. It is not known how much about the lessons of World War One they were aware, but of course any saboteur would have found placing a bomb on a mostly empty diesel tank would have had a better chance of success than in a cell of pure, smothering gas. If only the fire had started down there in the keel.

No matter, years later one book begat another book which begat a motion picture (and possibly another movie as of this writing) offering persuasive arguments for the "lone nut" theory, tending to forget even a suicidal maniac would have basically the same problem the British defenders were up against. The FBI eventually produced a report with the conclusion the fire was *not* the work of an athletic madman. Of the dozens of investigations, it was the opinion of the Feds that a terrorist simply would not work that way.[275] With this one could add the problem of how a Zeppelin could be destroyed in such a way as the witnesses testified it happened.

---

[273] Meyer, AIRSIPMEN/POLITICS, Pg. 214

[274] "Analytical Research of Covering Material of Zeppelin Airships LZ 127 *Graf Zeppelin* & LZ 129 *Hindenburg* " (1937) showed that the binding material of the layers of the LZ 127 consisted of cellulose acetate. The LZ 129 shows a different mixture, i.e. with "butter-acid".

[275] Addison Bain dug into the FBI's files on the accident at the National Center for Analysis of Violent Crime in Quantico, Virginia.

Seemingly oblivious to both, novelist Mike Mooney concluded a bomb had been set off in "the middle walkway."[276] If there was anyone listening who had attempted to shoot down Zeps during the Great War, they surely got a chuckle out of knowing smaller Zeps had suffered mid-cell artillery shell explosions and still sometimes made it back to base.

There were many more theories – and they continue to this day. Also as with other airship losses, lightning was discussed. "One immediate belief was that lightning destroyed the ship."[277] Since wind cannot break an airship not deliberately driven into it, this gave comfort to some. "The [laws of physics] declares that electricity has mass. When applied to lightning this opens the possibility that lightning exercises a hammer-likc power of beating on the ship's ribs."[278] A quick theory easily believed with so many explosive half-empty tanks of petroleum in the keel – until one remembers how many times the giant "Faraday Cage" was previously struck.

The *heck rund lauf wagen* riding on the track circle is visible forward of the aft *lande rad*. Vent hood for Cells I & II combined *gasschacht* is visible on the vertical fin; the fin's vent is visible at the aft upper corner of the red Nazi flag. Rain gutter is barely visible on the equator. From the Hepburn Walker Jr. collection.

Stan Washburn, a business partner of Anton Heinen in the *Air Yacht* venture, had been at Lakehurst with his car the fateful night. Washburn drove the injured Zeppelin senior officer Wittemann, who had first flown over in the ZR-3,[279] to the hospital. In 1996 Washburn was so kind as to write the author to tell of Wittemann's confession that the Zep officer had inadvertently started the fire by activating "new" electrical cell fullness indicator. These indicators would be checked and exonerated, but not just because there would have been no air in the cell. The identical electro-mechanical sending units were installed in the next hydrogen filled airship, which Wittemann commanded.

---

[276] Mooney told the Miami *Herald* for the 16 FEB 72 issue that he had found a sketch of a bomb.
[277] C. E. Rosendahl, "What Really Happened to the *Akron*," *Liberty*, Dec. 16, 1933, pg. 4.
[278] H.W. Blakeslee, "Blow by lightining or twisting by gusts might have doomed *Akron*, scicence shows," *Asbury Park Evening Press*, April 7, 1933. On April 19 the paper reported Dr. W. J. Humphries said *Akron* was struck.
[279] A. Wittemann, *Die Amerikafahrt des ZRIII*, Wiesbaden, Amsel-Verlag, GmbH, 1925.

As recently as March 2000, the Baltimore *Sun* claimed *gasoline* had ignited the hydrogen. Bill Rutan, who had heard the backfire, said he smelled gasoline. Supposedly used to start the diesel engines, the demonstrated real nemesis of airships had somehow leaked and ignited. Oddly, this is given some credence by the Fischer report. The last Navy man to ride a Zeppelin noted there was some 66 pounds of gasoline aboard, though he did not say what the ten gallons or so was used for. Mr. Rutan was so kind as to write the author with details. Sadly, Rutan's newspaper interview had the gasoline fire spreading *directly* to the hydrogen, never minding what it would encounter on its way. Engineman Bentele was still available for comment and reassured Dr. Bain through the Zeppelin Museum's Dr. Meighorner, that unlike the LZ-127, no gasoline was ever carried aboard LZ-129. The big diesels had been started with compressed air. [280]

A respected airshipman suggested diesel chars were the ignition source. Such discharges were rare, but not unknown, when reversing the engines. They had been known to burn small holes in engine car fabric and even main covering. But this theory sought to blame hydrogen – by having it hanging around below the equator (!) to be so ignited. Harold Dick wrote to Paul Litchfield: "The possibility of an engine backfire igniting the gas we have always discounted because of low temperature of the diesel exhaust. It is becoming increasingly difficult to even advance a theory as to what may have ignited the hydrogen." Sadly, no one was listening to Ralph Upson.

P. L. Teed had written 12 years earlier, "The danger resulting from a hydrogen leak …is not very great, for hydrogen weighs 1/14th as much as air, therefore the gas leaks from the bags rushes upwards, away from the engines and the crew, to escape from the top of the airship." Nor were conditions perfect for hydrogen ignition. The safety handbook points out, "Aside from mixture composition, the flammability limits also depend on a number of variables, such as ignition energy, initial pressure, initial temperature, and relative humidity…the flammability limits are widest with low humidity and narrower when humidity is high." The cell's pressure was only slightly above the air outside. The drizzly May night's humidity reading was 98%. Small wonder, according to accounts, Lehmann's dying words were: "I don't understand it…"

Unfortunately for the pursuit of truth, the investigators eventually narrowed the focus not to what is seen burning, but instead, to the invisible lifting gas. One way or the other, it *must* have been the gas; it didn't really matter how. Bob Gordon wrote, "The *Hindenburg* was destroyed while landing at Lakehurst, when exhaust flames from the motors or static electricity probably ignited hydrogen gas as it was being valved." Closure then could finally come simply, easily, when some rationalization for hydrogen's involvement was whipped into shape.

One report would define the limits of how far one could go in defining hydrogen's role in the fire. The New York Times declared "Wrecking of the *Hindenburg* Laid to 'Demon Proton' by Scientist," an interview in which Professor Otto Stern, "one of the world's most eminent atomic physicists" of the Carnegie Institute of Technology. "The proton, inner electrical core of atoms of hydrogen gas, with which the giant dirigible was filled, stood revealed as the evil spirit who burst loose from the outer shells of electrons that held him in check." Stern said the accident was no surprise. "On the contrary, it has been a miracle all along, and a tribute to German genius, that the happening of what amounts to the inevitable could have been avoided all through the previous great voyages of the *Hindenburg* and other Zeppelins." Yet the interviewer admitted Professor Stern was reluctant in applying the knowledge of the "demon" inside hydrogen directly

[280] The similar gasoline Maybachs on the American rigids were compressed-air start as well. Author's interviews with ZRS enginemen John Lust and Eugene "Pete" Calande explained how the engine would be stopped, a lever to reverse the cams was worked, and the engine compressed-air started in the opposite direction.

to the *Hindenburg* disaster, as Stern lit a cigar and patted his hydrogen tanks. Ignorant of the fact the pure gas inside the cell could not burn, an entire theory was created having the origin of the fire *inside* cell IV where Lau had seen a *reflection*. That was only 90° out of phase and a short distance from where port observers first saw the fire come over the top, so this scenario gave closure to some. For these people, pure gas had caught fire, and popped out the cover at the top of the ship. Somehow, the pure gas had burned the ship from the *inside*, outward.

Others realized pure gas could not burn, so for a time, the gas cell *valves* were put on trial. Known to have stuck open once, perhaps a valve had not reset properly during the last gas release – even if only the forward cells had been valved. Perhaps Cell III's stuck-open failure back in 1936 was still leaking? Years later Zep machinist German Zettel told Henry Cord Meyer he thought the valve was never properly repaired. There was a chorus of "maybe so's" from people who mistook the aluminum rich bright fire, showing through the mostly empty interior, to be hydrogen burning *inside* the ship. Von Meister would later state he saw hydrogen burning in the cell IV's vent shaft.

Remembering that air displacement was their livelihood, the masters of the rigid airship had fashioned a system of gas management that featured several safeguards. First were the electrical manometers Whittemann had suspected. These instruments could detect as little as a two percent change in cell status. The sensor's two percent variation design was no accident. A two percent loss of lift was considered the maximum allowable loss for good reason. Since no fixed weight of the airship's construction could be changed, a loss of even such a small percentage of such a huge balloon would mean payload would have to be dumped – perhaps in a hurry.

Control car officers insisted there had been no indication of gas loss, and they were watching the readouts. Dr. Bain later calculated that, to remain undetected, a stuck valve in one of the aft cells could leak no more than 1.44 cu feet/min in the allotted time. The gas shaft vent setup, witnessed to work even in the hangar, had the additional venturi effect of at least six knots of wind over the hoods, and that is conservative considering the greater altitude and the gust that ripped fabric and required throttle response. But even six knots would yield an air exchange rate in the gas shaft above the valve of 530 cu ft /min. At the extremely conservative lower limit of 4.1% hydrogen-in-air, a gas flow of 22 cu ft/min would have been required to yield a flammable mixture. That would still be located in the isolated shaft, safe from spreading a fire even if somehow ignited.

A similar analysis for cell IV shows a minimum value of two times the detectable flow for combustible concentrations to be available. It just wasn't possible so much could have escaped undetected, and the airshipmen knew it. Rosendahl would tell a newspaper reporter that he doubted a large loss could have gone unnoticed, and would write: "…it would have been detected by the gas pressure instruments indicating in the control car…" So the leaky valve theory was discounted then and there, but it would survive as an alternative if one did not like the other rationalization of how hydrogen had been involved in the fire. After all, it all happened so fast, surely the gas, with its very fast flame front speed, *must* have been heavily involved, right?

Lee Payne [281] relates how that often-quoted overall time reference was reached. "Later when the newsreel footage was timed with a stopwatch, it was found that the destruction of the *Hindenburg* took just thirty two seconds from the appearance of the first tiny flame until the final collapse to the ground." This somehow became a historical benchmark in spite of Rosendahl's

[281] L. Payne, LIGHTER THAN AIR: AN ILLUSTRATED HISTORY OF THE AIRSHIP, Orion Books, NY 1977

statement "...but unfortunately not a single photograph has turned up to show the first appearance of fire," let alone how much time elapsed before the fire reached topside where it was visible from starboard, or then how much time elapsed before the movie cameras got cranking. Nor is there any way to know which cut was timed; in every popular version, the different camera's films are *spliced together*. Payne also summed up what happened next:

"At the end of eighteen days' testimony from scores of witnesses, Eckener was asked to present his own personal conclusions. First he stated that Commander Rosendahl and two other airship experts standing at different parts of the field all agreed that a light or flame appeared on the top edge of the ship just forward of the rudder fin a few seconds before the explosion... One expert witness described it as the slowly burning flame of pure hydrogen and, Eckener concluded, at that position on the ship it could not have been anything else... The ship had made a sharp turn in approaching the landing field. The strain of the turn could have snapped a bracing wire, which in turn could have torn open a gasbag in the stern. The escaping gas collected at the top of the hull just beneath the outer cover. It was ignited by static electricity when the landing ropes touched the ground in the highly charged atmosphere following the electrical storm."

As the leading airship expert in the world, Eckener's comments were widely publicized, and became a mantra still chanted today. No one pointed out the three persons standing on different parts of the field were still all on the port side, or that a "slowly burning flame of pure hydrogen" is not just invisible, but 'pure' and 'slow' is impossible. Even the mater-of-fact record of the station log stating the "...fire originating at or near the stern" was no longer part of the story. Listening to those comments, one would not know the fire had started starboard aft, as witnessed and diagrammed by people in a position to see the origin, and only later to appear topside. One would not know that, to place an exact time for the origin, one would have to add several more seconds to Eckener's "several seconds before the explosion." With all due respect to Eckener, pure hydrogen makes *nothing like* a visible "slowly burning flame," or even a flame at all. The fire they saw most certainly *was* something else. His unrehearsed remarks were widely paraphrased and portions emphasized.

The first leap of faith in this scenario was damage caused by turning the ship. No one seemed to remember that the Americans had been criticized for building their airship sterns sections too weak to handle dynamic tail loads. Bruce Davis, son of US Navy Chief Airship Rigger "Shakey" Davis, was so kind as to write down how he and his father felt the new design of the LZ-129 was substantially weaker than the LZ-126 and 7. The Germans bragged their cruciform structure would support the upper vertical through even greater loads than the lower vertical; later we found out why. When Karl Arnstein's biography was published, it was revealed his 1935 suggestions had been followed. The LZ-129 tail had been further strengthened, benefiting from the ZRS bad experience and the wind tunnel tests the American Navy did not take seriously. But to these theorists determined to finger the cause of the 'leak' in the tail that was stronger than all others, and had routinely sought out fast storms, *could not seem to have stood a gentle turn*.

A US Navy observer on several LZ-129 flights had recorded his observations. "There seem to be no restrictions on using full rudder and elevator throw in order to effectively control the ship. *Harsh rudder and elevator angles as stated above are used when necessary.*"[282] Yet months later this pinnacle of the airshipman's art had somehow broken a stern wire just slowly turning to land in the relative calm *following* the storm!

[282] Emphasis added. C. Knox, LCDR USN: "Report on the *Hindenburg* Flights, 15 SEP 36

It was not generally known at the time, but the tail had once suffered more than just harsh turns. During the hasty launch for the carefully staged three-day propaganda flight on March 26, 1936, Lehmann found he had to risk a downwind takeoff rather than delay the highly anticipated appearances. Suddenly the tail broke loose of its cable while the nose was still secure. The bow got away from the handlers and rose to a 14 degree angle, bashing the fin and crunching the rudder. Goons raced around the Lowenthal field to grab cameras, but Goodyear employee Harold Dick managed to hide his. The damage photos were made public in his book with Dr. Robinson, who would summarize: "Thanks to the sturdy design and construction of the *Hindenburg*, the damage was entirely local, confined to the lower after portion of the lower fin, and lower portion of the rudder." No interior wires had broken at all.

Following the tail dragging at Mines Field in Los Angeles, the *Graf* continued across the country to a landing at Lakehurst. This embarrassing damage was local, but lead to Navy design change #2 in the ZRS ships, which ultimately lead to the LZ-129 tail being made stronger. White streaks caused by envelope glue in this long-stored negative in the Hepburn Walker, Jr. collection.

Getting the gas out of the cell so it could mix with air to become flammable is the cornerstone of blaming hydrogen in the fire's origin. Historians naturally tend to emphasize the broken wire theory in spite of the complete lack of evidence or precedent. In fact, just the opposite. Of LZ-127's close call in California, von Schiller said "It was much too warm, and the ship had not received enough hydrogen," so the ship had to make a dynamic takeoff. Dragging the tail in the dirt busted up the rudder's bottom – and yet the ship was still fit to cross the country through the same mountain pass that nearly ripped off the undamaged USS *Macon's* horizontal fin in 1934.

Robinson wrote, "Dr. Eckener… argued insistently that the sharp turn had overstressed the after hull structure, causing one of the bracing wires to break." With all due respect to the late Doc Robinson and other historians, the actual testimony does not sound very insistent at all. The 22 MAY 37 record reads: "Witness could imagine a shear or brace wire had become parted..." How sure was he of this speculation? Not very sure, since the record also reads: "Witness had observed that shear or brace wires on a number of occasions had been broken, but none had been broken on the *Hindenburg*."

There was good reason for this scenario to be difficult to swallow, as the record shows Eckener went on to explain: "The breaking has not occurred during the last years because the pretension necessary to be given these wires is at present much better known than formerly." Rosendahl appealed for reason, writing: "It is known that of the thousands of wires in an airship, once in awhile one has failed in flight, but in such cases, even hydrogen ships, often not even the slightest damage has resulted." The experienced airshipman had good reason for saying that. The hard steel wires, set to a stretch tension with a tensionometer, were not springs. Nor did they act like springs. Wires sometimes broke because they did *not* give like a spring. More akin to a bicycle spoke, a broken wire would likely just lay there.

Furthermore, Hugo Eckener had also imagined the supposedly broken wire "…had caused a rent in a gas cell." The lost art of the rigid airship does not allow a live demonstration today, but the gas cells, by their low-pressure nature, were not easily slashed. All but the smallest cells were no where near full. "They were quite a bit up, the cells," said Lau. The relatively limp, rubbery fabric would present a difficult target for a slasher. There is not a single example in the literature of even a stretched-tight cell at pressure height being cut by a snapping wire even in violent maneuvers. To the contrary, in the ZR-1 loss, just the opposite happened. Vladimir Pavlecka observed that the height-climber Zeps, on which ZR-1 was based, were particularly dependent on structural support provided by 100 % cell fullness during the entire mission except the return phase, when loads were tens of thousands of pounds less and gas would be vented to lose altitude.[283] Rick Zitarosa e-mailed "On some occasions, a ship would return 'exhausted.' That is, barely enough gas to keep it afloat, most of the water ballast and fuel gone, maybe with a soaked outer cover, some battle damage and a couple of engines out as added excitement. In this condition the Zeppelin would make a 'hard landing' and have to be dragged and carried into the hangar by the ground crew; on at least one occasion, the rear gondola had to be cut off to get the ship light enough to move!" Again, even among these very lightweight, hastily constructed airships maneuvered in desperate combat situations, there is no record of these wires slashing a cell.

Nonetheless, Hollywood would use this incident as an inspiration in the 1975 movie, showing how a careless crewman's stomping a wire had it whip back with such fury as to cut the heavy fin fabric. Incredibly, they did not show such an event happening during the landing. (But then the story concerned a mad bomber, so they could not very well use a slashing wire.)

Harold Dick would to his last days insist that a shear wire could not have cut a cell.[284] He said only a gas cell wire itself would have been capable of doing damage. These little-known wires' role in even a strenuous turn was never addressed. A surprising number of people who ridicule the ZRS stern design as lacking the essential cruciform of the "much stronger" German ships see no inconsistency in signing up to the broken-wire-doing damage scenario. Nor do they realize the deep ring and the cruciform design were both the work of the same man. Both investigation commissions would agree that broken shear wires varying effects *had never included serious damage*, but this would go unnoticed.

Again, many writers have imitated Robinson as he covered the next step of this bedrock of the hydrogen ignition theory. "The recoiling end, he theorized, slashed open one of the after gas cells, permitting the escape of a large amount of hydrogen…" Did Eckener *really* believe in two such unlikely occurrences? To the contrary, the actualy record reads: "There were two

[283] Lee Payne's conversations with Pavlecka, *Aerostation*, December 1999
[284] Dick/Robinson, pg. 148

interconnecting and remote possibilities. If it has been necessary to assume two remote possibilities occurring at the same time, then the probability of their occurring at the same time is not likely."

While there was no evidence for this second step either, this theory *further* imagined that, following a wire snapping and cutting a cell, the damage was extensive enough to cause a *significant* gas leak from the torn gasbag. Again a misconception has been repeated, that he though it was a *slight* leak. Eckener's testimony states flatly: "The loss of gas must have been appreciable.... Under those conditions, naturally the accumulated gas... must have been a very rich gas and not an explosive mixture of hydrogen and air – but more of a pure hydrogen." It was as if he did not know pure gas would not burn! Not satisfied, there was another backup story.

With time came speculation that the cells were just leaky, period. Rosendahl told a reporter, "The seepage of hydrogen from the cells would have amounted in one hour to one-fortieth of one per cent of gas in the cell...the gas that might have seeped out was very, very small." Forgotten was the successful *Graf Zeppelin's* much higher permeability rate while straining through electrical storms. Others, ignorant of the LZ-129's less porous cells, would insist no explanation of how the gas escaped was really necessary. Again, the outcome-based argument: gas *must* have been there; there was a fire, after all, there *must* have been a leak.

The airshipmen knew it would take a very large amount of gas to contain any realistic amount of energy, and could conclusively state no such loss was possible for *several* reasons. First were the aforementioned electrical manometer readouts that would have shown even a two-percent leak in the smallest cell. But before there had been these fancy instruments, the keel watch provided a constant vigilance against the loss of the airship's precious gas. As the rigger walked his beat, he was alert for escaping gas two ways. A leaky cell would "mushroom" – its bottom retracted to a higher point than the healthy cells. Nevil Shute, for example, recorded in his diary during R.100's Canadian voyage, "Gasbag 7 appears to be leaking as it was when we started and has risen a good bit..." Later he noted, "...7 has risen to about 3 feet below the E longitudinal."

Crewmen assured the inquisitors the LZ-129 cells were all evenly inflated. Nonetheless, there would be rumors that one cell was visibly higher than the others, as if gas had been lost. One story had one of the cells almost empty just before the fire. By contrast during the inquiry Lau was asked "Would you consider that they were up at a height greater than usual for a trip of this kind?" Lau answered, "Nothing occurred to me to be out of the ordinary." He added that accused Cell IV "...was at the same level as the other cells."

Shute wrote, "Hydrogen is fairly easily detected, however; it has a peculiar warning of its presence, more of a savor than a smell. I do not think it is toxic; a man gassed by hydrogen recovers quite quickly in clean air." Since hydrogen is odorless, this "flavor" may have been caused by glue or other emission odor from their cattle-gut cell construction. The 1940 hydrogen manual explains, "After use in a rubber envelope, hydrogen frequently has the odor of rubber." Odorless when pure, the lifting gas presented the most danger of displacing air in the lungs when fresh, before it could absorb some odors during later confinement in stinky cattle gut or rubber bags. The Great War literature contains an asphyxiation case at Nordholz, in which an L30 (LZ-62) rigger was overcome while working topside. (That's half as many men who were lost to engine fumes aboard the L53 [LZ-100] on 13 MAR 1918, alone.)

In the LZ-129's much improved gelatin-latex cells, to warn of small leaks, an odorant may have been added to the gas. This principle is in use today, as methyl mercaptan is added to odorless cooking gas in a few parts per billion. The odorant is obvious to humans with a normal sense of smell long before a leak can reach combustible proportions. Fast-moving hydrogen molecules mostly head skyward, but trace amounts go in all directions. The human olfactory is many times more sensitive than hand-held detection instruments. The odorant was an effective way of being aware of even excessive permeability losses long before a flammable concentration could exist.

All four men in the stern survived; all specifically answered "no" when asked if they smelled a hydrogen leak. Did they know what a leak smelled like, one might ask? After all, there had never been a fire before. Again we must remember that the men's livelihood – indeed, in dynamic flight, their very lives – depended on *keeping gas in the bags*. You can bet they were keenly aware if their ship was losing gas, since impact with the ground or their having to swim home was a much more realistic outcome than fire.

Eckener had in his 22 MAY 37 testimony "...made the following statement and set forth his thoughts therein, not because his having formed any definite conclusions as to what may have happened, but because some possibility might have been brought to light which might properly be further investigated." In other words, he was laying out some speculation that he thought worth checking out. Reporters, and later historians, missed this entire point. It follows, therefore, that the first part of Eckener's testimony actually speculated not about how it got there, but that "...assuming there was free gas aft..." thereby agreeing with the outcome-based adage, "there was a fire, there must have been a leak." He said so earlier: "Considering the type of fire the witnesses said they had observed, it must be concluded that there was free gas in the rear section of the ship."

How was that possible? In the report line that began "Witness could imagine..." Eckener was saying, "The gas from the torn cell escaped upward and filled up space between outer cover and cells in the rear part of the ship..." If it was going to be involved in setting the outer cover afire, this was a must. Problem was, the outer cover was no realistic barrier to the gas. Hydrogen's high rate of diffusion enables it to penetrate very dense materials, up to and including *cast iron*!

It was if he had just ignored the testimony of C. E. Earle, Chief Chemical Engineer, Bureau of Aeronautics, Navy Dept., Washington, DC. [285] Earle stated: "... diffusion of outer cover much greater than that of a cell...doubt that there would be *enough concentration to set off combustion...*" The newspaper article quoted Earle saying only tests on the fabric itself would be conclusive, and the Germans we quoted as having promised such tests. We must remember the 1940 manual stating "Because of its lightness hydrogen is very mobile and diffuses with great rapidity through other gases. As it has a high rate of diffusion it readily escapes through fine openings." This actually gives modern safety technicians some worry, as they write under 'safety considerations:' "Hydrogen readily migrates through small openings and through torturous paths."

Also forgotten was hydrogen's convection and very fast dispersal rate, almost three times as fast as methane in air. So, as Earle pointed out, any hydrogen leak would tend to sail upwards on a straight line and go right through the cover, even as it was disappearing into the interhull air.

---

[285] Emphasis added. At least one newspaper refers to Earle as the Bureau's "chief hydrogen expert" and in other LTA references he is called a "helium expert."

Closing Lehman's book, Rosendahl again emphasized "...nor could it be shown that there was free hydrogen or any dangerous mixture of hydrogen present."

Knowing this, someone came up with the idea of making hydrogen visible to the naked eye, by action of something that was visible. It was volunteered that the wind-driven fluttering outer cover was actually 'trapping hydrogen trying to get out.' Eckener, possibly surprised that what he suggested to be checked out was not being quoted as a factual explanation for the accident, later speculated that it would take something like 40 to 50 cubic meters of gas loss *per second* to flex the outer cover. No one addressed *where* the gas would actually go if somehow retained by the cover. No one seemed to remember one "fluttering at the top of the stern" observer was at the bow, not the stern, and the other had seen the flutter *in the prop wash*.

At a cell pressure of only one-inch of water, such a flow rate would have required a massive, most noticeable hole in a gas cell, if then. Modern audiences, treated to such turkeys as Hollywood's ZEPPELIN, would believe the film's hissing sound as gas escaped bullet holes in the cells, in spite of being obvious nonsense. One final thing that has gone unnoticed in the sea of noise: hydrogen leaking aft, making the tail "heavy" and thereby sagging, is incompatible with it being retained by the outer cover.

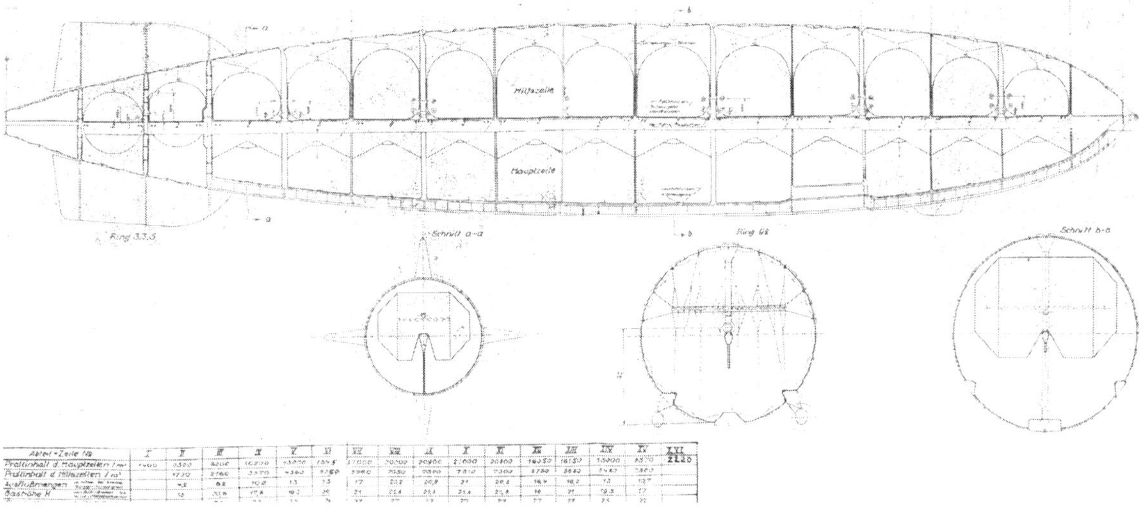

Original LZ-129 plan called for 14 helium filled *Hauptzellen* to enclose 14 hydrogen-filled *Hilfszellen*; cells XVI and I would be helium only, for a total of 30 gas cells. Cheap and renewable hydrogen would automatically be vented to obtain initial cruising altitude and later manually valved to compensate for fuel weight burned off. Engineman's rule of thumb was a basic consumption rate of 500 kg/hr, necessitating about 500 m3 to be vented to compensate. L-Z archives blueprint courtesy Addison Bain.

The entire notion that gas escaping from a cell, with less pressure than a person's exhaling breath, could somehow yield a rapid *back-and-forth* flutter motion in the reinforced and wired fabric, which on topside was supposedly visible to a ground observer hundreds of feet away, is the most asinine rationalization in the entire collection of literature. Yet it too would be repeated, and sworn to, for a generation.

Amid the "trapped gas" confusion only token effort was applied to explain away the vents' capability. If the interdeck hull vent was even discussed, it was decided since the station log showed only six to eight knots there 'had not been enough wind moving over the vents to evacuate the gas.' But ignoring the higher altitude, and the late gust, this was a confusion of the gas valve vent shaft versus the interhull vent system. The interhull vents were designed to exchange air during rapid altitude change, and were not dependent on ram air or slipstream for suction. They would have to be able to handle large changes in air volume even if the ship was being thrown *backwards* while rapidly changing altitude in a storm. If the vents were somehow clogged and the covering suddenly a substitute gas cell, the gas would have floated upwards to the highest point on the sloping stern's spine. If aft, that would be the top of the upper fin, also no where near the stern origin of the fire.

Eckener's "imaginings" had thus far asked for three very unlikely events to be checked out. Needless to say, the fact they had to happen in an exact order within a specific time was not emphasized. Yet even these three baseless speculations did not explain the fire. Oxy-hydrogen cannot burn without an ignition source. No one at the time was quoted as explaining how an ignition source hot enough to ignite oxy-hydrogen (if a perfect mix, 858°F) got *under* the cover, without first *igniting* the cover which lit off at 383°F.

So therefore came the second most often quoted supposition after 'there was a fire, there must have been a leak.' It goes like this: 'There was a leak, there must have been a spark.' Indeed, Eckener's testimony reads: "...it must have been ignited by some spark."

Some spark? What spark was that, one wonders? The report shows Eckener cautioning "The probability of a spark having been generated at the point was from past experience a remote one." Indeed it was. The steel mooring cable would have, of course, conducted any accumulated charge at the point it touched earth, but the steel cable never got anywhere near the mast during the flying moor. Just like he said, past experience would also show the spark would have been down there as the cable approached the ground.[286]

As later tests showed, the LZ-129's fabric was a very good insulator. This envelope of highly resistive fabric would have all but eliminated any chance of static electrical energy finding a place to enter. If there was some defect in this armor, the next thing the current would encounter would be the "Faraday Cage" of aluminum structure and steel wire bonded together to make one giant conductor. So the spark would have had to be inside.

The manila landing ropes would not conduct electricity when they first hit the ground. Eckener noted that, stating "It would appear queer that if the ignition was caused by such a static equalization spark the ignition did not occur at the moment the landing ropes were dropped, which ignition would have happened without a doubt had the ship been grounded by the throwing down of a conducting rope." They are not good conductors, having been kept coiled up and dry inside the ship's nose. Eckener also noted the fact they were hydroscopic and would be soaking up water, thereby becoming conductive. "Slowly the landing line became damp from the light drizzle still observable at the time. Therefore, it was believable and possible that the potential gradient between the ship and overlying air masses was then sufficient to generate these static sparks."

The manila ropes were connected to steel cables near their terminus. Had the manila lines been conductive, the spark would have developed down where they neared the soppy ground – not somewhere near the upper tail fin. Instead, like an unpowered extension cord being connected safely and later switched on, the circuit was complete to ground. About four minutes of soaking

---

[286] Author experienced this concept first hand while watching night landings aboard aircraft carriers *Roosevelt* and *Kitty Hawk*. It was possible to see a blue spark jump from the aircraft's arresting hook to the steel deck.

later, they became conductive, and they would pass any accumulated charge to ground. Trouble was, this would not have generated a spark because the circuit was complete *before* it became conductive; there was no gap for the current flow to have to jump.

Again, *The Literary Digest*: "If you want complete immunity from lightning, sail on a Zeppelin, where lightning cannot enter. The 803-foot long hull is composed of huge rings made of aluminum alloy. Every section of these rings and the longitudinal girders are connected so that the whole outer structure of the hull makes a single electric conductor." No one thought to dream up an idea of how the Faraday-cage structure, which had shrugged off lightning bolts, somehow failed to safely carry away this lesser charge. Neither Eckener nor anyone else at the hearing evidently reasoned this would take yet another unlikely happenstance.

There would have to be some sort of bad joint in the structure just at the point where the somehow undiluted oxy-hydrogen was lurking under the cover. The ropes become conductive, the charged frame discharges, except for, say, one bad rivet joint – which then overcomes the resistance of the oxy-hydrogen (it is an insulator) and arc-jumps across it, thereby igniting the mixture. Of course the *Digest* article was specific this would not happen when it stated flatly, "The ride the lightning has taken was harmless because it found no need for leaping a gap from one section of the ship's frame to another thus causing sparks." Surprisingly, someone has not yet dreamt up the idea of the supposedly strenuous turn breaking a structural component back aft, exposing raw aluminum and providing a gap that the discharge would have to jump across.

Even acceptance of this five-miracles-in-a-row scheme did not really explain the fire as it actually happened. Hydrogen burns six times faster than gasoline. A bubble of gas, if somehow trapped under the cover, remained in concentrated form and then reached by an ignition source, would still have to obey other natural laws. It would have collected in the upper fin. Less than half would have actually burned. Its straight-upward flame path would have made a local hole in the fabric, but not provided enough calorific value for long enough duration to make the fire self-sustaining from that point. An ugly maintenance headache, nothing more.

Though other authors have tried to paint a picture that such sparks ignited an oxy-hydrogen mix outside the ship, everyone realized, since experience showed, that exterior ignition would do no harm. Nor was it possible, as we shall see, that an exterior electrical event was somehow transmitted inside. No one ventured a guess how aluminum powder, which can be ignited with 0.00095 Btu of electrical energy – that is *two-tenths of a joule* – could possibly pass this static electricity through its matrix without its own ignition.

No, if hydrogen was to be blamed, it had to be *under* the cover and set the cover afire from below. Eckener's "imaginings" were of course directed at explaining the most popular origin of the fire as heard from the starboard witnesses. Revisiting that, one record reads: "Three independent observers standing at considerable distances from each other, who give the impression of being very reliable, have stated under oath that a light or flame appeared at the point of attachment of the stabilizing fin about half a minute before the explosion." All three were on the port side. Also, there always has been a big difference between *light* and *flame*, even though they are frequently interchanged ever since the Burning Bush of Exodus. He then said, "This flame appeared, according to one witness who must be considered an expert, like that of slowly burning hydrogen, hence pure hydrogen." (That *would* be slow, since pure hydrogen wouldn't burn, and if mixed with air and burning, could not be seen.)

Publicly, no one reminded Dr. Eckener he was first talking about supposedly seeing burning gas on the *outside*, which meant he could not subscribe to the popular concept that Lau had been seeing pure gas burn *inside* cell IV. Eckener was recorded, "He thought that Lau's seeing the fire burning within cell No. 4 was therefore at the second stage of the process of burning. Lau could not have seen or heard the relatively small fire that must have burned under the upper cover and also probably would have been unable to see the reflection in the cell."

With so many unlikely base elements having to be in place, and so many unlikely events having to happen in exactly the right sequence, it seems astonishing that, even the incomplete request for possibilities he was simply requesting to be checked out, these somehow became etched on stone tablets. Yet they are widely quoted as gospel today. The other Germans were not supporting Eckener's unprepared remarks. By June 10th a *New York Times* item reported "The German Commission reported that it considered impossible that there could have been a bad gas leakage with the resulting formation of masses of inflammable material around the Zeppelin. The report pointed out that special instruments on the airship would have recorded such an occurrence and that these instruments were always closely observed during landing maneuvers."

Damaged by nitrate decomposition and less than sharp from long exposure in the failing light, Fred Tupper captured this image of ZRS-4 revealing her companionway lights showing through the non-aluminumized keel walkway. Brighter light would of course shown through both sides. Printed from a long-unopened negative box in the Hepburn Walker, Jr. collection.

Finishing up Lehmann's book, Rosendahl assured the reader that the investigators would get to the bottom of it. "Scientific tests and experiments of many kinds are being continued in Germany and in the United States as well, in the effort to pin this mystery down to a definite cause." But the Germans would not be sharing what they found any time soon.

Most Americans would be satisfied with either of several incompatible explanations, as if a murderer who had cleverly destroyed evidence was still convicted, just because everyone seemed to know he was guilty. "There was a fire, there must have been a leak." When pressed for evidence, they could dodge by saying the fire had consumed the telltale component that failed. In the case of the "lone nut" theories, the device that ignited the hydrogen was of course destroyed. No matter that rumors about bits and pieces – a battery, a timer – all proved to be false leads.

Bent on enhancing the weak premise, Captain Hans von Schiller would be interpreted by Len Deighton that the broken wire end became a white-hot igniting tip, and if a tail wind had come along, it could have ignited trapped hydrogen. Neither man acknowledged even a perfect mixture required 858° F for ignition, or that the wind was going the other way. Author Tanaka still blamed the wire by suggesting, "The spark might have been caused by one of the metal wires that supported the gas cells snapping…"

Rosendahl wrote: "Theories advanced included snapped wires, broken structure, disintegrated propellers, lightning, sparks from a motor, St. Elmo's fire, the whole range of electrostatics, and even the phenomenon of catalysis." At once there was a chorus of "me too's" from everyone who thought they had something that fit any of the rationalizations.

Over the years of telling, another man who survived in the bow, Albert Grotzinger, would be "promoted" from cook to rigger (even chief rigger) and would have supposedly been mumbling something about having to fix a problem in the aft.[287] There had been hastily patched damages to Cells II or I, a backup rumor had it. "In Europe and America, Commissions were appointed, but the primal cause of the catastrophe was never definitely established. Thus the end of the era of Zeppelin airship aviation is shadowed in a veil of mystery…"[288] Finally, after many weeks of deliberations, the commissions' results were made public.

Both studies found no evidence of sabotage. The final report documents *do not* contain any conclusion as to the origin of the presumed leak that created the required hydrogen-air mixture. Both discussed Eckener's broken wire/slashed cell/hydrogen pooling, but admitted there was no evidence supporting any of it, and noted broken wires in other ships had in the past done no real damage. By leap of faith, the gas was presumably *just there*. Somehow. After all, what did it matter? There had been a *fire*; therefore, any idiot would have to realize there had to have been a *leak*. And it did not matter about hydrogen going through cast iron, the cover held it, somehow.

"The investigators concluded that the explosion (sic) was an accident, caused by hydrogen that escaped from one of the gas cells and was ignited by a spark."[289] There had been a leak, so there must have been a spark of some kind, any dummy can see. "Mention of the fluttering of the outer cover on the upper port stern, just prior to the conflagration, as testified by witnesses, is to be found in both documents. *Much importance was attached to this part of the testimony, since the fire appeared to originate in the vicinity of this flutter. It likewise helped both bodies of investigators to arrive at the most probable cause of the accident.*" [290] No matter several witnesses had seen cover flutter in the prop wash. It was a physical impossibility that the gas, at little more than atmospheric pressure, to have the power to *flex* outer side covering. It was even more impossible for any ground observer on either side to *even see* the topside covering. Yet *both* groups based their conclusions hydrogen present because people saw the outer cover flapping in the wind and prop wash!

Only the Germans made mention of where gas, if escaping its cell but somehow trapped by the cover, would have gone: into the vertical tail fin. They speculated the tail would be susceptible to

[287] To be certain, on Dr. Bain's third trip to Germany, he again asked Albert Grotzinger if he could remember *anybody* talking about damage aft… and the answer was an emphatic 'no'.

[288] L. Tittel, GRAF ZEPPELIN – HIS LIFE AND HIS WORK, Zeppelin Museum Friedrichshafen, translated by P. A. Schmidt, 1995

[289] S. Tanaka, THE DISASTER OF THE *HINDEBURG*, Madison Press, Toronto, 1993

[290] Emphasis added. R.W. Knight, Acting Chief, Air Transport Section:, Comparative Digest

steep potential electrical gradients, and that the nearly invisible flames had started "in the upper rear." They concluded "...as a hydrogen-air mixture burns nearly without color, under the given circumstances the first ignition could have happened on the upper part of the vertical stabilizing fin and could have quickly spread to the leading edge of the fin on the ship's body." Yet the Germans somehow had also agreed with the Americans the fire became visible first well *forward* of the fin. "Both bodies of investigators are in complete accord with regard to the first appearance of open flame, concluding it to have been on the top of the ship, forward of the vertical fin, over cells 4 and 5." How it started invisibly on or inside the fin only to become visible far forward is not explained.

Various ignition possibilities were suggested. The Americans used the words "brush discharge" while the German Commission preferred "equalization of tension between wet spots on the exterior and the ship's framework, the latter being better grounded than the covering after dropping the manila landing ropes." How a spark got inside the previously perfect Faraday cage is not offered even a token explanation. No matter that nothing was really proved, this was the closure everyone needed: a rationalization of how the pre-determined cause actually did the deed. In the court of public opinion, hydrogen was damned.

There would be various attempts to recreate the hardware and conditions to try and duplicate the accident. Mortal men would be challenged, and failed, to duplicate so many miraculous events in exactly the right order as imagined by Eckener:

1. A wire would have to break aft during a gentle turn, even though no LZ-129 wire had ever broken, even when the tail smacked the ground.

2. This wire made a huge gash in a cell, even though no Zeppelin breaking a wire had ever suffered even slight cell damage.

3. This significant amount of gas must evade detection either by the rigger's notice of uneven bottoms, the men in the aft smelling an odor, or the manometer's 2% sensitivity of change.

4. Hydrogen must then be retained by the outer cover not believed capable of containing it in combustible concentrations, while avoiding the vent system easily capable of ejecting it, while not floating into higher sections including the fin.

5. "Some spark" must come along and get by the most perfect electrical insulation of the cover, avoid the perfectly bonded structure, and ignite a waiting, accessible small amount of gas, that in spite of its low heat radiation and fast rate, burns a hole in the outer cover, while nature's lightest gas somehow burns downward and aft.

Hitler would declare it an "Act of God." Indeed, the Almighty *would* pretty much have to be involved for such a string of unlikely events to happen in the precise order required! It comes as no surprise that Karl Arnstein would tell his biographer he thought Eckener was covering up for the company. As Dale Topping wrote, "Göring did not want a finding of sabotage since this would make his Nazi police look bad, and the airship crew was "loaded with them." On the other hand, Eckener and Dürr could not afford to make their own company look bad by admitting any errors in design or operation of the airship. Their only way out was to conclude that the accident was an act of God, and that is what the report concluded."

There would be compelling reasons to reach this closure – not the least of which were financial. One newspaper report stated "The operators of the *Hindenburg*, however, at their own expense, had insured each passenger under a passenger accident policy which provides benefits in case of death or total permanent disability of 30,000 reichmarks, about $12,000.00..." The article suggests the Warsaw Convention's limit on liability offered no protection if the operators could not prove they took all possible precautions to avoid the accident.

Dragging the investigation on to explain how the gas escaped, then paused, then became ignited, then actually sated secondary ignition as Eckener had speculated, would delay closure. The "Act of God" need not be explained… nor could anyone be held liable. Ultimately, it came down to a matter of money. Payments were made, the dead were laid to rest. The airship itself had been insured for 6 million reichsmarks, a sum that was paid in full without question of what caused the accident.

The German investigative team returned to the fatherland the first week in June. The *New York Times* said "It will continue to investigate the cause of the explosion with the help of a number of German scientific bodies and will exchange views with a similar commission in the United States." Yet not another word seems to have been published here about how they were coming along. No matter, the public's attention moved on to other things. Ralph Upson aside, no one else in America seemed to have a problem with the conclusion, proved or not. We had helium; what happened to some swastika-emblazoned Zeppelin wasn't our problem.

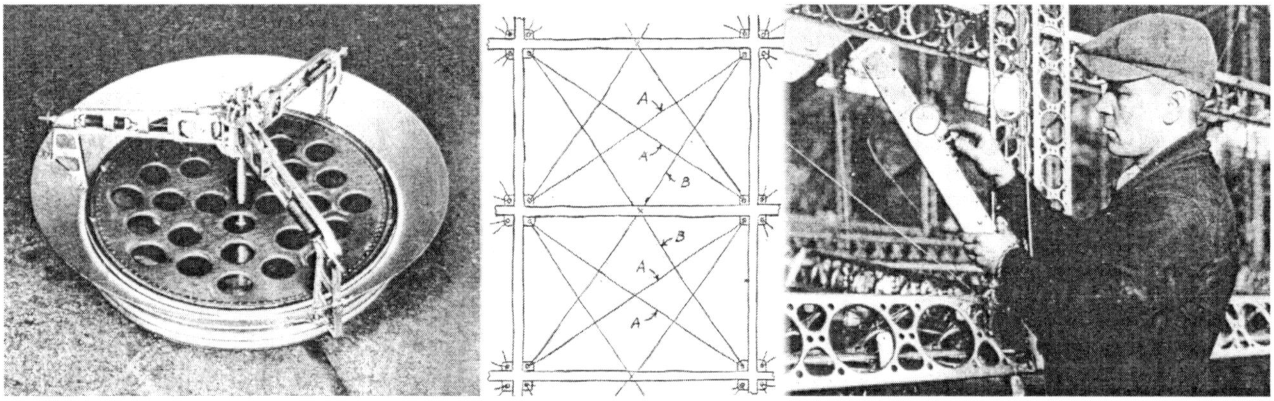

Spring-loaded automatic valves were set to open below the pressure that would damage the cells. "The necessary tension in this bracing induced heavy compressive loads in the peripheral girder..." wrote Mowforth. Installation and use of the tensiometer to balance the tens of thousands of bracing wires was among the most labor-intensive operations constructing a rigid airship, including the ZRS shown in these Goodyear and Lakehurst documents courtesy Dr. Dale Topping.

Suddenly the Germans seemed to have *always wanted* helium. If only they'd been given the helium they'd supposedly just been waiting to get, the fire couldn't have started! Authors would from that point on conclude an item on a wish list was an actual order, but the economics of helium inflation were very real. In a May 26th newspaper interview, Eckener told the Senate Military Affairs Committee that Germany must have helium gas at fair prices… stating the present $75 to $100 per 1000 cubic feet should be reduced to $2 to $3 per 1000 cu ft. Michael D. Regan discussed some later legal implications in "The Helium Controversy." [291] Bob Gordon had set a tone back then when he wrote, "The German government could have secured helium for both the *Graf Zeppelin* and the *Hindenburg*, but balked at the price."

---

[291] H. Stein, ed., AMERICAN CIVIL-MILITARY DECISIONS: A BOOK OF CASE STUDIES, The University of Alabama Press, Birmingham, 1963

Quietly, however, the men whose livelihood depended on solving the mystery *before* they completed the next airship could not depend on divine intervention not happening again. Perhaps they heard the Lord helps those who help themselves, because the Germans did not stop searching for the cause. Quietly, some behind closed doors, they were doing just that. For the next several months they labored, be there helium or no, for the ship then under construction.

That October, a mention of a "lantern lecture" given by German team electrical expert Prof. Dieckmann was run in London's *Aircraft Engineering,* dated the 12th. The Professor had told the *Lilienthal Gesellschaft* (German Aviation Research Association) that a coincidence of *five* individually harmless events conspired to create the disaster. By this time, the early emphatic denial that hydrogen could not have leaked seemed to have been forgotten. Dieckmann claimed that rain dampened the stern, an idea that would be repeated as an alternative explanation to the now monumental 'tail heaviness.' Without explaining how, hydrogen must have escaped; neither did he go in to how it was not noticed nor how the cover managed to retain it. Dieckmann named three other incremental causes. [1] "The execution of the landing maneuvers at the unusual height, so that the airship was in a highly charged electric stratum, and there was considerable difference of electric tension between the hull and the earth. [2] The thunderstorm, causing as experiments have shown, the electric tension in the atmosphere to undergo sudden changes. [3] The damping by rain of the landing ropes, intensifying their conducting properties."

Oddly, Dieckmann's previous remarks about how LZ-129 had been painted with a different dopant formula *were not mentioned*. Of course, the talk got little attention in America anyway. In 1938, Ray E. Brown, head of the helium program of the Navy's Bureau of Aeronautics, translated a report from Dieckmann. Brown added some information on his own, gleaned from people who had witnessed some testing. Various test sections of gas cells and coverings had been assembled and subjected to conditions simulating those of the fateful night. As the legends grew, "explosions" had taken place.

Gordon Vaeth, airship historian and author, noted that Dieckmann had mentioned the dopant change back in 1937. Reading Brown's appendix, he suspected a covering cover-up. "When the [*Hindenburg's*] outer cover material was painted with the original dope used on the *Graf Zeppelin*, no explosions were obtained under any conditions, but explosions resulted whenever the *Hindenburg* type of dope was used."

Had the Germans made a covering change that failed to protect the lifting gas as well as in LZ-127? If there was to be a connection between electrical activity on the outside of the cover that somehow got to the hydrogen without harming the cover, Dieckman's comment about dielectric differences had to be it. Vaeth commissioned a dielectric comparison study on his own,[292] but it was a dead end. His expert could find no dielectric differences between the LZ-127 and 129 outer cover materials tested. Yet even this detail would be left out when the matter was discussed, even as late at Vaeth's 1990 article in *Weatherwise* magazine.[293] Vaeth recognized Dieckman as the unsung hero of the affair. The earlier lecture called out five steps to fire; they just did not have anything to do with the five imagined steps of Eckener. No one noticed. Neither did any report mention the new blue primer in the role of electrical insulator.

Not that any of the above would have made *any* difference. Fritzsche wrote, "Given the largely symbolic value of the Zeppelins, it was no wonder that the Nazis discontinued the entire airship

[292] Vaeth via phonecon with the author
[293] Vaeth's article was summarized and partially run in *Buoyant Flight*, Vol. 38, No. 4, May/Jun 1991

project when the worldwide publicity surrounding the *Hindenburg* explosion [sic] in May 1937 cast doubt on Germany's technological infallibility and thus Germany's political pretensions. By 1937, in any case, German military might was better represented by the air force than by airships." The last rigid airship flight was made just before World War II quickly overshadowed Zeppelins. After the war, even though the Nazis were gone, no great revelations concerning the accident were volunteered. Eckener would die in 1954.

Rosendahl would during the rest of his life try to correct the misconceptions that the public had absorbed from the media. In his 1938 book WHAT ABOUT THE AIRSHIP? he would emphasize the airship had not 'blown up.' "It is clear the *Hindenburg* was not the victim of an explosion but rather of progressive burning after initial ignition." He chafed at the incompatible but constant insistence on broken wires, cut or leaky cells, stuck valves, mysterious tail heaviness, or some other kind of malfunction. Rosendahl would commission a film in 1946, *History of the Rigid Airship*, which opened emphatically with the Germans' supposed poor judgment on using hydrogen. Yet before it was done, the film admitted that "even with hydrogen" the Zeppelins had an enviable safety record. Elsewhere, Rosendahl wrote: "The *Hindenburg* arrived in perfectly normal condition, made a skillful and normal approach, and was being landed in an entirely normal way. In my opinion, there was nothing wrong whatsoever within the airship until the fire broke out. In other words, the *Hindenburg* was not lost as a result of any weakness or defect in the airship as a type..." "Rosy" died in 1977, some saying he was leaning toward the sabotage idea toward the last part of his life.

Designers of the stressed-skin or metalclad airship would later have Alcoa's much stronger 24S-T available as a building material. Nearly half again as strong as the ZMC-2's 17S-T, the new aluminum could be spot-welded, making even Hill's remarkable automated riveting machine obsolete. Carl B. Fritsche's 1938 query article "Why Has The Airship Made Greater Progress?" pointed out the airplane had made improvements in speed, range and performance due to better aerodynamics and better propulsion. He concluded the transition to all-metal airplanes was a logical path for airships as well: "The abandonment of fabrics in construction, utilizing the most modern alloys available...In fact, a fabric covered airship cannot be built strong enough to fly at 150 miles per hour and still carry a decent payload."[294]

The transition to all-metal airplanes was started even with those hook-on craft taken into the ZRS's bellies; the aviation metalsmith gave worry to the aviation carpenter's mates their skills were outdated. Even fabric covered control surfaces were disappearing as the war wound down. The Navy wanted no part of this seemingly logical transition already enjoyed by HTA. Rosendahl did not favor the metalclad design, and the unique ZMC-2 was not fully explored by the Navy. "Rosy" passed up an offer from the otherwise anti-airship Roosevelt administration to build a larger 3 million cubic foot airship purposely length-limited to favor Detroit Aircraft's squat design. One can only wonder what would have been possible if Litchfield had leaned toward accepting Detroit's offer of a partnership. As the new World War loomed, the metalclad could have been hydrogen inflated with little fear of fire. Ralph Upson went on to teach aerodynamics at NYU and work aircraft during WWII. The man who recognized the big problem with flammable airships was an adviser to Boeing when he died in 1968, climbing a mountain. One of his team at Detroit Aircraft, Valdimir Pavlecka, went on to distinguished service in HTA and powerplants, but always followed the dream of an improved metalclad. He told Lee Payne, "Today, with metal bonding, epoxy laminates, aluminum-magnesium-lithium alloys, carbon and

[294] Reprinted in *Gasbag Journal* No. 33, Sept. 1997

boron fibers, gaseous fuel turbines, and boundary layer removal, we can create a new generation of metal ships as far ahead of the ZMC-2 as today's jumbo jets are beyond the plywood airliners of the 1920s." In the decades since Pavlecka's death, the advantages have become more acute. (One could imagine a self-healing metalclad based on automotive tire selants that flow out into a puncture and harden in the air.)

A generation would be born into the "knowledge" that hydrogen must be supernaturally dangerous, since photos of the spectacular accident were accompanied by text that indicated no one was ever able to figure out how hydrogen started the fire. With the advent of household television, the mixed-together old newsreels and snippets of Herb Morrison's record would be reborn. Walter Cronkite would narrate a *20th Century* segment on rigid airships, which featured Rosendahl and his recollections. Many other TV shows would follow, and continue to this day. While there was endless speculation, no one could duplicate how hydrogen made its start, then stopped to be somehow ignited.

Not unlike the German technology sought after the Great War, the new rocket science would become all the rage in the 1950s. Airships quickly took a back seat, offering only the usual stable camera platform to observe the rocket's remote impact point. The incredible endurance flight of the "Snow Goose" – a record of powered flight that stands to this day – would be back page news. CDR Hunt and his ZPG-2 crew landed even as the manned space program was taking off, and there was no room on the front page.

The Navy went on to deflate its last blimp in 1962.

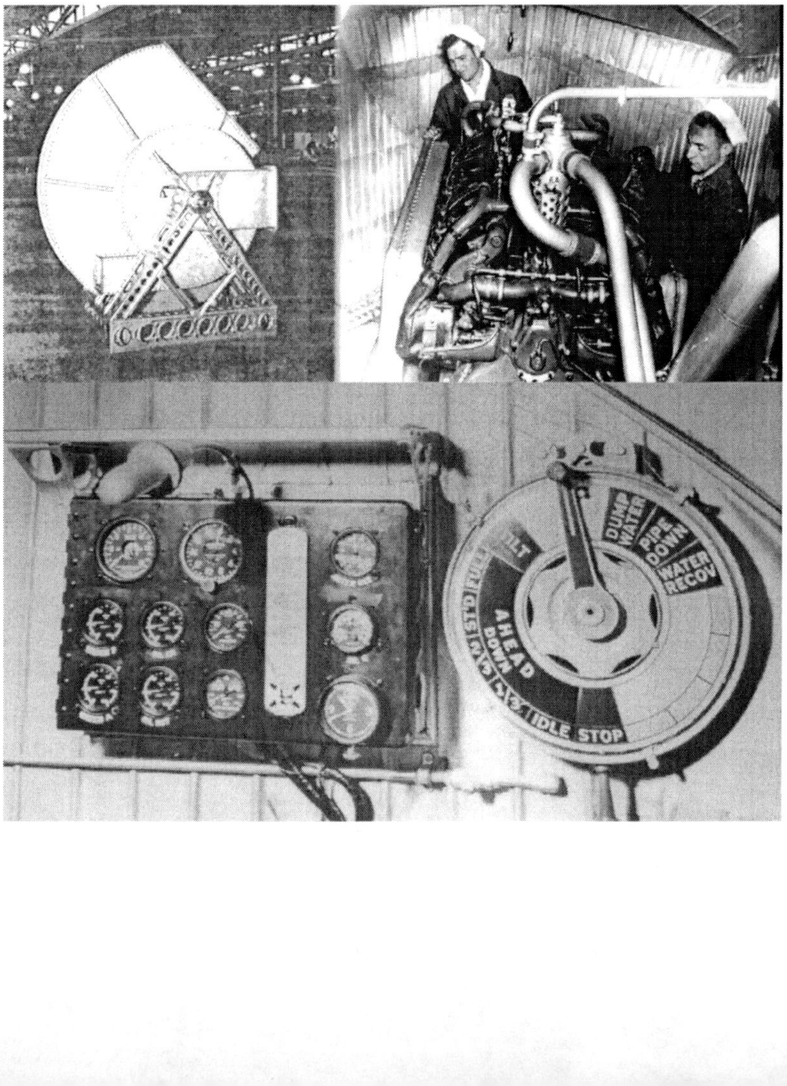

Huge belt-driven fans below the ZRS engine room deck plates (top, left) blew air across ducts covering the exhaust manifolds (top, right). This protected the rocker-arm oiling enginemen from burns and provided first stage cooling. Telegraphs wire-tugged commands from the bridge out to the engine room control panel to order the creation of ballast water or, for full power, recovery by-passing. Water pumps below the deck plates moved this disposable weight into forward or aft ballast bags, which allowed continuous trim adjustments – which were often necessary. LTAS, NARA and Akron's "World of Rubber" display photos.

Chapter Eight: Random Events, A New Idea

There was still one Navy blimp flying out of Lakehurst when young Addison Bain reported to the Marshall Space Flight Center in Huntsville, Alabama. Bain was employed first as a technician, later as an engineer designing ground based hydrogen systems and writing specifications for hydrogen equipment, spending some time working on hydrogen improvement problems in the lab. Transferred to Kennedy Space Center, Florida, in 1964, Bain worked on the *Saturn* LH_2 plant used to provide 2400 psig hydrogen for fuel cells. A few years later Bain was appointed by the KSC director, Dr. Kurt Debus, to represent KSC at the newly formed Aerospace Research Data Institute (ASRDI). Working with well-known safety expert Paul Ordin, the team headed by Dr. Frank Belles initiated the task of writing the hydrogen standard that was eventually published as the NASA hydrogen safety handbook.

Through the course of reading related material on hydrogen, the *Hindenburg* accident was part of the literature. Bain did not pay a lot of attention to the matter, but noted inconsistencies in the reports and details of the event itself.

In 1973, Bain and 34 other hydrogen fuel experts decided to form the International Association For Hydrogen Energy (IAHE). This exposure helped find more applications for hydrogen, including use as a fuel for all forms of transportation. In the background, there was that nagging problem that the *Hindenburg* fire neither looked like nor was described as a hydrogen fire. Surely some historian would work to correct such an obvious error someday, thought Bain. But no airship historian would ever challenge the status quo; none one questioned what had become a cornerstone of airship lore. If you had a problem with one explanation, there was another and another, all with the common thread that hydrogen had burned first, and destroyed the ship from the inside out. Indeed, after the last hydrogen filled blimp stopped flying in Europe, the legend of hydrogen and the LZ-129 grew with every retelling.

In 1990, Bain was assigned to NASA headquarters, Washington DC, just across a street from the National Air and Space Museum. The 25 foot model of the *Hindenburg* was on display in the lobby. Checking, Bain found that the Universal Studios crew that built it for the movie had not been able to get detailed information on the airship's construction. Bain did find enough information to show the ship's materials were highly flammable. More research, turning up references now mentioned here, led Bain to suspect it was not necessary for hydrogen to have played a major role in the fire. But the effort was stymied for lack of detailed information.

Meanwhile, this writer had been gently but firmly admonished for introducing an error when laying out Jim Shock's US NAVY PRESSURE AIRSHIPS 1915-1962 for publication. Following pilot Steve Adams's description of the air-to-helium valve during a flight on *Shamu,* this writer reached the hopeful but incorrect conclusion that in the event of envelope failure, helium could be directed into the ballonets to act as sort of a back-up envelope that might slow a fall. *Aerostation* editor Donald Woodward pointed out the error in his book review, and James Shock would have to endure the frustration of being blamed for what now seems like a silly assumption on the publisher's part.[295] Still, as James Joyce said, "Mistakes are the portals of discovery." Why *shouldn't* the ballonet work that way?

---

[295] The actual function of the air-to-helium valve is to allow air to be rammed into the helium envelope in the event of a catastrophic casualty. This allows the envelope to survive a hole that otherwise so lowers the pressure that the bag looses stiffness and the control cables become slack and useless.

When young people first learn about airships and their limitations operating in our atmospheric ocean, a common question asked is why one cannot simply compress or expand the lifting gas on board. It works for submarines! Oh that someone could build a lightweight, lightning fast compressor and a featherweight high-pressure storage tank.

Yet to this writer, it seemed that age old daydream could actually be realized in a hydrogen-powered airship. Expanding the fuel from its reduced form would allow lift to be borrowed, and burning it off would allow the displacement to be controlled. It fit together in the applicant's mind something like the device the uses the energy in a scuba tank's air expansion to spin an underwater propeller before the diver breathes the air. You don't have to carry the compressor, you just benefit from the energy it has created and stored. The idea would only work if 1) the fuel had an enormous expansion ratio and 2) very little energy by volume. It would be more efficient, of course, if the fuel had a good energy-to-weight ratio as well. But no readily available reference seemed to define hydrogen's Btu in terms of weight, or energy per pound.

A devoted "helium head" since reading Dr. R. K. Smith's *Akron/Macon* book in 1970, this writer chafed at the common misconception that airships were dangerous because the LZ-129 "blew up." Upset by the constant hand wringing about the dangers of a fuel we used on the space program with great success, this writer had no idea of the lack of visual evidence of hydrogen's involvement. The author thought a picture of the LZ-129 fire would make an excellent basis for a safety poster warning about becoming complacent about fuels around static electricity.[296] It was time to move on after 50 years, for sure. Otherwise, this writer was irritated by the constant harping about the LZ-129 loss while ignoring all the positive aspects of airship accomplishments. This writer just wished the whole subject would go away, and had never written anything about it. So the hydrogen fuel idea was stalled for lack of information, namely the Btu value of hydrogen by weight. This writer sought answers at the 1994 IAHE Conference, that year being held in Cocoa Beach, Florida, mythical home of the astronauts.

Among the delegates was Addison Bain, who answered the author's question, offering to get the value...but did the inquisitor know anything about airships? The following lunch started a period of intense activity. Bain shared his research and conclusion that the airship was covered with highly flammable materials. While "some spark" might have ignited some hydrogen, he suggested, the way the thing took off was unlike any hydrogen accident in history. To this writer, if he meant hydrogen had a *lesser* role in the fire, all the better. When Bain supplied figures that showed hydrogen has nearly three times the energy of the best gasoline by weight, this writer became committed to the idea of using it in an air vehicle whose performance was directly related to weight carried. The ballonet error provided a further inspiration. Since hydrogen contains so little energy by volume, the fuel would have to be carried in some form reducing its bulk. It should be possible to allow it to expand into another ballonet inside the helium cell, driving out the no-lift air. It could then be consumed in the fuel cells and engines as required to shrink its volume, as ops in our atmospheric ocean called for.

There would be time to pursue hydrogen-fueled airships later. First, Bain would need specific information about the composition of the LZ-129 covering in order to re-create it for tests. Surely the volume of materials our historical action team could bring to bear would quickly overwhelm such a novice airship researcher, this writer thought. Quickly scanning available records revealed nothing that detailed the fabric to the required degree. We got the word out to the airship history

---

[296] A mercifully short fire during a Shuttle Orbiter OMS pod servicing was traced to static electricity, and resulted in changes to the anti-static precautions taken during self contained suit (SCAPE) operations.

team, but preliminary results were not encouraging. Gordon Vaeth had closed a 1990 article on the subject by saying, "As to what dope was used on the new ship, there may no longer be anyone who knows." Five years later, "Sig" Geist, LTAS member whose father had helped build the LZ-129, would write, "Too bad our NASA man is asking about covering fabric only just now, a year after Emil Hubner died. Emil was a textile technician known for his expertise in applying the envelope to the framework of the LZ-129 and LZ-130, and he may have been assistance with.... dopes and lacquers."

Water-recovery radiators carried heat away up the ZRS sides while condensed water dripped back down into keel bags and tanks. Engine #7's heat was ducted forward to warm crew spaces. Engine cooling radiators mounted on the prop shaft housing on ZRS-4 (left) were moved under scoops on ZRS-5 (right, #8) as part of streamlining efforts that included the fairings seen. Roy Gibbens photos.

When AEROSTATION editor Donald Woodward first responded positively, he had promised research on the scuppers surrounding the ship's topside vents. Yet Woodward would die on January 2nd, 1996, before we could even call San Diego again. Luckily, senior team member Hepburn Walker Jr., WWII airshipman and LTA consultant, had some samples of LZ-129 fabric that had not burned, probably having been blown loose in the ruptures before the fire reached them. Walker had dug them up from the mooring circle where she fell prior to the circle's tracks being torn up to make blimp landing mats. Walker's kind lending of these priceless artifacts was a giant first step toward re-creating the material. Other members of our "helium head" community responded positively. Cheryl Ganz, editor of *The Zeppelin Collector*, lent samples from her personal collection, including one swatch from each of the two *Graf Zeppelins*. Kent O'Grady lent and arranged for still more from the British rigids, as Airship Heritage Trust lent samples including some believed to be from a Schütte-Lanz airship. Obviously only non-destructive testing would be possible, but much could be learned thereby.

The NASA Material Sciences Lab at the Kennedy Space Center first performed one series of tests that verified what Vaeth found earlier: dielectric attributes showed no significant differences between LZ-127 and 129. The infrared spectroscopy tests agreed: "LZ 127 - LZ 129 (Ganz), no notable differences." So LZ-129 could not have had some peculiar component that would have passed electrical energy inside in a manner different than her predecessor. Dieckmann *appeared* to have been barking up the wrong tree.

Working on a non-interference basis, technicians and scientists worked on the samples during what became known unofficially as "project H." It was difficult to find time between some calamities in the space business those months, since so much legitimate lab work kept pouring in. But the lab workers even volunteered some time to evaluate the sample material, toward the eventual goal of recreating the covering with great accuracy.

Testing of cover samples included chemical and physical analysis using the scanning electron microscope, X-ray energy dispersive spectroscopy, optical microscopy, as well as tests on flammability, electrostatics, surface and volume resistivity and corona discharge exposure. Returning from Germany, Bain found papers waiting. "About ninety (90) Infrared spectra were collected and compared, with considerable effort placed [on] analyzing fine details of the data. R-100 British sample, ...aluminized layer indicated mostly nitrocellulose characteristics. LZ-129 (Walker sample), ... fabric was cellulosic in nature, with a cellulose acetate butyrate dope plus possible nitrocellulose dope present on the fabric (difficult to confirm, cellulose itself absorbs in the nitro region) ...LZ-129 (Ganz), fabric with cellulose acetate butyrate dopant on the aluminized side, ...indications of possible nitrocellulose being present." There were maddening inconsistencies in the samples, understandable since five different fabric weights and mop-applied dope over the acres of covering were used on LZ-129 alone. One of the conclusions in the report addressed the chemical difference: "The butyrate acetate would probably be more flammable than the acetate alone..."

Meanwhile Bain was trying to obtain as much information as was available on the covering's manufacture and makeup. He located the references on the subject quoted earlier in this book. The company that manufactured the dopant, Herberts, responded, sending Bain a spectrograph of the dope makeup. Among other things, the graph showed a peak of 'Na' (Sodium) of about 2.8v on a vertical range of 4.0v. Sodium would explain where part of the yellow color came from in flames reported by witnesses. Armed with these reports and tests, Bain set out to recreate a section of *Hindenburg* outer cover. Still, there were many questions. Could one get the exact same cloth, that one vet claimed was "fine Egyptian linen?" Nitrate dope had not been used in years, could some one still mix up a batch? Could the hand-applied technique be duplicated? And, what about the iron oxide, said to have been painted on the inside?

As it turned out, the same questions were being asked across the Atlantic. The Zeppelin Company had committed to building a full size re-creation of the LZ-129's port side passenger section for the city's new museum, set to occupy the old city train station. The Zeppelin Company archives contained enough information to tell the company as they sought to re-create the outer covering… *that they really didn't want to recreate the outer covering*. The city's fire marshal vetoed the idea. Such combinations of highly flammable components simply would not be tolerated in today's environment of building safety.

While the "helium heads" had gladly lent their priceless artifacts for study, they couldn't be asked to donate them for destructive testing. Anyone in this country trying to get a chemical supply house to sell that witches' brew of explosive ingredients might find himself reported to the FBI as a terrorist making a bomb. It would take a trip to Germany by Addison Bain to find a collector and purchase a sample of real LZ-129 covering. Back to the lab long enough to verify its authenticity against the Walker samples, here was a piece of history that could be sacrificed for the good of science.

First came a basic flammability test. The sample still had a faint odor, reminding Bain of being near a Piper 'Cub.' A standard section of the precious material was placed above a Bunsen burner-like apparatus in a chamber. The flame was ignited under the sample in a test chamber using the "needle rake" method. The sample immediately caught fire with a vivid yellow flame. (Hoehling: "...Mace...watched '...a column of yellowish flame.'") The sample took only three seconds to curl like a dried leaf, and a few more seconds to be consumed. During its destruction, sparks of aluminum could be seen flying off the sample. Aluminum's ignition temperature of 959°F was probably not reached throughout the sample before it became a flake of carbon. Yet the samples' relatively slow burning over a constant gas flame seemed to take quite a long time, in relation to the seemingly much faster flame progression seen in the newsreel films. If a gas flame or petroleum fuel flame had been under the cover and set it alight, it certainly seemed the entire ship's destruction would have taken longer – a lot longer. It did not look or act like a hydrogen fire, the simple burning of cellulose wouldn't account for the ferocity either.

Then there was the question of the atmospheric electricity reported by the witnesses. It is a phenomena not completely understood even today.[297] Always blamed for setting a gas-air mix on fire, one would have to wonder if tons of highly flammable covering material could not have been ignited from the environment effects at the time. But just guessing spontaneous combustion, like oily rags in the garage, would hardly constitute any more proof than snapping wires and hydrogen's imagined U-turn.

As another e-mail correspondent would later put it into words, "Any nitro-explosive (and doped fabric is very similar to a nitro-explosive), needs an igniter to detonate it. For example, you can quite safely put TNT onto a fire and it will burn (just don't breathe the fumes). However, the energy level available from an electrical spark may well be enough to detonate it, even when the spark is generated by relatively low-energy static electricity... as it depends on the energy density, to take the material above it's self-ignition level, not the total energy available." The correspondence reminded the author of some playful Morton Thiokol employees who put a lighter to a small chunk of their company's solid propellant mix. It fizziled around on the floor. By contrast, he company takes great precautions to insure no stray electrical energy gets anywhere near their solid propellant products, lest they are unable to tell it's not their regular igniter, and head for the stratosphere.

At least one of the theories circulating had the *Hindenburg* having been struck by some form of hard-to-see lightning. Witnesses all agreed the storm had past and there were no strikes in the area, but theorists suggest a striking bolt had been all but invisible, and set the ship afire. They point out that lightning from *clear* skies has killed people. This is not a likely explanation, however, since as discussed, the hydrogen filled airships had repeatedly been struck by lightning in flight, with no more harm than holes burned in the fabric, with the lightning carrying on through the path of least resistance. No one at the time knew of the Ohio Insulator Co. tests.

Nonetheless, Bain worked with the NASA lab to employ their high-voltage test rig to check out this theory. A small strip of LZ-129 sample covering was shot with 30,000 volts on the NASA test stand. After burning a small hole, the current continued on the path of least resistance. "It is interesting to note that in all of the tests the fabric used on the zeppelin and that on the fabric-covered Duralumin airplane were *not* ignited by the (electrical) discharges. This goes to show

---

[297] Perhaps this ignorance is universal. In his book THE DAY AFTER ROSWELL, Col. P. J. Corso reports the *alien spacecraft* fell victim to a lightning storm that infamous New Mexican night!

that the fire hazard is negligible where a path of *high electrical conductivity* is provided."[298] The experimental results were consistent with real-world airship experience.

Suspicious of the atmospheric electrical activity reported many years earlier, Bain's initial research revealed the general lack of knowledge in this field. New discoveries were still coming along, as scientists had for example just identified upper atmospheric activity dubbed "sprites." Bain had composed and distributed a questionnaire to airline flight crews asking for observations of electrical activity during flight. There were reports of a little known manifestation known as *ball lightning* having entered an aircraft and traveled down a passenger corridor. Another incident found ball lightning falling into a bucket of water and bringing it to a boil! Bain diligently sought the help of experts in the area of electricity in the atmosphere.

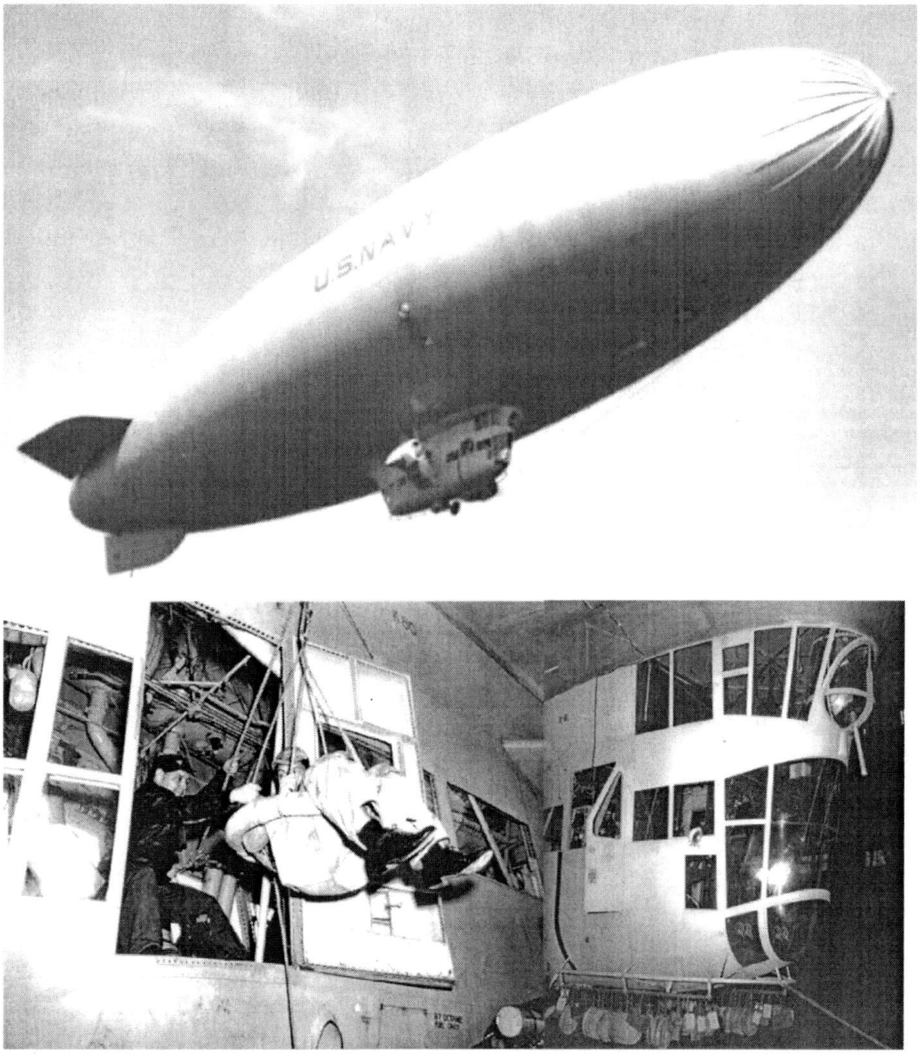

The Goodyear-designed K-type airship, the workhorse of World War II, performed anti-submarine warfare, escort, rescue, mine-hunting, R & D, and some twenty other types of missions impractical for other vehicles of the day. 134 of this type were built, more than all other US Navy airships combined. They ranged over the American coasts, South American east coast, and the Med. Some believe their combat successes have long been obscured by classification; K-34 (top) bombed the U-107 off Georgia in August of 1943. US Navy airships flew approximately half a million flying hours with an astonishing "all causes" availability of 86.8%. A dozen were lost to gasoline fires. Photos courtesy the NAA, Hepburn Walker, Jr. and James Shock.

The point about good conductivity turned out to be very important. Likewise, that 1936 *Literary Digest* article had inadvertently made an important point: "Makers of the ship followed the principles laid down by Faraday, who proved that electricity prefers to travel on the surface of an object. When lightning leaves a cloud to light on the airship, it remains on the surface until it can find a lower potential - usually another cloud. Then it hops off again. The ride the lightning has

[298] Austin, A. O. "Lightning Investigation as Applied to the Airplane," American Society of Mechanical Transactions, 55-2, 1933. Item courtesy Chuck Jacobs; italics added. Austin was chief scientist at Ohio Insulator. This conclusion probably should have noted that in the *Akron* model tests it was possible for lightning to set the exit point on fire, not unlike Brock & Buckingham ammunition that both blasted and ignited the fabric.

taken was harmless because if found no need for leaping a gap from one section of the ship's frame to another thus causing sparks." [299] The writer was also unaware of the *Akron* model test.

There was even more reason for atmospheric electricity to tend to hug the Zeppelin surface. Further tests showed the covering's aluminum particles in matrix exhibited electrically retentive properties associated with capacitors. This was no surprise to those familiar with capacitor manufacture. Some 'caps' are made simply by rolling very thin sheets of aluminum around a dielectric material. Further, references list the dopant material as not just combustible, but possessing *high dielectric strength*. These attributes added, then multiplied by 27,299 square meters [300]- *six and three quarter acres-* of electrically retentive surface area, made a capacitor the likes of which the world had never seen.

The consensus was that Professor Dieckmann had been on the right track when he concluded "The thunderstorm, causing as experiments have shown, the electric tension in the atmosphere to undergo sudden changes." Bain worked with Professor Uman,[301] who explained what would have happened during the prevailing electrical conditions. With LZ-129's altitude, the process of *charge induction* between the airship's capacitor-like skin and nearby cloud formations was established. In the 1975 movie, the artists toyed with the idea of electrical manifestations seen on the airship. Normally harmless phenomenon like St. Elmo's fire really had been experienced on Zeppelins since before Lehmann reported it sprouting from his fingertips in flight. J. D Barry wrote, "The term 'molecular electroluminescence' may be appropriate. I believe the phenomena are electrical in the sense that atmospheric electrical discharge activity is most likely responsible for their formation."[302] However, that fateful night a much more powerful form of energy had developed. As Dr. Gomp told Bain, the witness statements indicated a rare phenomenon called *corona discharge* formed on the airship, whose surface was really just a huge capacitor hovering at about 200 feet. Professor Alexander Dressler concluded that, to be visible, a charge would have to be at least 10,000 volts per square centimeter. To be visible at the given distances, and to have existed for at least a full minute's duration, that night's corona discharge may very well be unique in aeronautical history.

Dieckmann had concluded the charge on the LZ-129 probably reached 200,000 volts. It was known the areas of highest potential would normally be the bow and the stern. Assuming higher voltage charges could not exist in close proximity to the landing lines, the highest charge would then be farthest from 'ground,' hugging the stern. That is right where Lizzy Tobin saw the origin of the fire, aft of the rudder surface. (Ironically, it was the same point the lightning exiting the *Akron* model has set its fabric ablaze – though in the LZ-129 current flow was headed forward.)

The German commission had talked about electrical "tensions" equalizing. Isolated, each fabric panel would have its own unique charge owing to all the variables of distance, angle, humidity, and so on. A panel-to-panel event would have been caused by fields hugging the capacitive surface and cutting across one another to equalize potential, but did not necessarily have to penetrate the grain to reach the lower potential. However, a panel-to-ground event *would* have cut through the fabric in a path to earth potential. To test this thinking, another atique would have to be sacrificed to the God of science. Another precious strip of LZ-129 fabric was placed on the

---

[299] Emphasis added: "The *Hindenburg* Rides the Hurricane" also Reader's Digest, DEC 1936
[300] figure given on L-Z blueprint
[301] M. A. Uman, THE LIGHTNING DISCHARGE, Orlando Academic Press, Inc. 1987.
[302] J. D. Barry, BALL LIGHTNING AND BEAD LIGHTNING: EXTREME FORMS OF ATMOSPHERIC ELECTRICITY, New York: Plenum Press, 1980.

high voltage test stand. This time fabric was mounted so as to be *parallel* to the energy, so the arc moved *through* the grain, without being given the opportunity to burn a lighting-like hole and continue on its way. The voltage was set at 30 kV, much lower than the estimated 200 kV, and the power was applied. *The sample fizzed into flame and was annihilated!*

It was an epoch-making day. Addison Bain and his team had proved that hydrogen need not have been present in the ignition that destroyed the famous airship. Unlike the published rationalizations, the theory had been borne out in laboratory tests.

As it turned out, Bain discovered his were *not* the first tests that reached that conclusion! On his fact-finding trip to Germany, Addison Bain had toured the Zeppelin Company's manufacturing facilities. He saw the work being done to make the new Shuttle External Tank hydrogen tank domes, for which the company had the contract.[303] Mentioning he had been having trouble obtaining historical material, the Zeppelin Company opened their archives to him with access rarely before granted an American – if ever.[304]

Though the dusty archives would make him temporarily ill, Bain discovered some very interesting material. He found a series of hand-written reports to Dr. Eckener from an electrical engineer named Otto Beyersdorff. Hand written in somewhat technical German language, it was quite an effort to find someone in Florida who could translate the photocopies. But eventually the translations came through.

Beyersdorff had written on 28 June 1937:

"1. The actual cause of the fire was the extreme easy inflammability of the covering material brought about by discharges of an electrostatic nature..."

Another Beyersdorff letter says:
"... [T]he reported details about the fire made the causes of the origination so definitely clear - namely, the ignition of the covering caused by electrostatic discharge - I found it necessary to report as follows: The 'Werf' was kind enough to send me a sample of the covering material; with which I was able to perform tests in the laboratory matching the conditions of the accident and which proved the material to be extremely easy to inflame. An acquaintance of mine, an academic advisor, who took part in some of the tests, was amazed at the high degree of inflamability."

Beyersdorff cautions the covering material must be changed:
"2. By using the same material, an ignition has to always happen again, when the proper circumstances are present, as for example Lakehurst."

Beyersdorff pointed out it did not matter what gas was *inside*, the airship fabric would burn from the outside anyway: "...that an incombustible gas cannot prevent a burning of the covering as long as this one has the characteristics of the test results."

---

[303] Some people would find it ironic that the Zeppelin Company was helping build hydrogen containers!
[304] Years earlier Professor Meyer had encountered some suspicion when visiting the archives. Payment in advance was requested of him. It seemed some "ugly" American had just been there researching a disaster book and left a string of unpaid bills.

Dieckmann had evidently believed Beyersdorff. In a letter to Bain in May of 1997, Mrs. Dieckmann verified her late husband reached the same conclusion sixty years earlier but "...convinced [the] reason not to publish [the] Professor['s] tests [was] due to reputation and insurance reasons..." Hitler and the almighty pocketbook, a powerful one-two punch of an incentive if there ever was one. Since no one was asking the right questions, all the airship historians working until the end of time would never have stumbled on the answer. Dale Topping wrote that Karl Arnstein had a dim view of the LZ-129 investigations. "Historians can be objective but lack judgment in technical matters requiring interpretation, while those with a background in airships cannot help but be slanted by their own experiences." Indeed, not a single airship historian had ever even noticed the parallel between the interior glow noted by witnesses to LZ-129's fire and the WWI Zeppelin in combat. Hydrogen doesn't glow; inexplicable, it was simply ignored. When Eric Brothers discovered the long forgotten clipping showing the fabric's ignition on the *Akron* model tests, one wondered how this 'mystery' had gone on so long.

Most successful of the postwar nonrigids, the ZPG-2 (top, and at left the 2W) still holds records set both as stunts and during operational flights. Like the *Graf Zeppelin*, ZPG-2s crossed the US, ranged from the Artic to the Caribbean, and crisscrossed the Atlantic. Variants sought submarines, stood radar barrier and acted as flying labs. The largest nonrigid ever, the ZPG-3W (right) would have exceeded the ZPG-2's performance but was not given the chance. Courtesy Hepburn Walker, Jr., James Shock, and LTAS.

As it worked out, BUOYANT FLIGHT broke the news to the "helium head" community just before the May/June '97 issue of *Air & Space* hit the news stands. This writer was very proud of his small role in helping Addison Bain solve this old mystery, and was beaming that the story was heard outside the airship community. Here, then, is Dr. Addison Bain's thesis.

(Chapter Nine)
The Hindenburg Disaster: A compelling Theory of Probable Cause and Effect
By Dr. Addison Bain, Ph.D.

Study Abstract: This study examines the disaster of the airship Hindenburg, which occurred at Lakehurst, New Jersey, on May 6, 1937. For nearly 60 years, the prevailing hypothesis has been that the Hindenburg's hydrogen gas used for buoyancy was the basic design flaw. Two separate boards of inquiry each rationalized the premise of two sets of conditions to justify the cause, namely the presence of free hydrogen and the subsequent presence of an ignition source. The investigation process in each case proceeded down the path of rationalizing the most credible reason for free hydrogen to materialize and then to rationalize the most credible source of ignition. Although the airship wreckage was examined, nothing could be found to conclusively support any other rationalization. Limited experimental testing was done (such as gas cell conductivity) but nothing conclusive was reported at the time that would question the airship design. Eyewitness accounts and photographic coverage constituted the principal evidence for the investigation.

Numerous theories were postulated by outside sources as well as the American and German investigative teams. These were all categorized and reviewed. In the final reports, the Hindenburg envelope was never mentioned as being suspect. In a newspaper account at the time and then later, as an article in a magazine, a Ralph Upson, inventor of the metalclad airship, did question the use of fabrics for airships in hydrogen service. A Professor Max Dieckmann later conducted fabric test comparisons but this was oriented toward electrostatic conductivity.

The question this research effort is intended to address is based on the author's examination of the original film footage and other documentary evidence in an attempt to explain certain conspicuous observations as follows:

1. The Hindenburg did not explode, but burned very rapidly in omni-directional patterns.
2. The 240-ton airship maintained trim many seconds after the fire initiated.
3. Falling pieces of fabric were aflame and not self-extinguishing.
4. The inferno colorization is characteristic of a forest fire, not a hydrogen fire, as experienced by the author.

The purpose of the study was prompted by these suspicious events. It appeared hydrogen might not have been the initiating factor, thus perhaps, leading to a different theory than that established in 1937. The public may have been misled to assume the disaster as solely a hydrogen fire. If it was not, then the misconception maintains a negative image about the use of hydrogen. Finally, the purpose of this study was to conclude that the Hindenburg disaster was a result of the frailty of human engineering not unlike the Titanic, Space Shuttle Challenger and similar disasters.

The initial approach to the study (1990) was to conduct an exhaustive review of the literature and make contacts with airship experts and airship historians. The focus was on airship materials as the author was suspicious of the fabric covering, having learned that a cellulose nitrate dope with powdered aluminum was perhaps used on the Hindenburg. The former chemical is the basis of gunpowder and the latter a fuel component used in solid rocket motors.

During the fall of 1995 and throughout 1996, a number of unexpected events occurred which dramatically revealed sources of significant information and complemented the course of the study. Fabric samples were provided from an individual that had been stationed at Lakehurst, fabric samples were provided also by collectors (or purchased from them), and interviews were conducted with survivors and eyewitnesses. The NASA Materials Science Laboratories at the Kennedy Space Center offered analytical services. The pinnacle of the study occurred when the author was provided the unprecedented opportunity (for an American) to examine files at the Zeppelin Archive in Friedrichshafen, Germany. The Zeppelin works and new Zeppelin museum was visited.

The overall results indicate two primary conclusive aspects. First, the prevailing atmospheric conditions and the unorthodox method of landing at Lakehurst could prompt severe electrostatic discharge activity on the airship. This factor is consistent with the original conclusion concerning the ignition source. The unpredictable behavior and consequence of various forms of atmospheric electrical activity is embraced by modern experts of static electricity, the experiences of airline pilots and solid propellant rocket motor (SRM) scientist's expert in electrostatic discharge or ESD as they call it. Secondly, the lacquer doped fabric envelope of the airship was tested and found to be very flammable and could be successfully ignited using electrical arc sources. The materials used in the makeup of the envelope, other airship construction materials and the hydrogen used for buoyancy all could have contributed to the resulting conflagration. Evidence that further supports the conclusion includes examination of the design of LZ-130 (Hindenburg sister ship). This research revealed design changes (after the Hindenburg disaster) in an apparent attempt to counteract static buildup and reduce the flammability of the LZ-130 airship hull. Unpublished German tests, uncovered by the research, also substantiate the severe flammability of the Hindenburg envelope when subjected to electrostatic discharge. During 1998 a copy of the FBI report (1937, 337 pages) was acquired. The FBI basically followed up on inquiries surrounding potential sabotage. Nothing conclusive was found. The conclusion by Hoehling was characterized by the FBI as, "pure speculation". The "evidence" by Mooney proves false.

<u>Statement of the Problem and Purpose:</u> The Hindenburg accident at Lakehurst, New Jersey, represented an archetypal disaster of high technology--the Space Shuttle Challenger accident was another--which was documented with full media coverage that perpetuates within the public image. The Titanic encounter with an iceberg represented another of man's high technology efforts in which the ship's demise, in the cold temperature of the North Atlantic, continued to capture the public's attention for decades. The fundamental cause and effect of these events are all similar, each an example of the frailty of human engineering. Either through the lack of appropriate materials testing or selection, and technical communication, all the disasters can be attributed to a common phenomenon. That is, materials incompatibility with the operational environment. This factor is the basis of the study regarding the suspected demise of the Hindenburg. The research question: Is it plausible that an atmospheric corona activity occurring on, or through, the surface of the airship envelope provided sufficient energy to trigger ignition of the flammable covering materials, which in turn rendered a rapid flame propagation across the airship, with the subsequent sequential rent of each gas cell as they are progressively exposed to the flame energy, thus leading to the incremental release of hydrogen as an added fuel to the resulting conflagration? This would be an alternative to the widely held view of a "hydrogen explosion".

The author feels a brief review of the Titanic and Challenger disasters accent the purpose of the proposed study and provide the reader with an understanding of the author's inspiration to revisit the Hindenburg incident. In the case of the Titanic disaster, the presumption persisted for decades that simple pilot error, miscommunication among the crew, and lack of sophisticated modern-day electronic surveillance and detection equipment sufficed as the concluding explanation. Having been outfitted with the electronic gear mentioned, perhaps the Titanic would have sailed for years without its crew's knowledge that it was a floating disaster destined to happen at a time of appropriate environmental conditions. The largest ship built to date, luxurious, with all the apparent attention to engineering detail, virtually unsinkable, the ultimate in turn-of-the-century technology, yet the question remained to some as to why she sank so fast. Eyewitness accounts describe the event as a casual rubbing of the starboard hull with a wall of ice. "Designed" to take at least three days to sink in case of the worst-imaginable catastrophe (head-on collision with another ship), the Titanic slipped beneath the water's surface, carrying more than 1,500 passengers, in less than three hours. The submergence equation had a missing unknown.

In 1985 the Titanic remains were located and in 1991 samples of the steel hull were recovered and tested. Subjected to the Charpy V-notch test for ductility, it was found that the high sulfur steel was very brittle. Adding to the ductility problem was the environmental factor as the water temperature at the location of the disaster was one-degree (F) above freezing. Typically, certain classes of steels lose their strength as the surrounding temperature is decreased, most notably in cryogenic applications. For a mechanical/chemical engineer, the solution to the equation descriptive of a tragic event was satisfied.

The Space Shuttle, the most highly technological machine ever built, brings into play practically every engineering and scientific discipline. The enormous talent base this represents was fortunately available to rapidly solve the equation descriptive of the 1986 Challenger disaster. Nevertheless, before it was concluded that a solid rocket booster O-ring seal failed materially in the cold temperature environment existing prior to launch, the media conjectured that the onboard hydrogen fuel was at the root of the cause. Despite the fact that the Space Shuttle carries 6-1/2 times the hydrogen as was onboard the Hindenburg, it was proved not to be a causation factor, but did contribute to the conflagration effect, as did other onboard propellants. The role of hydrogen in the disaster still perpetuates the idea that hydrogen is too dangerous. For the Challenger, like the Hindenburg, hydrogen was necessary.

The Hindenburg incident, a minor accident compared to subsequent air disasters and lives lost, but a major event because of its spectacular impression and media coverage, continues to perpetuate the myth that hydrogen is unacceptably dangerous. The board of investigation that looked into the Hindenburg incident concluded, as many theorists since then have conjectured, that it was a simple case of hydrogen ignition. Other than sabotage theories, there are variations of the basic "conclusion" addressing the unequivocal source of free hydrogen and a number of probable ignition sources. Even today, books and articles on the subject allude to a common presumption that hydrogen was conveniently available (by a mechanical act of some sort) and the obvious culprit.

At the onset of the study, it appeared that most information would be from documentation and the only physical evidence would be from the available photography of the disaster. Subsequent breakthroughs occurred, however, which provided material samples, laboratory analyses, unpublished archival data, and the opportunity to work with individuals of various disciplines and experiences.

Introduction: The search for information included a review of book bibliographies, computer searches at libraries and NASA RECON, articles, reports, magazines, news clippings, trade journals, videos, photos, drawings, private collections, and extensive time at various archives, reviewing files. In view of the nature of the study focus, concentration was placed on information about airship materials, atmospheric effects, hydrogen characteristics, fire behavior, material compatibility and airship experiences. Identification of detail concerning the Hindenburg (LZ-129) was important. It later proved important to also obtain detail on LZ-127 (built and operated prior to LZ-129) and LZ-130 (built and operated after LZ-129). Although there is a number of books specifically on the Hindenburg published since 1937; there are also numerous airship-related books and other publications that include information on the Hindenburg. Much of the other data on relevant topics was excerpts from technical books, journals and reports.

Use was made of the NASA library on airships and related data as well as the National Advisory Committee for Aeronautics (NACA), American Society for Testing and Materials, and the National Fire Protection Association. The National Archives in Washington, DC, were visited to review and obtain copies of photos, videos and films. The National Records Center in Suitland, Maryland, was visited on three occasions in order to review and acquire data from the 14 boxes of information, mainly on the original investigation. The Smithsonian Air and Space Museum in Washington, DC, was visited on several occasions to go through the still photos and 18 file folders, an accumulation of data since 1936. In what is called the "rare book room" arrangements were made to read through the original transcript of the investigative board testimony. At the archive in College Park, Maryland, a copy of the unpublished 350-page digest of the original Air Commence investigation was obtained which proved to be a valuable source of information. The published air commerce bulletin (ACB Vol 9 No 2, August 15, 1937) typically used by historians is a truncated version of the American Board of Inquiry results. Where necessary, a copy of the full testimony was also purchased. The Lakehurst Historical Society in Lakehurst, New Jersey was visited as well as the Naval Air Station to accumulate information. The NACA historical file in Washington, DC, contains journals since the early 1900s and a review of that file provided many reports on airships. The Zeppelin Museum in Friedrichshafen, Germany, provided valuable information upon a review of reports, drawings, books, and photos. This information was in German but the archivarin provided very helpful assistance in locating relevant material. This material was later translated in the

US, with the help of consultants. At the onset the author was concerned about getting the information at the museum in view of the difficulty Universal Studios encountered in obtaining data for the movie, "Hindenburg" in 1975. The art director and unit manager were sent to Germany to gather data for use in building a scale model. "Little was found which would aid in the building of the model". A number of private collections, most notably the John Roby collection on Aviation Technical Literature, San Diego, California, provided a number of relevant books and reports that were purchased and reviewed. Many of these items were out-of-print materials. Reasonable success was made in using commercial book search specialist companies to obtain books not available in libraries.

Upon review and assessment of the overwhelming amount of data accumulated and the anticipated categorization of topics, a reference source was set up to conveniently file, track notes and organize the data. The appendix format was used on principal sources with sub-files instead of the conventional bibliography. Some 9,000 titles or abstracts on books and reports were reviewed. About 300 appeared relevant and were scrutinized for detail. It was concluded the only real fact finding of record was the original investigations and some follow-on work at the time. Most of the research and publications to date were done by historians and journalists with much duplication and overlap, and opinions without convincing evidence. Occasionally, however, a detail of significance was uncovered and noted as potentially relevant.

No other specific study of the nature of this research was identified. To complicate matters, specific detail on the Hindenburg outer cover was not to be found in a direct manner. Doping processes were discussed but in airship generalities or were addressed to other than the Zeppelin fleet. Specific books on LZ-129 allude to nitrate or acetate cellulose dope with aluminum powder but did not clearly admit formulation or application, a key aspect of the study. Clearly a need existed to attempt to develop empirical data. This observation further established the need for the research project. Other methods of data acquisition were required where information may not have been published such as from eyewitnesses or private sources.

A detailed analysis of the books, articles and other documentation, as a result of the literature review, provided relevant and useful information. However, the very specific detail needed regarding the fabric makeup and doping process used was not to be found. In some airships such as the Akron, the dope used was clearly stated as cellulose acetate. A text on Zeppelin type airships did provide a general but interesting and useful formulation of the typical doping process. Photos show methods of dope application for LZ-129, but no supporting detail was provided. Clearly the source of other information needed to be pursued. These included contacting members of the lighter-than-air (LTA) group, specific authors and historians, possible witnesses and direct contact with Zeppelin-related entities. Communication was established with prominent people having had, or in the process of, conducted study on Zeppelin airships and, more specifically, LZ-129 or related Graf Zeppelins. Telecons, letters, personal interviews were planned and the questionnaire technique implemented. A visit to Germany was arranged in cooperation with the Zeppelin Company to seek other data and set up potential interviews with survivors of the Hindenburg disaster. In the fall of 1995, a study relationship was established with an LTA member and noted airship author-historian. This event led to meeting another individual previously assigned at Lakehurst in the 1940 era. This individual not only had fabric artifacts that were from LZ-129 but also was willing to offer them for testing. The possibility of other material sources became apparent and this was pursued. During 1996, a number of events occurred that provided further data input to the study. A questionnaire was prepared outlining questions and statements of possible fact concerning design detail of LZ-129. Telecons were first made with selected recipients of the questionnaire to obtain their permission and establish a personal dialogue about the study. This process included contacts with the LTA, a noted author and historian, an eyewitness of the disaster and a Goodyear employee assigned to the Zeppelin Company during LZ-129 construction and flight service. On-site interviews were conducted with the Goodyear-Zeppelin employee in Wichita, Kansas, the retired fire chief and the historical society librarian at Lakehurst, New Jersey, mechanic and steward on board the ill-fated LZ-129 in Friedrichshafen, Germany. While in Germany, on several occasions, many days were spent researching files, reports, drawings and books at the Zeppelin archive. Data acquired required translation to English by consultants in the US, via paid contract, in-kind-service correspondence exchange or interviews. Contacts were made with static electricity and lightning experts and organizations such as the National Center for Atmospheric Research. Airline pilots were interviewed concerning any corona discharge they may have experienced aboard aircraft. A related questionnaire was processed.

The Fire Science Research Station in Missoula, Montana was visited to discuss forest fire behavior with fire specialists and to obtain reports on crown and plume-dominated fires. Discussions were held with security (NASA), law enforcement agents (ATF) and agents of the Federal Bureau of Investigation at the FBI National Center for Analysis and Violent Crime. These discussions were arranged to obtain information and opinions regarding explosive devices and typical bomb-terrorist behavior. An explosive's expert previously employed at China Lake, California, was interviewed. It is noted that sabotage was never proven; on the other hand, arguments apparently have not been systematically formulated to discredit the theory. Communication with a collector of airship materials in Chicago revealed not only the existence of fabric samples from LZ-129, but LZ-127 and LZ-130 as well. This find led to opening up the study to empirical data collection from the three airships for comparison. This would complement the literature information as to what design changes may have occurred from LZ-127 to LZ-129 and then LZ-130. Non-Hindenburg airship fabric samples were loaned from other sources such as the American Airship Heritage Foundation and also there was the purchase of items from other collectors. Not only fabric samples of the airship envelope were acquired but framing, gas cell material and such items as the wallpaper used for the typical LZ passenger lounge wall covering.
Authentication of the specimen origin would be a concern. Reliance was placed primarily on the collector's own need to validate source originality and value. However, laboratory analysis would be used to cross check like items from different sources to enhance sample validation. In one case, a gas cell specimen was claimed to have been from LZ-130 by a collector. Laboratory analysis matched a description of the cell from a translated German test report on the LZ-130 cell. This was a test case for credibility only, as the specific results were not relevant to the study objectives.

With initial material samples in-hand, a meeting was held with the Microchemical Analysis Branch of the NASA Materials Science Laboratory at the Kennedy Space Center to examine the samples and seek consideration of their assistance. The laboratory personnel provided in-kind service for the study project on a non-interference basis to space program priorities as they may occur. The investigation was prolonged unexpectedly due to some space program related problems that required the lab's attention and the introduction later of additional samples and work by the author. Rare samples loaned from collectors were examined closely to determine a selective site to remove a small representative specimen for analysis. A protocol was established in specimen identification and tracking and the order of the various analysis techniques to be employed. All samples were photographed and video taping used where appropriate. The instrumentation and techniques used are hereby listed basically in the order of the protocol established.

a. Optical microscopy observation to initially characterize the specimen make-up.

b. X-ray energy dispersive spectroscopy (EDS) in conjunction with the scanning electron microscope (SEM) was used to provide a qualitative and semiquantitative analysis of the samples to detect all elements in the periodic table with atomic weight higher than boron.

c. Infrared spectroscopy (IR) was used to identify organics present in the specimens. This proved to be a very difficult task as the technique is more of an art than direct science as in the case of EDS. About 90 spectra would be run over the course of many months in an attempt to isolate the specific organic make up of the samples provided.

d. Thermogravimetric analysis (TGA) was used to assay relative percentage of inorganic contents such as aluminum.

e. Resistance and conductance tests.

f. Static electricity susceptibility test.

g. Material ignition and flame sensitivity characterization.

h. Arc discharge exposure tests.

A video copy of the newsreel films of the disaster and still photos from Smithsonian Archives were put in digitized format using NASA electronic lab computers and the NASA quality assurance computer. Commercial computer and imagining laboratory sources were also used inconjunction with the NASA effort to develop photo colorization and produce special imaging and photo enhancement.

Results of the Study: The findings are categorized into topics starting with background type information to acquaint the reader with significant events of the disaster and introductory airship data. Brief discussions of prevailing theories are provided as well as key data from the archives. Information is limited to those aspects considered relevant to the study project objectives.

About Hydrogen: The author assumes the reader is not familiar with the characteristics of hydrogen. There are some features about hydrogen that are very important to the study and therefore a brief review of those features will be addressed first. The source for information comes from not only good references such as the Bureau of Standards and the NASA Hydrogen Handbook, but the years of the author's experience in dealing with hydrogen. For purposes of this study only the significant and relevant aspects of hydrogen behavior are addressed. The following key characteristics of hydrogen are summarized:

-Hydrogen, the lightest of all elements, at room temperature is very buoyant and diffuses through air much more rapidly than other gases
-Hydrogen, like methane, is a colorless, odorless, tasteless, and non-toxic gas hydrogen is flammable in air, can be ignited with a relatively small amount of energy (at stoichiometric) and has a relatively faster flame speed if its concentration is not very lean or very rich
-Hydrogen has a higher ignition temperature than common fuels and burns with a nearly invisible flame in daylight
-Hydrogen flames emit a low level of thermal radiation, thus reducing secondary ignitions and burns, and inhibition is strong when humidity is high

Airship Descriptions and Operations: The Hindenburg airship as it was christened was the 118th rigid airship constructed by the Zeppelin Company (Luftschiffbau Zeppelin GmbH) and was assigned the shop number, LZ-129 (registration D-LZ129). Two other airships need mentioning, as they are important to the study. The "Graf Zeppelin" or LZ-127, and the "Graf Zeppelin II" or LZ-130, airships that were likewise built at the same site in Friedrichshafen, Germany. The LZ-127 operated prior to the Hindenburg and the LZ-130 operated after the Hindenburg disaster. There are many good references that provide detail information about the design and construction of these airships, but for this study only relevant key aspects will be pointed out for orientation purposes.

Graf Zeppelin: LZ-127 was designed in January of 1927 with construction completed and its first flight in September 1928 (A18-168). Like typical Zeppelin design it was made up of an aluminum alloy girder framework, exterior engines and a fabric envelope. Seventeen interior gas cells using hydrogen provided the buoyancy. It had a capacity of 75,000 cubic meters, was 236.6 meters long and 30.5 meters in diameter. It had a crew of 36 and could accommodate 20 passengers. A unique feature was the incorporation of a fuel gas as primary fuel for the engines. This fuel called "Blaugas", named after the inventor Dr. Herman Blau, was a mixture of propylene, methane, ethane, acetylene, butylene and hydrogen, was slightly heavier than air, and was contained in twelve separate gas cells in the lower part of the hull below the hydrogen gas cells. Its consumption, therefore, had little effect on the ship's disposable lift. After being in service for almost a decade, accumulating over 17,000 flying hours, it was grounded after the LZ-129 disaster and later dismantled for needed war materials (March 1940).

Hindenburg: Not long after LZ-127 was in service work started on LZ-128 with a much larger payload capacity (40 passenger) and a volume of 155,000 cubic meters. According to LZ-128 drawing No. H 1361, August 1930, the airship would also use the fuel gas contained in eight lower cells.It was the disaster of the British rigid airship R101 at Beauvais, France on October 5, 1930, that led the Zeppelin works to consider the use of helium and to use Diesel engines running on heavy oil, instead of fuel gas, to reduce the fire risk. Work on LZ-128 was halted and the decision was made to redesign a larger airship, the LZ-129. To achieve the same payload characteristics the nominal capacity of 200,000 cubic meters was chosen, and therefore the LZ-129 was conceived in 1931. Because of the expense and limited availability of helium at the time a novel arrangement was designed for the LZ-129 gas cells. The sixteen gas cells were to be a double arrangement, that is, one cell of hydrogen inside of another cell of helium. The helium would provide primary buoyancy and the buoyant, but easily obtained, hydrogen could be vented as needed to maintain trim. The airship was 245 meters in length and 41.2 meters in diameter and would require a crew of 40. Fifty passengers could be accommodated (initially). Construction of LZ-129 was well underway when it was made known that helium would not be available from the United States. At the time the only major source of helium was in the "helium-rich" natural gas in the Texas/Oklahoma areas. Because of the NAZI war movement in Europe the US took the position of not providing helium under the premise the airships could possibly be used in the German war machine. Thus the Zeppelin works went on to reconfigure LZ-129 for only hydrogen service. For the 1937 season, the passenger section was modified to accommodate 72 people to take advantage of the total hydrogen lift. New types of gas cells were designed based on a Goodyear and Bureau of Standards development in the US. Rather than using the typical animal intestinal membrane (called goldbeaters skin) a cellulose film sandwiched between two cloth plies was used as the substrate. Several layers of mixed compositions of latex and gelatin were applied. This had the effect of making the cells translucent. The Zeppelin works claimed the permeability of hydrogen was held down to less than 100 cc per square meter per 24 hours. This was an improvement over the very expensive "goldbeaters" type of cell (one liter/sq. meter/24 hr). In addition to the gas shafts to vent hydrogen from the cells directly to the top of the airship, a ventilation system provided airflow in the space between the gas cells and outer cover. Any accumulation of hydrogen from cell permeability or inadvertent rent of the cell fabric would escape out through the specially designed vents. A number of references discuss the "good" vent system. Officer Bauer discussed the excellent ventilation at mooring and even while the airship was in the hangar. The envelope of the Zeppelin type airship typically was made up of a cotton fabric, with some use of linen and then doped with a formulation of various cellulose syrups, solvents, dyes and pigments. The general literature is not specific as to the formula used on LZ-129. H. Dick defines airship dope as a solution of cellulose nitrate or acetate in acetone in the book on the LZ-127 and LZ-129. More detail on this topic will be addressed in this chapter under Airship Envelope and Doping Procedures.LZ-129 first flight was March 4, 1936. It had logged more than 3,000 hours before the last flight (#63) to Lakehurst in May of 1937.

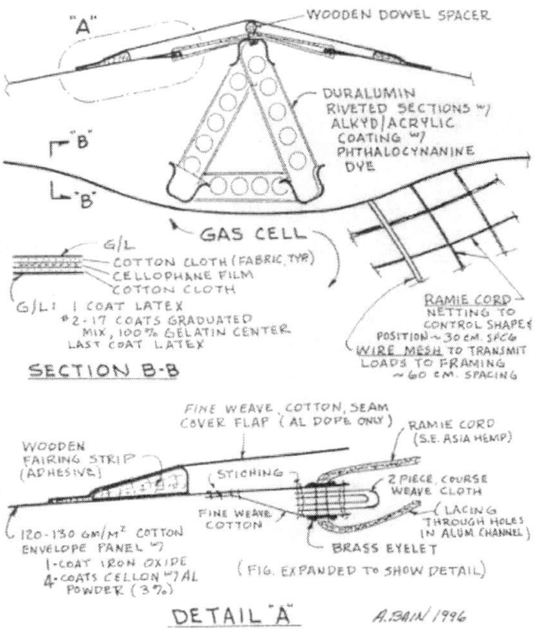

FIGURE K9

Graf Zeppelin (II): The LZ-130 was essentially much similar to LZ-129 in size and shape. It was planned that helium would be used. However for reasons stated before the US was not willing to provide helium. The LZ-130 drawings indicate that there were two configurations, one for helium and one for hydrogen service. The significance of this information is presented in the Design Modifications section under German Archive Data Findings of this chapter. The maiden voyage for the LZ-130 was a scheduled trip to Rio de Janeiro, Brazil for October 27, 1937. The first flight, however, did not occur until September 14, 1938. The delay was due to configuring the airship for hydrogen service and resolution of some technical problems, one with the supplier of the doping formulation for the envelope lacquer. The last flight was August 8, 1939. Like LZ-127, LZ-130 was dismantled in May 1940 after 30 flights. The use of the large airship for commercial business was postponed, so LZ-130 only saw limited flights for military applications. There were plans for a LZ-131 and LZ-132 but World War II and the rapidly competing airline enterprise forestalled the airship program in Friedrichshafen.

Hindenburg Landing Observations and Data: After a successful trip to Rio for the first flight of the 1937 season the Hindenburg departed Frankfurt, Germany, for the first season link to Lakehurst, New Jersey on Monday, May 3, 1937. Because of unfavorable head winds encountered en route to the US the scheduled arrival time of 7AM (Eastern Daylight Saving Time) was deferred to 7PM the same day, May 6, 1937.

Landing Meteorological Conditions: During the afternoon of May 6, cumulonimbus and cumulus clouds developed locally and a cold front moved slowly over the air station. Several heavy showers occurred between 6 and 7 PM with accompanying thunder. The storm was moving north and it was expected that by the time the airship arrived the storm would have moved away from the air station. Conditions at the time of approach were: ceiling 2,000 to 3,000 feet; very light rainfall; signs of clearing to the west; barometric pressure 29.72; temperature 60 deg F; relative humidity 98; surface wind light, variable and shifting. The wind at the top of the weather tower (96 feet) was west 6 knots. Occasional lightning was visible in the south and southwest, but none at the field at the time of landing.

Airship Status: Landing: The landing made on this occasion has been described as a high landing or flying mooring a favorite technique of the US Navy but unorthodox for the Hindenburg. The Germans preferred to approach the landing party close to the ground so the lines could be taken in hand and the airship walked up to the mast and be secured.
Altitude: The altitude of the airship was about 180 feet above the ground. It rose to 200 feet, at which point the bow port landing rope checked its rise. Thereafter, it descended to about 135 to 150 feet. The bow port, 2 inch diameter, manila trail rope had been coupled with the ground winch system. The bow starboard trail rope was being handled by the landing party in preparation to coupling with the winch. The main bow steel cable had been let out about 50 feet but never reached the ground. The airship was still 700 feet from the mooring mast. The 413-foot long bow ropes had been let out at 7:21 PM. The airship was stationary, and then a light gust of wind shifting from port side begins to swing it to starboard.
Gas Cell Status: The gas cells could be vented from the control car as a group or individually. Venting cells 3 through 14 could be done simultaneously and was called, "valving at the wheel". Cells 1 and 2, as well as cells 15 and 16 were interconnected and each set was controlled by a separate mechanism. Officer Bauer testified the only valving during the trip prior to landing was near Nova Scotia. Officer Ziegler indicated the highest elevation was about 1,400 meters. A review of the airship log shows the airship was typically at 200 to 350 meters in altitude except for one time at 1,400 meters, 41 hours out of Frankfurt. Map coordinate analysis and PV=nRT calculations based on cell design data, temperature and barometric pressure data show that the airship gas cells would require venting starting east of Newfoundland where the highest elevation was approached. This analysis was important to determine if, and how much, venting was required during the trip and the probable status of the stern cells, particularly cell 1 and 2 at the time of landing. Further discussion about these cells and their probable role in the accident will be addressed in the Summary of the Study.

Initial Observations: Fire Appearance: From the air station perspective a fire broke out at the upper stern area of the airship at 7:25 PM or about four minutes after the release of the bow ropes. From the Air Commerce Bulletin report, "Numerous expert and lay witnesses on the field testified as to where they first observed the fire on the airship. There was great diversity in this testimony for reasons that are very apparent. Among the most important of these reasons were the extreme rapidity with which the fire spread, the different positions of the witnesses with respect to the ship, the size of the ship, more than one-sixth of a mile in length, and an over-all height equivalent to a 12-story building, and the fact that at the time of the fire it was still daylight. It is estimated that the interval between the first glimpse of flame and the impact of the main body of the ship with the ground was 32 seconds. The great majority of the ground witnesses who testified as to the first appearance of fire were at the port side of the ship.".
An unpublished photo shows the fire in the stern area. Photographic experts indicate it appears to be a triple exposure resulting from a long shutter opening and movement by the camera operator. Computer adjustments were not successful to enhance or refine the image. Clearly though the upper fin and the

region of cell one and two is engulfed in fire and there is a burn area in the top surface of the cover forward of the vertical fin. In another unpublished photo the top of the vertical fin near the fin vent is not burning. Starting on page 413 of the original testimony, May 18, 1937, helmsman Helmut Lau testimony is documented. This particular witness is referred to because of his location on the airship at the time the fire broke out and his attention to the events observed. His location was in the bottom fin area. He describes seeing a bright reflection forward of cell 4 up toward the top and on the starboard side. "At first red and yellow and there was smoke in it." Cell 4 disappeared by the heat, then 3 and 5. " I saw the reflection of the fire through the cell wall ... first burning out of the fire, was on the starboard side above and, up." Lau also relates to the sounds he heard. Of significance is the cell bursting out on top, the strong detonation, heard after the fire and cell 4 disappearing, and then cell 3 and 5. Initially he heard a sound like a "frump" as one lights a gas kitchen range. This is what he first observes causing him to look up and then the fire reflection catching his eye. It is understandable the difficulty to describe exactly the sequence and relationship of events considering the circumstances. But the author suggests the first sound described by many "that shook forward through the ship", causing the dislodgment of the waste water tanks (noted on the historical photos, may have been what Lau witnessed as the bursting of the rear cells 1 and 2 causing the forward motion. The strong detonation he refers to later is consistent with the other witness reports and photos showing what appears to be cell 6 and/or 7 over expanding causing the well documented upward fire plume, commonly but inaccurately labeled, "the explosion".

Starting on page 758 of the original testimony, May 21, 1937, Ms. Elizabeth Tobin, resident of the Lakehurst air station, testimony is documented. The significance of Ms. Tobin's observation is that she was located facing toward the stern of the ship and had a starboard view. She indicates seeing a small flame in back of the fin at first, "it was very tiny and it ran right along the top of ship until it got amidships ... a continual line of flame, very small and saw-toothed, orange colored ... same color as the large flames which burst out of tail surfaces, a peculiar yellow orange color. Prior to first flame she had heard no unusual sound of any kind and everything seemed to be going about as usual". She also did not observe any sort of reflection internal to the ship. The author's contact with people who knew Ms. Tobin indicates her as a credible witness. The author, using a scale model of the Hindenburg and scaling out relative distances and location of Ms. Tobin, observed that at her vantage point she could see the "far side" of the vertical fin and hull interface. A flame in this area would not have been seen by the port observers. Also she was looking into the passing storm front and had the advantage of the relative dark background, contrary to the people on the port side, near the mast, or over by the hanger. In this case they were looking into the cloud clearing "sunset" in the west.

Witness Werner, an engineer, could see a part of the starboard side and indicates he saw the fire forward of the upper fin on the spherical curvature of the starboard side and that the flames started there. His attention was drawn to the stern, first seeing, "small curls of black smoke". From his vantage point the flame appeared near the center. He was sitting on the helium tanks (over 2,000 feet from the rear of the ship and only slightly off center to starboard side). Static Electricity Appearance. Although the boards of inquiry later concluded that some form of static electricity was involved in the incident, none was seen by any of the witnesses who testified at the inquiry. But in 1964 Dr. Douglas H. Robinson, a noted airship historian and writer, met with a Professor Mark Heald of Princeton, New Jersey. Heald was at the station with his wife and son and observed the incident but saw an unusual activity before the fire broke out. As Robinson says, "Heald...undoubtedly saw St. Elmo's Fire flickering along the airship's back a good minute before the fire broke out ". An account is noted of a witness reporting a "real light bluish" indication from starboard aft. Witness Kirkman alludes to a neon light type of activity from his starboard vantage point at 400 yards distance, but tends to characterize this observation as a tiny match flame that smolders prior to the breakout of the flames.

<u>Flutter of Outer Cover:</u> Witness R. H. Ward, in charge of the port bow landing party, testified he saw a fluttering of the outer cover port topside of cell 5. "It was a wave motion ... as if gas was escaping". This observation took on a lot of interest by Dr. Hugo Eckener (Director of the Zeppelin Co.) since loose hydrogen was unequivocally assumed the cause of the disaster. Lt. R.W. Antrim also saw a flutter but "after the port engine... it was very loose and fluttering". Eckener testifies it would take 40 to 50 cubic

meters of gas per second to cause a cover flutter. The author notes that the cell pressure is one-inch water pressure and the cell volume is down to about 7800 cubic meters (Lau info, LZ dwg). It is suggested it would take a sizeable opening to support Eckener's statement and the resulting flow (loss of gas rate) would have been noticed on the control car instrumentation, which it was not (Watch Officer, A. Sammt in charge of ship trim control). Airship expert H. Dick indicates that on the first flight of LZ-129 cover flutter was a problem and additional doping or antiflutter wires would be required The author recalls from the literature mention of flutter as a routine problem especially when the cover got wet. Also noticed in film footage from a vantage point above and to the port side almost all panels are fluttering during flight. This of course is from the aerodynamics of flight but illustrates the problem. Commander C.E. Rosendahl (In charge of the Lakehurst Naval Air Station, and an airship officer) says, "It seems to me that the fluttering of the cover...could easily have been caused by the passage along the hull of the ship of the light gust of wind which struck the ship on the port side... It is a matter of common knowledge that airship fabric which becomes slack in places in damp weather will flutter". In the original testimony (page 1199), Gunther Bock, Director, Deutsche Verersuckanstalt fuer Luftfahrt, says, " He thought the fluttering of the outer cover was caused by the air whirls caused by the fire within the ship ... after there was a flame, even a small one". The author points out that the observers, Ward and Antrim were up near the mast (well in front of the airship) on the port side to witness the flutter and the fire may have started on the starboard side (Lau, Tobin) then progressed over to the port side. It is noted that chief engineer, Sauter, on board in the lower fin, testified that the rear engines were running at full speed about 30 seconds before the fire broke out. Witness Kirkman testifies that at the time of first seeing a glow the engines were revving very fast. It is not clear if there is a relationship.

Heaviness of the Stern: There was much discussion recorded in the testimony about the apparent tail heaviness of the airship just prior to the intended landing. This avenue of discussion was of course to offer evidence that hydrogen was escaping from the stern, the origin of the fire. According to Rosendahl, "Had there been an appreciable leakage of hydrogen at the stern, it seems to me that it would have been detected by the gas-pressure instruments indicating in the control car, or by the loss of buoyancy in the stern, a condition which in the existing light wind would have been apparent to the personnel at the instruments and the elevator wheel in the control car. The stern, however, did not drop but actually rose somewhat while the ship hung on the landing ropes." Other accounts indicate that Captain Pruss usually approached the landing mast with a slight nose up attitude. Second Officer H. Bauer provided notes showing an angle of attack (inclination) of 3/4 of a degree positive (up at the bow) just prior to two sequences, one for 15 seconds, then one for 5 seconds, of releasing gas from cell 11 through 16 (forward cells). This action along with dropping of water ballast at frame 77 (stern) brought the airship in trim, while lowering the altitude 30 meters, which it held for six minutes prior to the fire breakout. Second Officer W. Ziegler in charge of the valving system and gas level meters in the control car testified that any gradual loss of gas in the rear cells would have been noticed. He further stated a rigger would hear escaping gas because of its whistling noise and it would have been heard at a distance of at least ten meters. The four crewmen in the stern-fin area reported that no unusual sounds were heard prior to the incident. During the hearing the crew witnesses were asked if they had smelled any hydrogen, they all reported they did not, but knew it had an odor. Pure hydrogen is odorless but per J. G. Vaeth, a garlic or onion type odorant was used in flammable airship gases. During the author's interview in Germany with E. Bentele, mechanic on board at the time, he pointed to his kitchen gas stove and indicated the smell to be not like the natural gas, but different.

Discussion of Prevailing Theories: Immediately after the accident two boards of inquiry were established. Under the Secretary of Commerce a US board was authorized. A German investigating commission was authorized by the Reichsminister of Aeronautics and the team arrived at Lakehurst May 13. Both groups independently advanced two basic theories, sabotage and accidental causes. These two theories will be summarized in this section, drawing information directly from both final reports of the investigation. Only certain aspects of the reports will be highlighted to focus on data relevant to this study.

Sabotage: Sabotage was examined under two classifications; the first -- external, including the use of incendiary bullet, high-powered electric ray and the dropping of ignition composition from an airplane;

the second -- internal, including the placement within the ship of a bomb or other infernal device. In view of the large number of people present at the time of the accident and that there were no reports of any unusual activity, any external cause was dismissed. Both investigating commissions and other experts investigated the accident site for suspicious clues. The surviving crewmembers were thoroughly questioned as to whether they had noticed anything out of the ordinary in this connection. The sabotage theory was dismissed due to lack of evidence. Since the disaster a number of movies, books and articles have been published in an attempt to dramatize the execution of an act of violence. However, to date, no physical proof has been offered, only subjective ideas to form the basis of the particular story line. The author suggests these theories are technically flawed, contain misquotes from the original testimony of record and have a problem with the timing of the accident. The timing aspect has been discussed with the FBI, NASA security and other law enforcement entities. Typical terrorist behavior would appear inconsistent with the circumstances at the time of the accident. The placement of a device in a cell of almost pure hydrogen (not flammable) is not logical. The placement of a device, say near an engine fuel tank is logical but does not equate to the location of the first sign of fire some distance from the tanks. Books written by Mooney and Hoehling have based their sabotage theory on a bomb or infernal device placed in, on, or near cell 4, the location of the first sign of fire by crewmember, Lau. Also germane to the argument is the sound Lau reported hearing prior to looking up at cell 4. The term "pop" is used by Hoehling, claiming that is how Lau described the sound. This term was used by the interrogator Major Schroeder essentially leading the witness, not Lau, as Lau states,"More a "frwump"; it was not a sharp detonation". (page 423 of transcript). Dr. Hugo Eckener concludes that the first sound originated in the area of the second gas cell from the rear (cell 2) and that flames broke out in that area. It is noted that cell1 and 2 were interconnected. Eckener testifies, regarding Lau's statements that, " I think that that was the second stage of the process of burning. Lau could, from his position at the time, not look vertically to the top of the ship through the gas shaft. He could only look forward and up at an angle, and he could have only noticed something of the fire as the flames beat down into the gas cell, and if I remember correctly as to what Mr. Lau said, that practically simultaneously with noticing the reflection of fire through the lower part of the gas cell the total gas cell was consumed, that therefore must have been the moment when the fire burning on top destroyed the top part of the cell and caused the escape of gas" (page 886 of transcript).

Accidental Causes: It is very important to note that both investigating teams in considering the possibility of accidental cause proceeded down the path that unequivocally hydrogen, and only hydrogen, was the problem. Therefore the teams set out to examine two factors. "There must be present (a) a combustible mixture of hydrogen and oxygen of the air and (b) sufficient heat to ignite such mixture." The question then became one of rationalizing the mechanism of creating free hydrogen and the source of ignition. The difficulty was of course to establish the two events simultaneously, complicated by the necessity of having a proper mix of combustible gas and the correct ignition energy at the precise location. This difficulty is not addressed in the reports. The following highlights the topics considered.

Free Hydrogen Source: Cell Diffusion, Valve Failure, Decreased Ventilation, Entry of Piece of Propeller, Fracture of Hull Wire, Major Structural Failure

Ignition of the Mixture: Mechanical; Chemical; Thermodynamic; Electrical (equipment spark, resonance, inductance); Electrostatics (ball lightning, brush discharge, St.Elmo's fire)

Conclusion: Both investigating groups concluded that the ignition source was from an atmospheric electrical disturbance such as a brush discharge. The German commission favored the theory that the discharge was caused by equalization of tension (static) between wet spots on the exterior and the ship's framework. Neither investigative report documents any conclusion as to the origin of the presumed hydrogen - air mixture. The idea of a wire break causing a rent of a gas cell has been the favored theory over the years by airship experts. There is discontinuity, however, whether it was a bracing, shear, or cell netting wire. In reading through the extensive literature on airship operations and accident accounts the author found no reference to wire breaking and subsequent cell damage. There are accounts of the chafing of a cell on airships requiring subsequent patching while in flight. Per Rosendahl, "It is known that of the

thousands of wires in an airship, once in a while one has failed in flight, but in such cases, even in hydrogen ships, often not even the slightest damage has resulted." Eckener in his 1937 testimony indicated experiences with broken wires, but not on the Hindenburg. The "sharp turn" at landing is the favorite explanation. Yet on March 26, 1936 the mishandled takeoff caused significant damage to the lower fin. No one identifies any broken wires as a result of this accident. In an interview with H. Dick, he mentioned he had gone over this many times as to which structural wiring system may have been at fault. The cells were loose except at the top, thus maybe the cell wire netting/support system at the top was a better one to blame. From Captain Lehmann, " To ignite or explode the hydrogen in the ship from other than premeditated or deliberate causes, would require escaping hydrogen or a free mixture of hydrogen and air, and a spark or flame as the igniting agency. But as the world knows, definite evidence to pin the cause of the fire conclusively on any of these agencies was lacking nor could it be shown that there was free hydrogen or any dangerous mixture of hydrogen and air present." From Stacey, " It was the Third Reich's policy, however, to never admit to any weakness. The idea of saboteurs bringing down the mighty Hindenburg was not acceptable to the German government, nor was the idea that a flaw in the design of the ship, or even simple human error, could have been responsible. Therefore, Hitler declared the disaster "an act of God." From Rosendahl, quoting Eckener, " As matters stand, there is not one possibility, but two, which might be interconnected. I would like to - because these possibilities are remote possibilities – I would be inclined to say that the possibility which I am about to state is an unlikely coincidence, because if it is necessary to assume two remote possibilities occurring at the same time, then the probability that these two remote possibilities concur is unlikely. But still, since the whole occurrence is extremely mysterious, we must proceed along these lines as the only sure point to start". Rosendahl then states, " The theory involving the coincidence of two remote possibilities advanced by Doctor Eckener is practically the same as the conclusion reached by the official board of investigation."
<u>American Archive Data Findings:</u>There are three significant findings to address from the photographic and witness testimonial evidence.

<u>Airship Trim:</u> The first few sequences of the fire photos show the airship still in trim. These photos were taken after several seconds of the first indication of fire or observations of a static electricity activity. Any serious loss of hydrogen in the stern as a result of a major cell damage should have immediately caused a noticeable descending of the tail section. This situation is also aggravated by the inadvertent release of the forward wastewater tanks and water ballast as seen in the photos. Clearly hydrogen is part of the burning process at this point as the cells melt or burn from the heat. The assumption of a large escape of hydrogen, as noted in the cover flutter argument, appears flawed.

<u>Flame Propagation:</u> From the review of the still photo sequence and a review of the video (newsreel film) frame by frame reveals that a very rapid propagation of flame in many directions had occurred. This is not characteristic of hydrogen gas burning due to its buoyancy. The outer cover simply was not resisting the almost non-radiant hydrogen flame. Knowing the distance between the hull frame rings and noticing the time lapse it appears the flame front travel was on the order of 5 to15 meters per second (variations were noted but the cover and the doping thickness/weights were also variable). Also, it is interesting to point out the observation that large pieces of burning fabric fell to the ground and were not self-extinguishing. These observations suggest that the fabric and its protective lacquer were quite flammable and some type of chemical reaction accelerated the fire progression along the hull. Hydrogen contributed to the forest fire inferno above the airship.

<u>Fire Colorization:</u> In the archive file there were two sources of data on the witness account of the fire appearance. One source has the record of a survey of about two hundred people (landing party) describing what they saw. The other is from the testimony digest. The consensus overall characterizes the flame as a white-yellow-orange-red, "bright" fire. The literature as well as the author's personal interview with witnesses and surviving crewmembers indicates the flame as very "hot" even at considerable distances. This is not a characteristic of a hydrogen flame where the wavelengths are mostly in the ultraviolet region (4340 angstroms and beyond) with some in the infrared region. Thus a hydrogen flame is not only difficult to see, if at all, it exhibits a very low radiation. In a visit to the Fire Sciences Laboratory in Missoula, Montana the author discussed the characteristics of a typical forest fire. They in

effect are large cellulose based fires and give out the characteristic yellow to reddish color and exhibit intense radiation for some distance. The whitish color content of the Hindenburg fire expressed by observers was perhaps due to the presence of the aluminum powder in the dope. Based on the above information a color composite photo of the airship fire was computer generated.

German Archive Data Findings: The archive information that was available included airship drawings, related correspondence and reports. Subsequent to the Hindenburg fire there apparently were a number of concerns in the design of the LZ -130 that led to a re-evaluation of some of its design features. The following outlines key findings.

Doping Process Concerns: In the June and July 1937 timeframe, considerable correspondence was transmitted between the Zeppelin works and the supplier of the doping compound, commonly referred to as "cellon". This cellon appears to have been a trade name and underwent various formulations to change its properties. The highlight of the correspondence focuses on improving waterproofing, reduction of cover fluttering and the general "safety" concerns of the resulting cover coating. Of note, in a June 30, 1937 letter, is that the lacquers produced by the Worwag and Atlas-Ago companies are processed of cellulose acetobutyrate. This cellulose is produced from woodcell material and has no inner structure. The material is selected over the cotton-based origin for economical reasons.

Instrumentation Concerns: As a result of the board of inquiry on the Hindenburg incident and the focus on the possible problem with the gas cell pressure measuring device a number of tests were conducted. The concern was whether an air-hydrogen mix might have accumulated within the measuring device that had an electrical rheostat. It was found that a modification was not required and the device as used on LZ -129 would be acceptable for LZ -130 in hydrogen service. Thus, any concerns about these electrical devices were alleviated.

Design Modifications: While LZ-130 was under construction it was not clear whether helium would be available; therefore the drawings depict a helium configuration and a hydrogen configuration. The primary difference, for hydrogen, was one of adding more ventilation and modification of the gas shaft vent hood. For hydrogen service it is specifically noted that a smooth metal cover be placed over the existing wood frame of the leading edge of the hood. The author notes that this precaution would reduce the possibility of a static discharge (such as St. Elmo's fire) at the upper discharge point from where the venting of hydrogen from either the gas cells or internal sources may occur. The LZ -130 drawings depict a common (helium or hydrogen service) change to the metalization of the final coats of cover dope for solar ray reflectivity and protection of the cotton cover. A 2% aluminum-bronze was specified. It is noted that for LZ –129 a 3% pure aluminum powder was used. In view of this finding the author investigated the properties of the bronze vs. aluminum as described in the literature. Basically the bronze should reduce the flammability/chemical reactivity risk and the high copper content should improve the cover conductivity. More detail is provided in the section on Airship Envelope and Doping Procedures. Another aspect that the correspondence addresses is the topic of graphitization. The letters simply imply that such a process is approved and is to be incorporated for LZ -130. Per H. Dick it was not common knowledge that for LZ -130 the ramie lacing cord used to attach the envelope panels to the frame was prestretched and run through dope containing graphite to make it conductive. This had not been done for LZ -129. An expensive and radical modification to the LZ -130 was the engine change from a "pusher" (historically used on the Zeppelins) to the "tractor" mode. It is noted that the engine exhaust gases and particulate were highly ionized. It is conceivable that during engine reversal at landing the action of the propeller wash on the swirl effect of the positively charged ions would be eliminated for a tractor configuration.

Fabric Testing: Perhaps one of the most significant findings in the file was a report reflecting tests conducted on the fabric with the typical dope coatings as used on LZ -129. The tests were done by Otto Beyersdorff, "Elektro Ingenieur". In two letters (to Dr. Eckener, chairman, Luftschiffbau Zeppelin) he makes the following comments: June 28, 1937. "Time and again I made comments to your company about the following important facts:

1. The actual cause of the fire was the extreme easy inflammability of the covering material

brought about by discharges of electrostatic nature; this caused the hydrogen to ignite.
2. By using the same material, an ignition has to always happen again, when the proper circumstances are present, as for example in Lakehurst."
July 13, 1937. "Because the reported details about the fire made the causes of the origination so definitely clear - namely, the ignition of the covering caused by an electrostatic discharge - I found it necessary to report as follows: The "Werft" was kind enough to send me a sample of the covering material; with which I was able to perform tests in the laboratory matching the conditions of the accident and which proved the material; to be extremely easy to inflame. An acquaintance of mine, an academic advisor, who took part in some of the tests, was amazed at the degree of inflammability. In further letters to the "Werft" ... I emphasized the latter circumstance time and again and mentioned that an incombustible gas cannot prevent a burning of the covering as long as this one has the characteristics of the test results."

Airship Envelope and Doping Procedures: Airship Envelope: The envelope for the rigid airship defines the shape of the ship. This leads at once to the requirement that the outer cover shall be and remain taut. Further, it must be weatherproof so as neither to admit rain to the inside nor get waterlogged itself. It should also, as an extension of its waterproofness, be water-repellent. It must, for retention of its own strength and for protection of the gas bags, be opaque to harmfully active light; and, principally to avoid superheating of the gas, must reflect as much incident radiation as possible. Added to this, it must have the lowest possible specific weight. (From J. W. Dyer) Linen or cotton was used in the LZ ships. Fabric panels sewn from material grades of 90 to 130 grams per square meter is indicated on the LZ -129 and LZ -130 drawings and located according to strength requirements. Cotton of course is the purest form of cellulose in nature. It contains more than 90% cellulose and has a self-ignition temperature of 464 deg F. Cellulose has approximately 60% of the energy content of bituminous coal. In addition to the fabric with its ramie cord lacing, wooden fairing strips and dowels were used for bracing and spacing. In the smaller airship (LZ -127) over 72 miles of cord had been used. According to Fritsche, "The fabric covering still presents the most serious problem in airship construction and operation." This statement was made to the Royal Aeronautical Society just two months prior to the British airship, R -101, disaster at Beauvais, France in October 1930. The disaster occurred during a thunderstorm. The cover was cotton, coated with iron oxide and aluminum powder pigmented cellulose nitrate. It is noted all the cover appeared to have burned.

Doping Practice: The function of the doping is to provide the tautness, waterproofing and reduction of solar radiation characteristics desired for the envelope. The dopes used in the LZ ships consisted essentially of cellulose nitrate or cellulose acetate. These cellulose syrups were mixed with suitable solvents and binders such as butyl or amyl acetate, alcohol, acetone, methyl ethyl ketone and castor oil. Light absorbing pigments such as iron oxide are introduced for the protection of the textile usually as the first coat. Succeeding coats of dope are applied and contain powdered aluminum principally for solar ray reflectivity. It was found that the pigmentation and metallic powder also assisted in reducing the deterioration of the fabric as well as improve other desired characteristics for the envelope. The dope mix was applied using the mop and bucket procedure as shown in many photos.

Cellulose Dope: Dopes made from cellulose acetate were given preference, even though they were more expensive than dopes made from cellulose nitrate. The production of the acetate required large quantities of acetic acid and acetic anhydride, chemicals hard to produce and obtain especially during the war years when such chemicals were important war materials, including medicinals. For this reason the readily available and low-priced nitrate dope was used wherever possible. The nitrate-based dope also was less affected by humidity. It was therefore used for some parts of the cover, but not used throughout because of its greater inflammability over the acetate based dope.

Dope Pigmentation: Light absorbing and solar radiation reflectivity based pigments were used that also enhanced the cotton cover longevity. The British extensively employed a dope used for the first coat consisting of a red iron oxide. The iron oxide coat was also used on the LZ -129 but only on the upper surface above the ship's equator. Such an iron oxide is used today in the make-up of rocket propellants. Its purpose is to act as a burning-rate catalyst.

An increase in the surface area of the catalyst significantly changes the burning rate (performance) of the composite propellant. During the early days of balloons and airships the solar superheating of the hydrogen filled containment was recognized by operators, perhaps unaware of the relatively high thermal conductivity of the hydrogen gas (over seven times that of air). It was this effect that prompted Enrico Forlanini from Milan, Italy, in 1909 to employ aluminum powder, "In order that the metallic layer would reflect sunlight and thus prevent heating of the gas." (H. Van Dyk translation from German Aviation Vol 26, 1997). The use of powdered aluminum in the doping process had many advantages even though it is hazardous to produce, handle and use. For example, during manufacture it is important not to allow the exposed aluminum surfaces, unoxidized at that point, to be exposed to air. Under this condition enough heat can be spontaneously generated to cause ignition of the aluminum. To prevent this a controlled amount of oxygen (3 to 5%) is provided to control the surface oxidation. The resulting oxide film then serves as an impervious layer to inhibit further oxidation in the presence of air. The self-ignition temperature of aluminum powder is 959 deg F. When a mechanism occurs that breaks the very resistive oxide film of aluminum surfaces serious consequences can result. The author was a member of the team that investigated the Brooklyn Hospital fire and explosion involving a liquid oxygen delivery truck. The breaking of an internal baffle exposing a fresh surface of aluminum led to a rapid reaction with the oxygen. Since that event the industry removed all such vehicles from oxygen service.

The author has been part of investigative teams concerned with gaseous oxygen events with aluminum. These events provide the foundation that atmospheric ozone may contribute to reactions with aluminum-impregnated surfaces, if adequate mechanisms are in play. The presence of the more reactive ozone not only reduces the ignition temperature but also causes the materials to burn vigorously. Noted in, "The Science of High Explosives", is the statement, " While aluminum has been used considerably in both commercial and military explosives, it has one important property that has not been generally recognized or appreciated, namely its ability to react explosively and powerfully with water under suitable high-temperature initiation." The reaction generates 2.5 times the heat as the same weight of natural gas. Metals are used in high-energy propellants to increase the total heat release during combustion. The most commonly used metal is aluminum, because of its high heat of combustion and combustion efficiency. The solid rocket boosters (SRB) of the Space Shuttle use a mix of 16% aluminum powder as the fuel. As mentioned earlier, iron oxide is also used in the propellant mix. In the steel industry, when mixed with iron oxide, aluminum powder forms thermite, employed to weld heavy sections of metal. Burning thermite can reach a temperature well over 5,000 deg F. It was noted earlier that aluminum bronze powder is called out on the drawings for the hull portion of the LZ -130. This is typically an alloy of 90% copper and 10% aluminum.

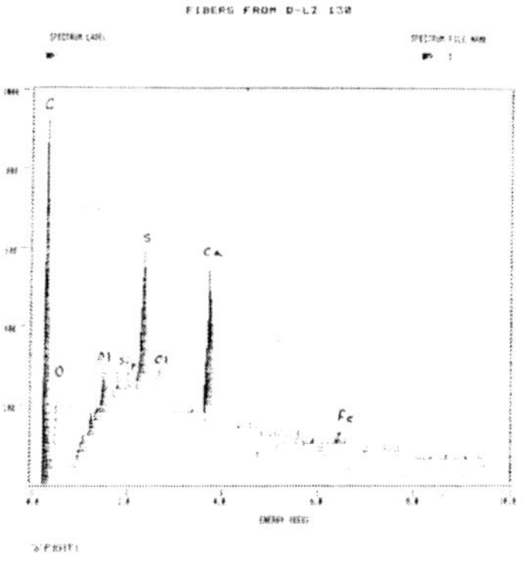

Scanning Electron Microscope (SEM)
X-ray Energy Dispersive Spectroscopy (EDS)

FIGURE K21

SEM/EDS ANALYSIS (LZ -130)

Flammability Concerns: The literature has extensive references to the flammability concerns, problems and testing of various materials related to the use of fabric coverings and their coatings. In 1921 a Mr. Dyer spoke of the importance of fireproofing. He suggested if acetate dope was used with "plenty" of pigment in it, appreciable fireproofness was secured. Dr. Guy Barr conducted tests to compare facility of ignition and flame propagation for acetyl and nitro - cellulose based dopes for cotton fabrics for aircraft, after the experiences of the destruction of RAF aircraft wings by fire. The interesting aspect of his testing was that it was done under airflow to represent the conditions in flight. He presumed that the flame propagation would increase as the aircraft speed increased. Tests proved to the contrary. At speeds above 20 miles-per-hour the flame would self extinguish; suggesting the fault had been with fabric saturated with petrol. This led to concerns on the danger of fire when aircraft were not in motion. The NACA TM 421 document discusses various procedures and formulations in using acetyl and nitro based cellulose dopes and states; "the question of inflammability of doped fabrics is one of great importance". The TM published in 1927, goes on to discuss experiments to fireproof the fabric by first immersing in ammonium phosphate and aluminum - sulfate prior to application of the protective dope. This process did diminish the rate of burning, but rendered the dope film less adherent, and concludes; "no real advantages can thus be obtained". In view of the use of aluminum bronze powder in the coating of LZ -130 a Bureau of Mines study on flammability tests of metal powders was reviewed. It indicated that aluminum samples would burn, whereas the copper (bronze) samples were not affected. In the book on airships by Jackson two interesting photos show airships L 31 and L 34 going down in flames. What is noticed is that in both cases the airship is caught in a nose down orientation with the entire bow engulfed in flames flowing back over the airship. Clearly hydrogen was involved in the fire at this stage but the apparently flammable envelope was a significant contributor. The author was not able to confirm, but translated German letters discuss the use of test panels on the Hindenburg. Apparently other materials were being tested, but their make-up and possible location on the airship was not identified. It was known that different gas cell designs were being tested on the LZ -130, as the author acquired a sample of the material from a German collector and it was comparable to a test report.

Related Atmospheric Electrical Phenomena: Historical Accounts. Quoting from Lehmann's book; "Tongues of fire are licking around the muzzles of my machine-guns, and around my head, too. And when I spread my hand, little flames spurt out of my fingertips.", describes an experience aboard the airship LZ 98. The name St. Elmo's fire is attached to this very common phenomenon. Recorded in such historical books as "Moby Dick" and "Hakluyt's Voyages" sea-going men first observed this atmospheric electrical activity. St. Elmo is a modification of St. Erasmus, the patron saint of sailors of the Mediterranean. The Hindenburg incident boards of inquiry concluded that the source of ignition was in fact a brush discharge, or St. Elmo's fire. Airships in flight acquire an electrostatic charge due to impact of dust and moisture particles in the atmosphere. This effect is noticeable at 40 kilometers per hour, increasing approximately as the 16th power of the speed. Studies by Dieckmann (electrostatic expert at the hearings) show that airships could acquire a significant level of electrostatic charge as influenced by nearby cloud formations (induction). One of his papers in German shows induced voltages as high as 180,000 volts. More than a century after Benjamin Franklin used a kite to demonstrate the electrical nature of lightning, Popular Science (1897) recreated the experiments using sophisticated electrometers. Among the findings: thunderstorms were not the only condition yielding marked electrical effects, sparks were obtained in cloudless skies. Also voltages increased as the kite rose.
A NACA report, released in July 1924, cautions: envelope materials should offer no resistance to the passage of electric current or act as a dielectric. In order to prevent the airship from becoming too highly charged with electricity, it is desirable for all conducting masses to have projecting points in considerable number throughout the entire length of the airship, except in the vicinity of valves or where gas can escape. It is noted that in the case of the Hindenburg, after the last coat of dope was applied the surface was lightly sanded to reduce aerodynamic drag. At the time of the Hindenburg disaster the bow manila ropes had been in touch with the ground for four minutes. Based on testing by the National Bureau of Standards of a section of the rope under various humidity's determined the "airship" would be 90 percent discharged in a period of 0.6 second to 170 seconds. Discussions with Professor Uman (lightning expert) about the configuration of the airship indicated that the Hindenburg frame was simply a lightning rod 200 feet in the air.

Types of Phenomena: There are various forms of atmospheric electricity and their characteristics are varied. The five types; ball lightning, corona discharge, point discharge, space charge, and spark discharge are categorized based on a recent consensus of experts in the field. Clearly, according to the literature on the topic of atmospheric electrical phenomenon, there remain many unknowns about the nature of this activity. Even today scientists are uncovering new types of "flashes" such as sprites and blue jets that appear connected to thunderstorms but they typically occur in the upper atmosphere. The eye witness accounts of the events just prior to the breakout of the fire on the Hindenburg (neon light, blue glow, saw-toothed flame, etc.) allude to some sort of atmospheric electrical activity. The question is whether a type of activity either in its initial form or some subsequent manifestation was of a sufficient energy level and dwell time to react with any of the airship materials. As will be discussed in the section on chemical kinetics it appears that other cases of reactions of materials similar to the airship with static type electricity required little energy or time. he author in private communication with a noted NASA static electricity expert (Dr. Gomph) pointed out that in view of the distance of the eye witnesses from the airship having observed the "blue glow" indicated, "that's a lot of coulombs". J. D. Barry goes in to some detail about experiments with what he terms ball lightning. He and others describe small luminous spheres (blue, white). Temperatures are deduced to be in the range of 6,200 to 10,000 degrees Kelvin. Energy levels (Joules) of ten to the fourth and sixth power are noted for balls with diameter of 10 to 15 centimeters. Many references address the controversial ball lightning, but Hellman claims they last from a second up to several minutes, some fade out slowly, others disappear abruptly, and have scorched wood and burned through wires. Dieckmann testified that one has to assume ball lightning within its surface contains a very high temperature, carries a high electrical charge and can have a duration of 15 seconds. He goes on to say that he does not intend to claim ball lightning caused the catastrophe of the Hindenburg, "but I want to narrate to you how under the assumption of such a ball lightning one could make the phenomena coincide with the statements of the witnesses". Also he indicates the scenario of one or more balls of lightning forming at the rear of the ship and passed relatively slowly forward, being very hot, "might scorch the outer cover".

Airline Pilot Experiences: During the course of the study the author learned from a number of pilots of aircraft that the corona discharge (St. Elmo's) and a form of ball lightning were quite common experiences while flying near thunderstorms. Experiences basically were described as a glow (St. Elmo's) first appearing around the cockpit window frame then manifesting itself into a ball entering the cockpit and "dancing" along the floor to the rear of the aircraft. One account, with two witnesses, describes an event that occurred during no thunderstorm but under cloudy conditions where a "ball", blue-green in color, bounced on top of the wing leaving a black scuffed appearance. A 1965 United Airline Report documents the accounts of numerous pilots, "with many thousands of hours of flying experience" about their experience with lightning. They almost unanimously agreed that there were two distinct types. One is while flying in precipitation near freezing conditions. Corona (St. Elmo's) is observed for several seconds then it discharges. The second variety occurs abruptly with no warning. It is most likely near thunderstorms. Pilots believe there is a slow buildup type of discharge from flying through the precipitation (no lightning). A brilliant flash and a boom can be heard throughout the airplane. It is noted that electrostatic dischargers (carbon-impregnated wick) are commonly fixed to the trailing edge of wing and tail surfaces to reduce the charge buildup.

Airship Specimen Acquisition and Utilization: Acquisition: The knowledge that airship artifacts (specimens) were available became known to the author in November of 1995. Of particular note was the availability of LZ -129 fabric samples from Hepburn Walker, a retired Navy person who was stationed at Lakehurst during the time when the rail track was being removed and pieces of the airship were uncovered that had been imbedded in the soil near the accident site. This led to a meeting with personnel of the NASA Materials Science Laboratory at the Kennedy Space Center (KSC) to seek their assistance in testing of the fabric samples (under the public Technology Exchange Program). The people were very cooperative and interested in supporting the test project. The work would have to be done on a non-interference basis with the basic lab priorities in support of KSC activities. A test project protocol was discussed and planned and the type of proposed equipment to be used identified. Later in February of 1996 the author met with Cheryl Ganz of the Zeppelin Collectors Club and not only were artifacts of the Hindenburg offered for study, but also fabric samples of LZ –127 and LZ -130. This immediately led the author to consider having the NASA labs do comparative testing. During the course of the NASA testing and word of the work being done having spread within the airship community, more samples were offered. One such source was Kent O'Grady, from his contacts with The Airship Heritage Trust. During the author's visit to Germany in July 1996 more fabric samples were acquired, such as from collector Deiter Weil and others. Utilization. Since the artifacts from people willing to share their collection were rare and needed to be returned intact, it was imperative the lab testing be conducted in a non-destructive mode. Items purchased by the author could be used for any purpose. The approach for these items was to conduct the same non-destructive tests as done on the collectors' items to assure correlation of the doping chemistry for the two sources. Also a tie to the collectors authentication was needed. This procedure allowed selective destructive testing at the end of the test protocol.

Airship Materials Testing and Evaluation: A synopsis of the NASA findings is covered in this section.

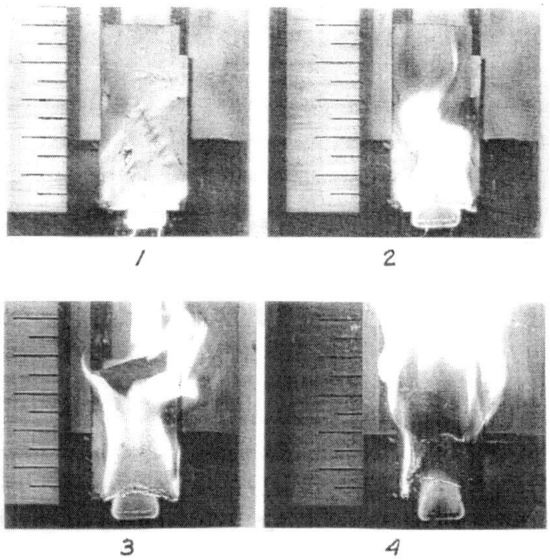

1. The LZ-129 frame was an alloy of aluminum with a minor amount of copper. This is consistent with the literature and description for Duralumin. The frame appears to have been coated with aluminum paint and a cover coating of lacquer of alkyd/acrylic origin. The lacquer contained a green blue phthalocynanine dye. This process is consistent with the need to mitigate frame corrosion, a serious problem on previous airships.
2. Analysis of the fabric paint from the three airships (LZ-127, LZ-129 and LZ-130) varied. LZ-127 dopant contained aluminum and cellulose acetate. LZ-129 dopant contained aluminum and cellulose acetate butyrate. LZ-130 specimen was similar to LZ-127 except for a chloride and a high level of sulfur and calcium. This suggests a compound such as calcium sulfamate, a flame proofing agent used in the textile industry. No significant amount of copper was present in the LZ-130 sample. This was bothersome since the LZ-130 drawing showed the use of bronze in the dopant for the hull section. Tracing back to the origin of the sample resulted in the fact the sample came from the engine car not the hull.
3. About seventy infrared spectra were collected and compared. The IR work is more of an art and requires extensive background experience in interpretation of the results. In the case of LZ-129, where the concentration of effort was applied, the spectra showed variations in the absorbance at the same relative wave numbers. It was concluded this may be the result of the method of dopant application, namely the mop and bucket process. The lab report indicates butyrate acetate would be more flammable than the acetate alone, and cellulosic material when wet becomes more flammable. Other airship fabric samples were analyzed as a curiosity. It was found that in fact the British used the cellulose nitrate with iron oxide.
4. The needle rake method was used to determine the flammability of the LZ doped fabric representative of LZ-129. Specimen #5 as it was labeled was crosschecked with known LZ –129 material to assure commonality. The test found the material flame property to be "non-extinguishing". Sequence of the test as recorded on video.
5. Surface and volume resistivity tests were done per ASTM D257. The results indicate readings of 8.22 to 9.71 x 10 to the tenth power, ohms per square cm, for surface resistivity and 4.61 to 7.07 x 10 to the thirteenth power, ohms-cm, for volume resistivity. The interpretation by the investigator is that the LZ-129 doped fabric panels exhibited high capacitance, resulting in a high voltage static buildup relative to the airship frame (at ground potential).

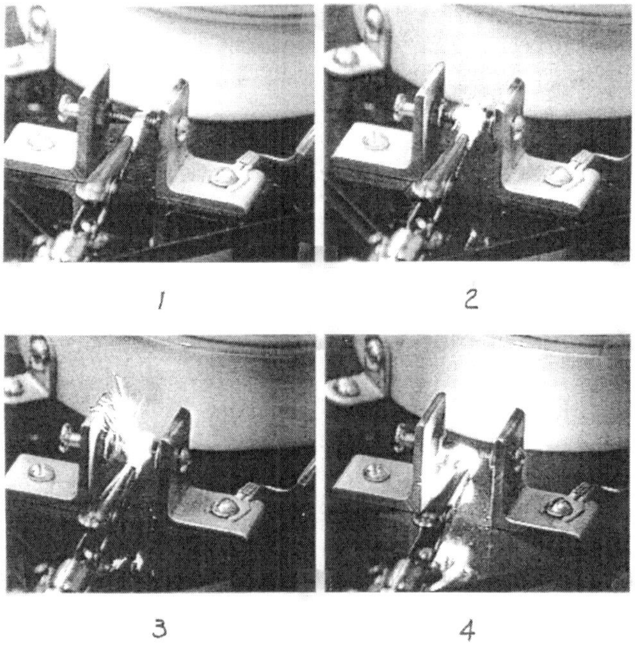

6. A multi-stage Tesla coil was used for the electrical arc tests. The gap electrode voltage was 30,000 volts at high current and 100,000 volts at low current. The LZ-129 representative sample(s) were first placed in a perpendicular orientation. The arc burned a hole through the material, but ignition of the sample only occurred if the sample was moved back and forth across the arc. In a parallel orientation the sample(s) would ignite almost immediately.
7. The lab notes cellulose butyrate acetate is derived of purified cellulose with acetic and butyric anhydrides with sulfuric acid as catalyst and glacial acetic acid as solvent. It

is noted for its excellent weathering properties, low water absorption, low heat conductivity, resistance to oil and grease and is nontoxic. However, it is also combustible and has a high dielectric strength.

Follow-on tests at other labs produced the following information: 1. Having identified the make-up of the LZ -129 fabric, materials were acquired to make up representative specimens for additional testing. At the Fire Science lab at Florida Tech the thermite reaction was demonstrated. This confirmed in more detail what was observed when the NASA video was run at slow motion (during the flammability test). 2. At the Florida Solar Energy Center, using their IR spectrograph, the butyl constituent of the dopant was confirmed. The electron microscope revealed the wicking through of the first coat of aluminum dope mixing in random areas of the iron oxide dope, thus providing thermite "hot spots". 3. A piece of airship frame with the blue coating, courtesy of the Zeppelin Company, was examined. The surface resistance measured "infinity"; thus the coating was non-conductive. 4. A two-foot by three-foot panel was constructed, using an aluminum frame and 100% cotton at 100 grams per sq. meter. Brass eyelets at 40 mm (LZ -129 specs) were installed on folded over sewn edges and the panel was laced to the lacquer-coated frame. Iron oxide and aluminum dopes were prepared using the same LZ formulation. It is noted that the acquisition of the chemicals had to be obtained through a certified lab and the aluminum powder from a different supplier. One chemical company would not sell the chemicals "together". A 200k volt Van DeGraff generator was purchased. The test scenario was to elevate the panel, attach a ground to the frame for use to ground the frame at a point in time, and position the generator under the panel. Still and video cameras were to be used to document the test. Unfortunately after two days of trying the generator would not cooperate, and the test was cancelled because of schedule conflicts. (The author hopes to resume this at a later date when finances permit).

Airship Materials Testing and Evaluation (Germany): In addition to the fabric testing by Beyersdorff, reported on earlier, from the German archive source there are three other findings to address. 1. Dr. Max Dieckmann, Munich Technical Institute professor and manager of Experimental Station for Aviation under the German Air Ministry, conducted tests at his institute in 1937 following the accident. He assumed that five conditions were required to initiate the fire: a) Inflammable mixture of gas under the cover at the rear of the airship. b) It was raining and the dampest part was at the rear. c) A high landing created greater potential gradient than if at low landing. d) A secondary thunderstorm was occurring creating a gradient undergoing very quick and very great changes. e) Landing lines became wetter and more conductive. His test consisted of a ten-foot square sample of the outer cover mounted horizontally. A rubber bag with a gas mix was placed under the cover with a piece of metal inside the bag. The cover was dampened. An artificial atmospheric electrical field was established. Ignition of the Hydrogen-air mixture always occurred. He notes that had the airship been metallically grounded the catastrophe could have occurred sooner. (The manila ropes had touched the ground four minutes before the fire was observed, and the metal cable intended to be attached to the mooring mast had been extended only about 50 feet). He further states that no ignition occurred during his tests with material coated with the Graf Zeppelin dope, but did so when using the Hindenburg type of dope. 2. In March 1997 information was sent to the author from the Zeppelin Museum concerning the coatings of LZ -127 and LZ -129. The information was a copy of the spectrographic charts from the Herberts Company the producer of the cellon "at the time". In the case of LZ -127 the coating was a "cellulose compound upon acetate acid base". For LZ -129, "cellulose compound with butyric acid constituent". Charts show essentially equivalent peaks for aluminum, potassium and iron. Calcium and sodium are shown in the case of 127, but for 129 no calcium and a higher peak for sodium. Small amounts of sodium can impart a yellow color to a flame. The Herbert's report confirms a report translated for the author entitled, "Analytical Research of Covering Material of Zeppelin Airships LZ 127 Graf Zeppelin & LZ 129 Hindenburg" (1937) depicting the cellulose acetate for LZ -127 and LZ -129 with butter-acid (butyrate acid). 3. In May 1997 Ms. Edith Dieckmann wrote to the author concerning tests her husband had done. Apparently he used a model. "When the cover of the Graf Zeppelin became wet and was electrically loaded and the holding rope fell down -- nothing happened. The same research with the cover from Hindenburg: The airship (model) burned at once." She goes on to relate that Rosendahl during a visit to their place in Munich "was convinced like Max Dieckmann. But at that time there have been two reasons not to publish it: Reputation (Hitler Government!) and also insurance reasons." The author notes it is not clear if the model tests were a separate experiment from the ten-foot square cover mock-up and gas mix in a rubber bag experiment as discussed earlier.

Significance of Related Chemical Kinetics: Discussions with a number of chemists brought up the possibility of three chemical kinetic reactions that may have come into play during the initial events of the disaster. These are postulated because of the chemical ingredients employed as part of the fabric coatings. The contribution of each reaction, individually or in some sort of combination, is of interest. The cellulose, wood strips, resins and like materials that made up the airship envelope and observed fire characteristics suggest the event was not much unlike a typical forest fire. Thus a discussion on crown fire behavior seems appropriate.

The Case for Ozone: Ozone, generated naturally when electrical arc charges pass through air, is one of the most powerful oxidants known. This form of oxygen has three atoms per molecule and is a short lived unstable gas that decomposes rapidly, especially in water. As the extra oxygen atoms split off from the ozone molecule they adhere to existing chemical compounds, typically changing them to more innocuous substances. The process of ozonation is a standard practice for water disinfection and applications in food processing. However, ozone being unstable cannot be stored, it must be generated and applied on-site. During the NASA lab tests where the airship fabric was subjected to the electrical arc, the distinctive aroma of ozone was present. As indicated earlier any mechanism whereby the ozone could make contact with the particulate aluminum may result in a reaction of its own. The dwell time of the event would of course be a function of the electrical discharge dwell time, which in itself may have masked the ozone contribution to the combined reaction mechanism. Barry indicates that Thornton postulated that ball lightning was predominantly composed of ozone and could rapidly decompose to oxygen with an explosive decay.

The Case for Thermite: The make-up of the base coat of iron oxide on the airship outer fabric with subsequent coats of aluminum powder based lacquers, suggests the possible condition for a aluminothermic process to take place. Commonly called the Thermite Process employed to weld steel or used as an igniting agent for incendiary bombs. Thermite, typically a mixture of aluminum powder and half an equivalent amount of iron oxide chemically reacts when ignited by a fuse or magnesium ribbon. (Chambers Dictionary pg 897) The aluminum is almost instantaneously oxidized in the process and temperatures of over 5,000 degrees F have been observed. The basic arrangement is an intimate mixture of the two ingredients with an over layer of powdered aluminum and sodium peroxide. In the case of LZ-129 the iron oxide film (topside only) amounted to 4.62 grams per square meter (gm/sq-m). The first aluminum coat contained slightly over 6 gm/sq-m of powder. Energy dispersive spectrography of the iron oxide side of the fabric showed aluminum bleed through, indicative of some level of intimate mix of the compounds. It is recalled the paint coats were applied using the mop and bucket technique, or brushing in of the solvent based lacquers. Another 12 gm/sq-m of aluminum based cellon was placed on top of the initial coats, somewhat characterizing the thermite preparation. The cotton fabric substrate varied from 90 to 130 grams per square meter depending on the location on the airship. Slow-motion video playback of the NASA fabric flame tests showed a vivid white sputtering reaction of the metallic residue. The thermite reaction can lead to what is known as a ballotechnic reaction. This has been pointed out by a scientist doing such work while at the Los Alamos National Laboratory. He points out the oxidation-reduction reaction between iron oxide and aluminum flake when mixed intimately occurs on a rapid time scale, typically microseconds. The energy release is approximately 3 kj/gm. There is associated with it a very bright orange light and shock wave.

The Case for Energetic Materials: The 1985 Pershing missile catastrophe placed renewed interest in the behavior of electrostatic discharge (ESD) activities around solid rocket motors (SRM). This event and other related ESD activity in energetic materials has prompted investigations that are reported in the CPIA publications and AIAA Journal of Propulsion and Power. It has been found that surface ESD phenomena in aluminum powder based propellants served as a mechanism for induced inadvertent ignition and sustained combustion of the aluminum based compound. According to chemists at the Air Force Rocket Lab at Edwards AFB the manufacturing process for the aluminum particles has been modified. The commonly used flake version of the powder has been changed to a spherical configuration.

This significantly reduces the sensitivity. A German report discussing the LZ-129 doping process indicates the flake version was used, intended to increase the surface sunlight reflectivity. The same report indicates that the aluminum powder was polished with an oily lubricant, then washed with gasoline. This may have removed the highly resistive oxide film on the aluminum surface, not allowing time for repassivation, thus increasing the sensitivity to ESD or other chemical reactive mechanisms. The Edwards AFB personnel and other interests in the study point out that the ESD and aluminum reaction is initially slow (on a micro second basis), but then exponentially accelerates to an order of magnitude of 1,000 inches per second. Timing the LZ-129 movie frames documenting the fire and calculating the flame spread across the envelope an estimate of 15 to 20 meters per second can be observed. The two sets of data are surprisingly comparable. From the AIAA paper on ESD hazards of energetic materials the following is extracted: " A severe hazard arises when energetic materials... are charged to a potential where breakdown occurs... which reduce the impedance of the energetic material and result in a rapid current increase. This can lead to arcing, establishment and growth of discharge paths, followed by catastrophic discharge. The resulting pressure and temperature increase in the very narrow discharge path may induce ignition and combustion". The same paper addresses the breakdown threshold as 10 to the tenth power, ohms-m. The NASA lab's resistivity test of the LZ fabric measured 10 to the thirteenth power, or one thousand times the "safe" level. A CPIA paper notes: " Flake aluminum particles, when compared to more spherical aluminum particles, tend to enhance local breakdown and reduce the critical electric field, Ebk." Environmental parameters such as temperature, humidity, charging time, etc., can significantly alter the Ebk. Classic dielectric breakdown is obtained when Ebk is exceeded. As current continues to flow, the discharge channel grows in size, and it is heated by the electrical energy deposited by the discharge. This heating will cause a chemical reaction to occur. The Air Force Research Laboratory proprietary computer program was used to run various thermal calculations based on Hindenburg coating data. From a twelve page report to the author the following significant comments are extracted: "Flame temperature of the cellulose acetate butyrate/castor oil/aluminum paint is 3,347 deg F, probably hot enough to allow effective flake aluminum combustion. This high temperature flame characteristic with aluminum in a seemingly nonenergetic binder appears incongruous. What is happening is that the high oxygen content of the cellulose butyrate acetate binder material can become sequestered by the aluminum producing very high flame temperatures producing aluminum oxide and large amounts of byproduct graphite. What this means is that the "safer" Hindenburg paint might be almost equally dangerous in producing a self-propagating fire as a nitrocellulose composition. Adiabatic flame temperature is so high that burning of the Hindenburg skin would likely be led by aluminum paint self-combustion followed closely by air oxidized burning of the remaining preheated fabric and iron oxide paint layer." "A US Army researcher once noted that electrostatic ignition of solid propellant required not a high voltage surface electric charge, but an abrupt electric discharge by electric breakdown through highly resistive energetic material to an electrical ground. This concept is readily applicable to the Hindenburg since the highly resistive outer fabric and paint structure were supported by lightweight metal tubing (frame) grounded to an airfield anchor post connected through an electrically conductive nose tether line. If the Hindenburg had a rather uniform electrical charge over its skin, it probably would not have ignited except for the electrical grounding line allowing an abrupt strong electric current flow through the aluminized paint layer."

<u>The Crown Fire Phenomenon:</u> Clearly hydrogen, once the gas cells were breached, contributed to the airship fire. However, most of this activity would occur well above the airship surface. It is interesting to examine the nature of forest fires as discussed in a number of papers by the Montana Fire Science lab. The phenomena of a form of crown fire called, "plume-dominated fires" is of interest. "Plume-dominated fires are associated with relatively low wind speeds, usually less than 20 miles per hour. They can cause unexpected and serious problems in at least two different ways. The first, results when the size and intensity of the fire produce extreme turbulence at the base of the convection column. The resulting pulsing and turbulence increases fire intensity and heat transfer to adjacent fuel and hence accelerates fire spread. The process feeds on itself and accelerates as the convection column grows." The second way in which a plume-dominated fire can be extremely dangerous occurs when a downburst of cold air descends from the convection column and blows outward along the surface at high speeds in all directions. These winds can be extremely strong and very dangerous because of their sudden occurrence. If they encounter

the fire, which is very likely, the fire becomes wind-driven for a short but very violent period in which extreme spread rates may occur. Downburst is initiated by evaporative cooling of precipitation that cools surrounding air." Examination of the photo sequences of the Hindenburg fire show a definite hull shrinkage for a brief moment at amidships suggesting an inner hull partial vacuum perhaps attributed to the fire convection column. Then as the stern descends to the ground the bow ascends and a fire plume is evident. It is noted that the airship has an axial walkway through the center of the hull, a configuration that perhaps assisted in the fire process, in view of the rapid melting of the gas cells. The plume appears to be predominate on the starboard side.

Summary of thc Study: This section will provide a discussion of the author's conclusions with supporting findings, a scenario of probable events, summary statement and recommendations.

Conclusions and Supporting Justification

Conclusion 1: Atmospheric and airship conditions at Lakehurst were conducive to formation of a significant electrostatic activity on the airship. Justification: a) The buildup of electrical charge on an airship from dust and water particles during flight was a common occurrence. b) The induced electrical charge on an airship is a function of altitude and presence of cloud formations. c) The weather conditions at the time of landing and the attempt to make a "high" landing were key factors in the retention of high static electrical charge on the airship.

Conclusion 2: The Hindenburg envelope design was incompatible with the environment encountered at Lakehurst at the time of the incident. Justification: a) The grounding ropes appear to have had a role in establishing the airship frame at earth ground potential. The method of attaching the fabric cover panels to the frame and the non-conductive materials employed permitted the cover to retain whatever charge was accumulated. Dissipation of the charge on the cover could not occur commensurate with the frame discharge, thereby a high voltage gradient was created between the frame and cover. b) Laboratory tests conclude that the fabric make-up was highly resistive, allowing an extremely high threshold of static electricity to accumulate. c) The airship cover was not uniform, electrical tension is the highest at the bow or at the stern area. The aft strboard engine had a significant amount of ionic exhaust particulate, a gust of wind occurred from port side shifting the airship starboard. The breakdown static discharge on the fabric panel surface occurs when a certain capacitance threshold is reached. External ionic (exhaust) behavior is known to influence initiation of electron flow.

Conclusion 3: The envelope fabric and doping process employed on the Hindenburg was sensitive to arc ignition and very susceptible to promoting flame propagation. Justification: a) Laboratory tests confirm that the dopant materials used on the envelope cover panels were made up of chemicals that are known to be highly reactive. There are several known reactions that could have occurred singularly or in some combination. b) Laboratory tests confirm that the fabric and dopant combination used was sensitive to arc ignition at levels significantly lower than predicted levels at the time of the disaster. c) Tests based on research of similar conditions of electrostatic discharge with similar materials show that the metallic-based dopant is sensitive to arc ignition or electron flow. The rate of reaction/flame propagation is shown to accelerate at speeds compatible with the visual data of the airship disaster. d) Laboratory tests confirm that the fabric and dopant combination was very flammable. e) The extensive modifications done to the sister airship of the Hindenburg, the LZ -130, after the disaster are indicative of the knowledge the Zeppelin engineers had of the design flaws of the envelope fabric and doping process used with the Hindenburg.

Conclusion 4: The premise of free hydrogen within the airship hull is not necessary to justify the cause of fire initiation. Justification: a) The preponderance of evidence characterizing the fire color, heat and behavior indicate that indeed hydrogen was not a necessary fueling agent. b) The circumstances just prior to the fire (airship trim, crew observations) and extensive and adequate ventilation of the airship discount the concept of free hydrogen as a probable condition for the cause of fire initiation.

Conclusion 5: Hydrogen, as well as airship materials of construction contributed to the resulting conflagration. Justification: a) The materials of construction, such as cotton/linen, wood, ramie cord, adhesives, celluloid, thread, etc., were moderately flammable. The additives of the metallic-based dopant were highly reactive and served as the predominant fire agent and accelerator. b) The remaining hydrogen at the end of the flight came into play as each gas cell was subsequently destroyed.

Scenario of Probable Events: Briefly, the following account of the disaster is proposed:
The origin of activity starts in the upper-starboard-stern quadrant as a chain of reactions of electron flow is relieved from the voltage potential pressure. One or more locations on the somewhat irregular cover surface rapidly break down either from cover to frame electrostatic forces or triggered from a wave of ionic engine exhaust gases, or both. The resulting electrostatic discharge, or discharges, initiate a reaction with the chemical nature of the doped fabric that accelerates across the panels in a number of directions setting up rapid combustion of the outer envelope. The intense heat flux heats up cell 1 and 2 to the point that the relief valve, either slow in reacting to the gas build up pressure, or unable to handle the flow of gas bursts the cell structure sending a lateral shock through the airship. The wind, having shifted, assists in the flow of fire toward the amidships more to the starboard side, rapidly burning through spot locations such as above cell 4. Burn through occurs on the top, on the top port side, on the upper fin and on other nearby surface areas almost instantaneously. Cells 3 and 5 disintegrate. Partial burning of the fuel-rich hydrogen has already entered into the fire equation. For the same reason cells 1 and 2 burst, the much larger cells 6 and 7 burst sending up a mushroom plume of fire and burning the structure while breaking the frame structure in two. The inadvertent loss of ballast toward the bow causes the airship to nose upward while the amidships and stern descend. The bow is rapidly engulfed in fire as it races around the airship in an omni-directional pattern. The entire event takes place in less than a minute.

Summary Statement: The Hindenburg did not explode; it did, tragically, catch fire but burned very rapidly, not in an upward 'Bunsen burner" mode (characteristic of hydrogen), but in an engulfment omni-directional mode. It was this aspect that prompted the author's curiosity that led to an exhaustive investigation. Also observation initially of photo footage and a review of eyewitness reports suggested other problems with accepting hydrogen as the sole contributor to the disaster. For example, the 240-ton airship maintained static trim for many seconds after fire initiation. The stern of the airship should have descended rapidly if, in fact, there was a major loss of hydrogen. Sheets of flaming fabric fell to the ground and did not self-extinguish Eyewitness accounts of the fire colorization describe it mostly as yellow-reddish. The hydrogen flame is nearly invisible. Unquestionably, hydrogen is an essential element to humanity. Other than its ubiquitous presence in the makeup of the universe, including planets, plants and animals is the fact that in the last few decades hydrogen has received serious attention as a future fuel and energy carrier. Touted as the forever fuel, it is always available, environmentally benign and safe as evidenced by its industrial safety record. However, the Hindenburg incident continues to instill fear in the public, as it is glamorized by movie films, videos, and still more books on the subject every year. On the other hand, pro-hydrogen advocates tend to diminish the "apparent" cause-and-effect role of hydrogen in the Hindenburg tragedy. These situations are inappropriately misleading current and future generations. To illustrate, the following sample quotations are referenced: - March 1995, Los Angeles Times: "The 36 who died were killed in the fall, not from burning hydrogen, which simply floated away." 1993, the Madison Press: "But there was one flaw in the giant airship's construction: the sixteen gas cells that filled its enormous frame were inflated with over seven million cubic feet of highly flammable hydrogen gas." 1969, Clean Air Now: "All deaths were caused by either burning diesel fuel or from people jumping to their deaths." 1937, Universal Newspaper: "The inferno of blazing hydrogen gas cremates all living beings in its path as the twisted skeleton of the balloon settles quickly to the ground."

The study should not only be of professional interest to the engineering (materials science) academia, but provide a basis to share with the public and the community of hydrogen experts, accurate information, and perhaps a more realistic account of circumstances surrounding the Hindenburg disaster. Contrary to some sources, the Hindenburg was not designed for helium exclusively. The idea was to use the nonflammable helium as a fixed buoyant gas and separate hydrogen-filled gas cells (internal to the helium cell) as the variable buoyant gas. The expensive helium could be retained and the hydrogen volume adjusted as needed for static lift. Despite the arguments to reduce the fire hazard by adopting helium or other nonflammable mixtures, the Hindenburg operation may still have become a casualty of overconfidence. It is this aspect the author wishes to focus on and bring to the professional community, namely the relationship of materials compatibility and environmental exposure. The development of advanced airships is currently underway in Germany and is under study by other countries. Whether hydrogen would ever be used again is questionable, but not unrealistic. The concern with its use is one of perception more than technical. Many advances have been made in hydrogen technologies and materials of construction. Although there are currently vast reserves of helium, the preferred lifting medium, it, like the petroleum reserves, is finite.

Recommendations: 1. Future airships should not be designed with flammable materials contiguous to potential sources of ignition. 2. Hydrogen should continue to be considered for utilization where its properties suggest benefits, remembering that it is flammable, but not eliminating it because of irrational fears of explosion.

James H. Collier, veteran of ZR-2 and survivor of ZR-1's wreck, saved this large piece of LZ-129 fabric blown clear via the rupture in the cell V/VI area. Some edges show tearing, other are singed. This laced piece was right at the equator; half has iron oxide as the first coat. Addison Bain acquired this and other artifacts from Dave Collier, who, as a boy, had gone aboard the LZ-129. Author's photo.

Chapter Ten: The Long Road Back, Then and Now

By the November '97 issue, *Popular Science* magazine had run the story, with a flaming LZ-129 on the cover. Mariette DiCristinia, who interviewed Addison Bain for the article, included a mention of the active e-mail debate in progress. Unhappily they also ran a graphic of the final LZ-129 configuration while using the caption from its original dual-gas design. This writer sent in a correction, which they ran in the next issue. However, the article pointed out that Bain's thesis had at least one vocal, knowledgeable opponent. The hydrogen-as-culprit belief had been entrenched like dogma, and no iconoclast was going to sweep such a religion away so easily. In one message, an opposing viewpoint took exception to the "neon blue glow" having actually been visible corona discharge. He posted to the e-mail list, "My handy college chemistry book describes a hydrogen flame as 'an almost invisible blue flame'... The same starboard witness [presumably Heald] reported that within less than a minute, a big burst of burning hydrogen then appeared about one third of the ship's length from the stern -- consistent with port side witnesses."

One could contrast this with the bulletin published by Kennedy Space Center's Technology Programs and Commercialization Office: "Firefighters who deal with hydrogen or alcohol fires typically have to rely on the antiquated 'broom method' to locate the invisible flames. This involves holding out a corn straw broom, waving it around an area and hoping it catches fire first." Indeed, the Material Safety Data Sheet warns, "In addition hydrogen burns with an almost invisible flame, and personnel may be injured by the flame because it is difficult to visually detect." In 1996, two engineers at Stennis Space Center developed a new fire imager system. "The fire imager...locates the burning flames." Firefighters have *not* developed a technique for defining location of invisible flames based on secondary ignitions of local materials. If a local material itself would burn as it pleased, a hydrogen fire therein could not be defined as a result.

That was only the beginning of the flaming rhetoric. Another argument was made that, even if it was not in on the initial ignition, hydrogen was certainly responsible for the speed of the ship's destruction. Given hydrogen's nature, however, this simply is not so. As Carl Wagener[305] put it in an e-mail posting, "The extraordinary complexity of a fire springs from these three components [in the fire triangle] which are themselves affected by a multitude of variables. It is therefore impossible to describe a fire quantitatively or to predict its course exhaustively... The fuel itself influences the fire situation in several ways. Parameters of primary importance affecting the behavior of fuel in a fire include position in the environment, "built-in" state, form (shape, thickness, surface characteristics, distribution, density, etc.) and finally physical and chemical properties (ignition and flash points, thermal conductivity, specific heat and heat of combustion, etc.). This variety of parameters explains why the fire performance of a fuel is only partially characterized by its chemical and physical properties and is therefore not an intrinsic property of the material - contrary to frequently stated opinions."

One might visualize part of this concept by the example of trying to light a newspaper with a match. If a match is lit on top of a rolled-up newspaper, little more than scorching can be expected. If the same newspaper is crumpled into a puffy column and the same match is lit under an exposed edge at the bottom, the paper is history. In both cases, the T_{ig}, (ignition temperature), the fuel, the available oxygen, and the ignition source would be identical, but the results would

---

[305] Courtesy Carl Wagener, Chief Scientist, Hermit Technologies, Inc.

be dramatically different. Flag-burning protesters never try to burn sheathed flags for two reasons; even staged acts of protest have to obey the laws of combustion.

Perhaps all the inflammatory remarks could be curtailed if there was some sort of simulation where properties of hydrogen and the other materials could be fed in so their behaviors in the fire might be tracked. One US Air Force Research Laboratory proprietary computer program was used to run thermal calculations based on parameters from the *Hindenburg* coating. "Flame temperature of the cellulose butyrate / castor oil / aluminum paint is 3,347°F, probably hot enough to allow effective flake aluminum combustion. This high temperature flame characteristic with aluminum in a seemingly non-energetic binder appears incongruous. What is happening is that the high oxygen content of the cellulose butyrate acetate binder material can become sequestered by the aluminum producing very high flame temperatures producing aluminum oxide and large amounts of the byproduct graphite. What this means is that the "safer" [airship] paint might be almost equally dangerous in producing a self-propagating fire as a nitrocellulose composition. Adiabatic flame temperature is so high that burning of the *Hindenburg* skin would likely be lead by the aluminum paint self-combustion followed closely by air oxidized burning of the remaining preheated fabric and iron oxide layer."

The Air Force report seemed to support the author's statement in the *Buoyant Flight* series that most any of the rigid airships, including our helium rigids, would have wound up the same way as LZ-129 had their fabric caught fire. More study, however, proved this was misleading. The other rigids would have burned much the same way, true, and would have no better chance for survival. But only the LZ-129 had a fatal flaw that made her vulnerable to the environmental conditions that fateful night, especially sensitive to ignition of the covering.

Likewise, one can understand why the highly flammable outer covering had repeatedly survived ignition sources that seemed hot enough to instigate combustion. Lightning bolts burnt holes, and chars from the engines bored holes. Yet, as history shows, if a sustained electrical current is applied long enough, outer cover ignition occurs. In December of 1916, the Navy ground crew was attempting to dock L24 into one of the Tondern sheds. A wind gust whipped the airship upward into an overhead electric light, smashing it in such a way the naked conductors intruding into the fabric started it burning.

Without the proper conditions of conductivity along the fabric grain, lightning strikes and engine cinders were no more than a maintenance headache for workers, assigned to climb, ironically, the 80-foot Magirus fireman's ladders purchased for the purpose. Finally published, the late Dale Topping had noted Arnstein's looking into electrostatic conditions being out of the ordinary the night of May 6th. He disagreed with Klaus Pruss, son of the Captain: "...noting that because of electrical storms in the area, all rubber factories in New Jersey were closed down on that day because there was too much atmospheric static electricity for employees to work safely (a static spark could ignite the factories' carbon black and rubber compounds). This was a very uncommon situation."

Meanwhile Bain's research was reviewed by Professor Veziroglu, of the University of Miami, prior to acceptance by the IHEA committee. The story was published in the International Hydrogen Energy Association journal, in collaboration with Professor Van Vorst of UCLA. The documentation had called out the fact that red iron oxide had supposedly been painted on the underside of the fabric above the ship's equator. The top-only coat was added to slow down ultraviolet degradation of the fragile cell fabric. None of the LZ-129 samples obtained had such an undercoat, but Kent O'Grady's working to get British samples and documentation showed the R.101 had red iron oxide as the first coat on the entire fabric cover. If it were true that LZ-129 had even half its covering painted with this stuff, there would have been additional problems with the fabric. In "Propellants Manufacture, Hazards, and Testing", authors Boyars and Lager discuss additives that can improve the burning of aluminum fuel. "A 10% increase is obtained by adding 2% ferric oxide." So if things weren't bad enough, the complete formula for a solid rocket motor was supplied from the LZ-129's equator to the top spine, bow to stern. In March 1998, S.R. Skaggs, Ph.D., wrote Dr. Bain of research at the Los Alamos National Laboratory that showed aluminum flake and iron oxide would also yield an oxidation-reduction reaction. While much is still classified, Skaggs verified the bright orange color reported, together with rapid surface propagation, was consistent with what are called ballotechnic reactions. The corona discharges were hot enough to initiate this reaction, which at three megajoules per kilogram of reactants, yield *more* energy than similar *pyrotechnic* reactions.

If powerful explosive reactions were not bad enough, there was more. "Aluminum powders are used in explosives...[but] when mixed with iron oxide...forms *thermite*, employed to weld heavy sections of metal. The temperature of burning thermite can reach 5,000° F. (2,760° C.)"[306] Common uses for thermite include destructive strips glued to classified equipment, as well as beam-shaped charges placed between steel railroad rails to weld them together. Oh, that a section from the *upper* cover had been available from the collectors for testing.

The project got a lucky break when Addison Bain visited the Florida home of a collector who claimed to have a piece of the *Hindenburg* cover "that must have been from the red swastika flag part" of the vertical stabilizers. Instead, the roughly foot-square doped cloth was actually a long sought section of *upper* hull covering. Held by a relative of a man on the ground crew, the veteran had picked it up and kept it sealed for many years. It could have been from one of the pieces seen raining down after the midship cell blowout, captured in the famous photos. Bain met their price, and it was back to the old stomping grounds for more study.

First off, it was obvious the literature had been wrong about iron oxide being on the inside. A moment's thought would have revealed the described process would have taken special scaffolding inside the ship, and some scheme to get around all the steel wiring, to paint on the underside. Rather, as the sample showed, the iron oxide was actually *just the first coat*. Layers of the acetate-based dope were then painted on top of the iron oxide. This was an important discovery. Under the microscope, it was seen that in dozens of places on the small piece, the more acidic outer coats had bled though and allowed the aluminum power to contact the iron oxide. Wherever and whenever this happened, thermite had been formed.

The "Analytical Research of Covering Material of Zeppelin Airships LZ 127 *Graf Zeppelin* & LZ 129 *Hindenburg*" from 1937 said the alkyd/acrylic phthalocyanine pigment added to deter corrosion of the aluminum structure was flammable, exacerbated since "it is dissolved by the cellon-lacquer of the covering material and adheres to the material." Thus, even the final coat of dope, which contained the heaviest concentration of aluminum dust, was so acidic as to soak through the other dope coats, which were already into the iron oxide coat. These had already soaked through the fabric so as to melt into the girder's primer paint. There was no way to obtain the exact result of those melded sections for testing. However, parts not in direct contact with the melted-paint girders were represented by the samples. In the final analysis, the aluminum-in-

---

[306] J. H. Meidl, *Flammable Hazardous Materials*, MacMillian Publishing, 1978.

cellulose substrate that was the upper outer cover is not unlike the aluminum hydroxyl polybutadiene used in *Polaris missiles*! Roughly similar to the Solid Rocket Motor stuffing we use to launch the Space Shuttle Orbiter today, the Zeppelin covering and SRM compounds are surprisingly comparable. The recipes are similar even down to small percentages of iron oxide, used as a catalyst in SRMs. The cast material contains both fuel *and* oxidizer, engineered to create a continuous combustion that *cannot be extinguished.* Like the Trident missile filmed bursting from underwater to burn until in outer space, the SRM provides its own oxygen.

As any firefighter will tell you, water only makes these metal fires worse, it cannot cool them and break the fire triangle. The only way to fit such a Class 'D' fire is with PkP powder, which attacks the fire chemically. A big bubble of pure hydrogen or helium would not snuff them out. Even enclosing it in a vacuum would make no difference, hence their use as missile strap-on boosters and satellite orbit insertion motors. How are these Solid Rocket Motors ignited? Electrical ignition, a'la a bolt of lightning, is provided by a pyrotechnic igniter. One can easily understand why the tallest structures for many miles on the coast of Central Florida are the lightning rods atop Pads 39 A and B. Manned space flight might be set back for decades if a stray bolt out of the blue ignited one Solid Rocket Booster (SRB).

Sadly, safety lessons are usually written in blood, and the SRM manual is no exception. A 1985 *Pershing* missile catastrophe placed renewed interest in the behavior of electrostatic discharge around Solid Rocket Motors. Robert Geisler, a SRM expert at Edwards AFB, explained the accident involving a rocket segment whose chemical makeup was very similar to the LZ-129 covering. Workers removing a shipping cover on a cold and dry German day inadvertently created a large charge of static electricity. The invisible charge found its way across the open grain SRM casting. Later tests showed the static discharge had ignited the aluminum powder in the propellant mix. The SRM segment, designed to burn at 25 meters per second, did so, killing the workers. In December of 1987, 5 men were killed and a facility utterly destroyed when, during a core removal, electrostatic discharges ignited an explosion of the ammonium perchlorate propellant. In May of 1988, a massive ammonium perchlorate explosion killed two and injured 350 as it destroyed not only the facility but also the surrounding community. The French SRM program had accidents; the Russians lost a nuke sub in 1968, the K-219, when missile propellant exploded. The high reliability of Solid Rocket Motor ignition - as well as other fatal accidents that resulted from static electricity causing accidental ignition - is a function of the ease by which their aluminum fuel is set alight *electrically.* In a 1943 US Bureau of Mines Report of Investigations, [307] various metal powders were subjected to electrical ignition tests. It took only *two tenths of a joule* (1 joule = 0.00095 British Thermal Units) to ignite aluminum powder. After sometimes fatal lessons, a microscopic examination of the rocket fuel lead to it being made more safe by making it more difficult to ignite accidentally. The aluminum powder fuel is now manufactured in tiny uniform *spherical* shapes. Geisler suggested the Germans would have had only non-uniform geometry particles, meaning microscopically tiny flakes; these are especially sensitive to electrical ignition.[308] Verification came when translated reports show the Germans polished their aluminum flake-powder with an oily lubricant, then washed it with gasoline. This certainly did not help make it safer, because the process may have at least partially removed the protective oxide layer on the flakes.

Burying the fuel in a matrix of nonconductive chemicals did offer some protection against accidental ignition of the powder. However, if the voltage (electrical pressure) is great enough to

[307] "Inflammability and Explosibility of Metal Powders," USBoM, October, 1943

[308] The French SRB program independently suffered similar problems and reached the same conclusions.

overcome the overall resistance of the SRM material, ignition occurs. Unlike simply raising the material to its ignition temperature, the electrical energy sets off a chain reaction. As an AIAA paper explained, "A severe hazard arises when...propulsion systems are charged... Discharge processes... reduce the impedance of the energetic material and result in a rapid current increase. This can lead to arcing, establishment and growth of discharge paths, followed by catastrophic discharge. The resulting pressure and temperature increase in the very narrow discharge path may induce ignition and combustion." [309] Geisler verified the close proximity of the iron oxide would have made a voracious thermite-reaction fire once the fabric had been electrically ignited. In 1998, Bain's work was critiqued by the Electrostatic Discharge Association

Those charged with solving the puzzle 60 years earlier did not have the benefit of computer simulations, but they did have some idea how the materials worked. As Gordon Vaeth had revealed, Professor Dieckman had at first casually admitted the dopant had been changed in between LZ-127 and 129. The NASA lab test results had said "…a cellulose acetate butyrate dope plus possible nitrocellulose dope present on the fabric." Bain's research included correspondence with the Herberts Company; they had supplied the 'cellon' – actually, cellulose butyrate acetate - used in the LZ-129 mix. Herberts verified cellon contained butyric acid, or "butter acid." Missing from both the *Grafs*, it appears this chemical element butyric acid was a key ingredient in the mystery.

Adopted for the new ship, the cellon was an improvement over earlier paints because it made the finished product more resistant to water absorption.[310] That was quite an important consideration, given the airship's more than *six acres* of surface area. Even an ounce saving of water retention per square meter would multiply into fewer control headaches. On one South American trip, the LZ-127 had entered a seemingly dry cloudbank but suddenly became tons heavier. After a near disaster, the cloud passed and the blistering tropical sun rapidly evaporated the weight and the ship became much lighter. With the limited control allowed by dumping ballast and venting gas, such episodes added 'better water shedding' to the wish list as well, even at greater cost.

A non-scale comparison of LZ-127 (left) and 129 shows *Graf's* off-center lower walkway, atop the fuel gas cells, vs. the *Hindenburg's* axial corridor, designed to allow access to what was supposed to be the inner cells. The smaller *Graf* had about the same range owing to the ingenious balance of lift and motive fuel. Courtesy Stefan Bauer and Walter Lion.

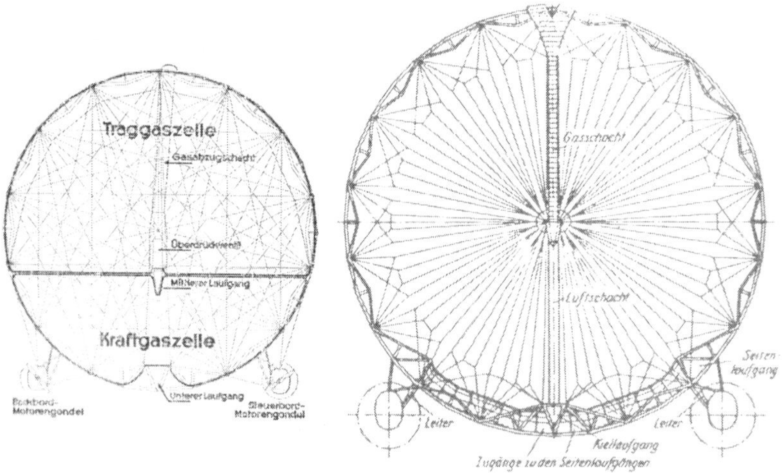

"Analytical Research of Covering Material of Zeppelin Airships LZ 127 *Graf Zeppelin* & LZ 129 *Hindenburg* " showed the LZ-127 layers consisted of cellulose acetate. The LZ 129 shows a different mixture, i.e. with "butter-acid," a key ingredient of cellon. However, the more water-

---

[309] "Current Assessment Methodology for Electrostatic Discharge Hazards of Energetic Materials," AIAA Journal of Propulsion & Power, Vol. 7 #6, 1991.

[310] The covering's hydroscopic nature is shared with the Space Shuttle Orbiters. Both the Low and High Temperature Reusable Surface Insulation must be re-waterproofed after coatings are burned off during the mission.

resistant formula had the effect of *greatly increasing* the overall impedance (electrical resistance) of the completed fabric covering. With only a fraction of today's knowledge of atmospheric electricity, it would not be fair to judge the Zeppelin builders for this seemingly innocent side effect of such a desirable change. Indeed, it might have been noticed, and ignored, or even *welcomed*. Remembering that 1936 *Literary Digest* article which said: "Makers of the ship followed the principles laid down by Faraday, who proved that electricity prefers to travel on the surface of an object. When lightning leaves a cloud to light on the airship, it remains on the surface until it can find a lower potential - usually another cloud. Then it hops off again." One could easily have reached the conclusion the increased impedance was *desirable*, adding an extra layer of shielding to protect the components below.

Likewise, the art of building an enormous "Faraday Cage" from more than ten miles of girders depended on clean aluminum joints that were riveted tightly together for perfect continuity. It's likely no one would have thought important that once the alkyd/acrylic phthalocyanine pigment protective coating was applied to seal this beautiful work against galvanic and other corrosion, it was also adding increased resistance between the outer cover and the structure. As stated, Austin's "Lightning Investigation as Applied to the Airplane" says "This goes to show that the fire hazard is negligible where a path of *high electrical conductivity* is provided." Here then was the *Hindenburg's* fatal flaw: the outer covering, holding a charge on its capacitor-like surface, offered too great a resistance to electric current flow. It could build up a charge that the non-conducting ramie cord could not carry away. Two capacitors glued together with more capacitive material would not create a conductor. Even if the entire mess was wet through to the ramie cord, the new blue primer would not allow the charge to carry away. Once it had accumulated enough charge to overcome its overall impedance, the covering was an accident waiting to happen.

Informed, Dieckmann's translated report could never be read the same way again. He had noted the problem was directly related to a change in the fabric's doping. His widow explained he knew what had happened, but calling out the specifics would have gotten him in trouble with Hitler's government. Armed with this new knowledge, we can look at Dieckmann's report as if he's *encoded* the truth for us to find at a later date. Oddly, the first indication that something amiss as a result of the dopant change had actually been published even as the author met Bain in 1994. In the German language, this astonishing statement was made public: "Dieckmann found the new hull material 'cellon' had no electric conductivity, ...could have been the reason for the *Hindenburg* disaster... considered a graphite painting on the inner side of the hull to prevent arise of voltage differences..." [311] Yet that book was generally unknown in the U.S., and no English translation was available.

Unaware of this published revelation, Bain continued to work with the electrical experts on the problem's solution. The lessons of the Solid Rocket Motor accident revealed the explosively flammable material was judged relatively safe to handle if it exhibited less than 10×10^{10} ohms per unit measure. Greater than that resistance would make it vulnerable to electrostatic discharge ignition. The KSC lab's electrical tests showed the LZ-129 outer cover measured 10×10^{13} per unit square, or *one thousand times greater* than the "safe" level. A lower landing, or clear weather, or even calmer conditions might have postponed the accident. But, a newspaper would summarize that, on May 6th, 1937, it all boiled down to *the wrong dope*!

---

[311] M. Bauer & J. Duggan, LZ130 *GRAF ZEPPELIN* UND DAS ENDE DER VERKEHRSLUFTSCHIFFAHRT, Robert Gessler Gmbh & Co. KG, Friedrichshafen, 1994.

The Bain explanation got some additional press, including the New York *Times* and the London *Times*, and even a mention in *Scientific American*. Bain co-authored a report for the German Hydrogen Association. A well-connected Chicago newspaper heard of the study and ran a story as if it were the first to break the news in the U.S. "IT WAS THE PAINT!" their headline shouted. It's hard to beat such an oversimplification, but we actually borrowed a Sherwin-Williams television commercial's lament for our title here. If only the Germans of 1937 had been given access to all this modern advanced test equipment, one might daydream. Less the politics, they might have been able to use the LZ-130 and return to transatlantic passenger service. Incredibly, as Bain would discover on his next trip to Germany, the airshipers *had* figured it out.

In a report dated Oct 28, 1937, a Dr. Lueder of the Wireless Telegraph and Atmospheric Electrical Experimental Station in Grafeling, Germany told the airshipmen their results. "The basic cause of the catastrophic fire is ultimately to be regarded as the poor conductivity of the aluminum paint coating of the outer skin and the good insulation of the blue anti-corrosion paint of the framework. The danger has increased presumably in recent years by the progress of the paint manufacturing industry, which has been to an increasing extent continually concerned with producing the best possible water shedding - and thus highly insulating - paints. The considerable resistance reduction of the landing tow ropes was made possible because the hemp ropes, each of which was attached to a wire rope, when thrown out during side winds and rain spray could become moist in their entire length. If the landing tow ropes had been attached to the airship's hatches directly, then the total conductivity value of the tow lines probably could not have reached the required amount for discharge, or only then after the ship had been hauled down to a lower height, when the field strength would not have been sufficient."

So ended the burning mystery forever. The Germans had figured out every aspect of the accident's true cause just a couple of weeks after the accident investigation team got back to the Fatherland. The highest voltage charge, collecting at the stern, waited until the bonded frame became grounded through the nose lines. "A reinforcing strip of balloon cloth about 4 inches wide is sewn to the inside of the cover panel and centered over the longitudinal wire..." said the rigid's manual. We can guess the best path to the most attractive ground was not through the ramie cord and through the insulating primer, but probably through the structure wiring, sown into the fabric to reduce flutter. The steel wires would not have had the girder's blue primer applied, and their ends were certainly electrically bonded to the structure. Just as lightning bolts are crooked, it is unlikely nature could provide a perfect straight-through shot from highest voltage to the best ground. The discharge undoubtedly cut across a *length* of fabric grain. This initiated the rapid breakdown, and started the bright conflagration that has had such a profound effect on our civilization.

The Germans knew it all along! Different dope, a more conductive primer, standard low landing, or even just clear skies; any one thing being different, and the waiting passengers would have boarded the *Hindenburg* and been borne by hydrogen back to Germany. *Graf Zeppelin* would have been turned around for another trip to South America, and transatlantic – possibly even transpacific - air travel would have been more luxurious for many years afterward. So the Zeppeliners were faced with applying the lessons learned to the next airship, to be certain no such thing could ever happen again. There are two sayings in the aviation business: You start with a bag full of luck and an empty bag of experience. The trick is to fill the bag of experience before you empty the bag of luck. LZ-129 fit the second truism, in which good judgment comes from experience. Unfortunately, the experience usually comes from bad judgment. Sadly, under that regime, no one was going to admit a mistake in design or construction.

Harold Dick wrote, "On May 6th, 1937, the construction of the LZ-130 was far advanced. The basic hull structure was complete and work was being done on the fins and rudders... Then immediately after Dr. Eckener's return from the United States, work commenced on altering the LZ-130 for helium operations." Only overtime was canceled, regular work continued. A 1938 world aviation wrap-up report said "...the accident which caused her destruction in 1937 did not deter the Zeppelin Company from its program of new building."

Inflated with hydrogen and doubtless the safest rigid airship ever built, LZ-130 *Graf Zeppelin* was history's largest sacrifice to save face for a political regime. (Inset) The twin two-blade *luft schraube* on engines *eins* & *zwei* were later replaced with experimental ground-adjustable three blade units. The jet-engine-like tail cone housed the *Kühlluft* for cooling the water-recovery and engine cooling radiators. Courtesy Hepburn Walker, Jr. and John Mellberg.

Too late to make it a fireproof metalclad, the Germans would leave no stone unturned in their quest to be sure LZ-130 could not suffer the same fate as its predecessor. Armed with the information not released to the public, the basic fabric itself was first to be scrutinized. The NASA lab's spectrograph showed a dramatic change for the LZ-130 fabric: the readout showed high peaks of sulfur and calcium not present in the earlier ship's samples. Calcium sulfamate is a *textile fireproofing agent.* Next, the chemical mix would have to be improved, offering the assembly more conductance. The NASA lab IR tests suggested changes were made in the dopant: "LZ 130 (Ganz), ...does appcar slightly different, perhaps the acetate without the butyl component..." i.e., no more *butter acid.* Yet, one more item proved to be the most telling of all.

LZ-130 blueprints Bain was shown at Fredrichshaften specified *bronze* to be added to the cover mix. How would bronze justify its extra weight even in an airship designated for the reduced lift of twice-as-heavy helium? That same BoM study showed that, even in oxygen, bronze *did not burn at all.* However, bronze-aluminum was more than just fire resistant; it was a master stroke, much more advantageous than the aforementioned consideration of painting conducting graphite on the fabric's initial coat. Remember, aluminum suffers from accidental electrical ignition from extremely small current flow. The 1943 US Bureau of Mines Report of Investigations reported that copper, bronze's major component, could not be electrically ignited with factors of ten times more energy than ignited aluminum. The lab's rig did not have enough power to ignite the copper in their test setup that had showed that, for example, steel could be electrically ignited and burned at 0.76 cm/s. Papers in the Karl Arnstein Collection[312] include a letter answering Knut Eckener's request for the Alcoa version of bronze-aluminum. A note to the Doctor dated 11-22-38 says they have obtained necessary papers and will ship "Aluminum bronze powder, Alcoa #422, extra fine," noted to be "2 lbs. used for pigment in varnish…" Now not only more fire resistant, there was no chance of the cover ever being ignited by the most powerful currents imaginable arcing with the grain. With bronze-aluminum mixed in their final coats, and the probable deletion of the "cellon" or at least the "butter acid" from the dopant mix, the individual LZ-130 panels would have been better conductors, and less capacitive-retentive. Yet, there was still a problem about conductivity in the cover's lacing to the framework. Would they have to paint a coat with graphite as first suggested to overcome the conductivity problem?

Instead, Harold Dick witnessed what seemed an odd new process. Workers doping the ramie cord had *graphite* mixed into the slurry. "…the lacing cord was prestreched and run through dope as before, but the dope for the LZ-130 contained graphite to make it conductive. This would have hardly been necessary if the static-discharge theory was a cover up."[313] But this had made little sense before the bronze-aluminum adoption was discovered. Bain's electrical experts verified the highly capacitive LZ-129 panels would *not* have carried their charges to the frame, even with graphite added to their lacings. Finally, the blue primer coat would have to be added before the hole drilling, or removed from the lacing holes prior to the ramie cord penetration.

With graphite added to LZ-130's ramie cord, then with the cord used to attach the bronze-aluminum covering to the structure, and no blue primer or a conductive primer in the holes, a complete current path was created. The newly conductive lacings tying the now easily-conducting panels to the structure *electrically bonded* the fabric together so the covering was more like, as well as part of, the "Faraday Cage" formed by the riveted structure. The slightest difference in electrical potential would have been conducted away immediately.

Obviously taking no chances the German Commission theory was possible, that fire could have started invisibly inside the vertical fin where loose gas would have logically accumulated, an entire line of vents topped the new vertical fin. With their entire future riding on this one airship, there obviously was not complete confidence that static charge was itself the singular cause of the outer covering's ignition. Dale Topping wrote, "Klaus Pruss, son of the captain, noted later that the electrostatic conditions were no worse at Lakehurst that day than had been encountered by the *Graf* and the *Hindenburg* before." Obviously high landings were going to be allowed for this time; what seems to have been the "last straw" was addressed, in the form of one very noticeable change to the power cars.

New Zealand's Lighter-Than-Air Institute[314] observed, "Another new feature was the use of ZF-designed 2:1 spur-reduction gearboxes fitted to the LOF-6 engines - these drove two pairs of two-bladed Heine laminated timber props mounted on double hubs, giving the appearance of four-bladed propellers. These were found to give a greater propulsive efficiency and would also generate less ionization. Later GX-Heine three-bladed laminated timber propellers were

---

[312] University of Akron Collection, courtesy Eric Brothers.
[313] Dick / Robinson, Pg. 149
[314] Lighter-than-Air Institute, PO Box 74-420 Market Rd. PO, Auckland, New Zealand.

developed - these initially experienced vibration problems which threatened to damage both prop and engine. This was eventually corrected and they were fitted to the rear engines of the ship experimentally. Had operations continued all engines would probably have been so re-equipped."

LZ-130 plans specify the metal vent hoods along the top spine (right) were to be seamless, a logical evolution when visible electrical manifestations – not always as harmless as St. Elmo's fire – were known to develop around discontinuities on a hull body. The twin dual vents, each hood topping the vent tower and the interhull area, appear identical to LZ-129 (left) in photos. Visible atop the LZ-130 at the anti-flutter wires sewn into the panels via a strip of reinforcing cloth. Photos courtesy James Shock.

Why, after 118 Zeppelins had enjoyed the efficiency of pusher style cars, were the LZ-130's turned around? Zeppelins up to an including the LZ-129 tried various generations of water scoops to take on ballast weight from bodies of water encountered in route, and of course there were the successful rain gutters as well. The published explanation cites the needs of the engine's water recovery equipment, yet von Schiller wrote *before* the LZ-129 accident, "As with the use of liquid fuel each kilogram of gasoline or oil costs one cubic meter of hydrogen to restore equilibrium, methods are today being devised to overcome even this difficulty… we shall soon be able to condense the water-vapor in the [exhaust stream] air." No easy task, since unlike the American airship's use of hydrogen-rich gasoline which produced water, the diesel had no hydrogen and would yield not a drop. The American systems took advantage of the exhaust heat rising, the water's weight allowed a gravity feed back into tanks near the steadily emptying fuel tanks. The ZRS ships had the radiators mounted from engines almost all the way to the equator, and the later Mark IV model developed at Lakehurst had actually achieved better than 100% efficiency. Keeping this bulky plumbing inside the car did not seem to make good sense, when the water would have to be pumped back up into the keel. Did this arrangement have an additional design goal?

Under a photo of the new car, the VDI paper hints there was possibly a special electrostatic relationship to the cause of the loss of the LZ-129. The LTA Institute continued: "The propulsive arrangements for LZ-130 involved several departures from previous Zeppelin Company practice, in view of the anticipated helium inflation, then the reversion back to hydrogen. For the first time the engines were to drive tractor propellers and the engine cars were to contain apparatus for the recovery of ballast water from the engine exhausts. While the arrangement was stated to have aerodynamic advantages, it may have also been influenced by electrical factors. The propellers, particularly in dry air, generated an ionized slipstream. Since at that time the cause of the *Hindenburg's* loss was still being investigated (with static discharge a prime suspect), placing the propellers ahead of substantial metallic structures, namely the engine cars, could help in absorbing some of this ionization, allowing it to be harmlessly dispersed into the airship's framework. Thus any charge carried towards the craft's fins by the slipstream would be minimized; as the LZ-129 fire started in the stern area it would a logically prudent arrangement." Since von Schiller had written of their knowledge that electrical charges would be carried away from the ship in the gasoline engine's water-rich exhaust, this might be the final piece of the puzzle. Nothing specific has been published definitely fingering the engine ion discharge as the 'last straw' that pushed the voltage potential over the threshold to overcome the resistance and ignite the fabric. Yet, seems to be nothing else that demanded the unique and time-consuming power car design reversal and inclusive plumbing. Indeed, the change was worked out and built entirely within the period the LZ-130 was designated never to fly without helium.

As a result, the word 'luxury' took on new meaning as Dick wrote in a letter back home, "there is no question but that the *Hindenburg* was a luxury airship for the ship had such an excess of lift when inflated with hydrogen. The designers were certainly not careful to hold down the weight as they might have been and that was naturally reflected in the LZ-130. The operators wasted perhaps even more lift than the designers did by carrying excessively large operating loads. All of this is ended now and we shall get an idea as to what these people can do with all the experience in designing and operating when they really have to get down to business." Meyer explained what happened following an intense effort on both sides of the Atlantic to get helium. "With this American effort, in early September 1937, Congress amended the Helium Act, which President Roosevelt promptly signed. Once Helium was [to be] available, the Zeppelin Company turned to convert LZ 130 to its new limitations. Not only did the inert gas cause a 7 percent loss of buoyancy, it also required about six tons of additional operating equipment. Most of the savings in weight had to come off the net payload of the airship. This adjustment was made by cutting the passengers to 40 and crew by ten; by limiting structural equipment weight; and by somewhat reducing reserves and fuel water and passengers supplies." Her new pressure height would typically be 1700 feet, causing severe operating restrictions.

The Germans had laid in quite an expensive infrastructure to process the helium, and developed a plan to partially compensate for the lost lift. "Electric helium preheat was to be used to help in conserving helium; an 18° rise would take some 14 hours to return to ambient, which compensated for much of the fuel consumed in this time." [315] The Mannesmann company manufactured eighteen-foot long, 2.75 ton high-pressure cylinders made of ½" thick steel. The German freighter *Dessau* arrived in Galveston to off-load hundreds of these pressure "bottles" ready to be filled with helium.

Eckener foresaw helium consumption at about 1,800 cubic meters per trip, owing to popping valves and overall diffusion. He wisely recognized the gas would also get away when trying to purify it, estimating this loss to be about 7% of the volume handled, working out to about 1000 cu. m. per trip. His experience also allowed for unscheduled losses through rips, tears and unusual high temperatures, estimating about 4,000 cu. m. per trip. Multiplied by 30 trips, the loss rate came to an estimated 204,000 cubic meters per season. The optimist Eckener tried to put a positive spin on it by writing "...a sufficient profit is still realized, to prove that helium inflated airships are economically decidedly more advantageous than the former hydrogen inflated ships in respect of which a loss in gas worth 16,000 RM had to be taken into account." [316] Eckener was

---

[315] *Gasbag Journal* No. 12, June 1992

[316] Dr. Hugo Eckener, "The Zeppelin Crisis," reprinted in *Gasbag Journal* No. 21, September 1994

of course side-stepping that LZ-129 had 4 tons of gutters, plumbing, and tanks with which he once picked up 8 tons of ballast from a single mid-Atlantic storm.

When Professor Henry Meyer revealed his findings of German disdain for the dual helium-hydrogen design of 1932, most of the LTA community glossed over the implication: *The Zeppelin service couldn't operate at a profit with helium.* The tone of Eckener's late 1937 writing assumed it could be done *if* the price of helium could be lowered to a more reasonable level. How much subsidy he could expect, just so as his government did not have to end a highly visible venture on a failure note, was not guaranteed. Some people are convinced the fledgling nationalized airlines of the day, joined by Pan American Airways, were pressuring the Americans not to sell helium at any price.[317] Then, oblivious to what it would do to German-American relations, let alone the rigid airship, Hitler marched his troops into neighboring Austria on March 13, 1938. The very next morning Congress acted: "...the sale of helium gas to the German Reich shall be prohibited."[318] A signature on the helium shipping permit that had been considered a formality, not part of the foreign policy process – that of Interior Secretary Ickes – was withheld. Even a personal plea by Eckener himself fell on deaf ears. In a letter to Dr. Arnstein dated July 11th, M. Benson wanted to avoid any misunderstanding that his efforts to secure helium in Canada might result in a tie-up with Canadian or British interests. Canadian helium was not practical to extract; there just was not going to be any from another source.

However, back home, work was continuing on another engineering marvel for the LZ-130. Ballast-water recovery would have been useful all along, since hydrogen may be cheap, but it isn't ubiquitous or free. A system to recover water from the exhaust was even part of HMA No.1, the *Mayfly*. Stokes wrote, "The system comprised extension of the engine exhaust by 400 ft of thin section pipe along the keel and then forward again to the engine car, in the manner of the keel condenser current in steam launches, terminating in a water separator and storage tank. Trial in the shed gave water recovery of 52% of fuel weight, doubtless enhanced with the speed of flight improving heat transfer. Another trial was carried out in 1917, at Kingsnorth experimental R.N.A.S. station. As related by Wing Commander Cave-Brown-Cave, the contemporary doyen of British aircraft engine development. With a 240 hp engine, probably an "Eagle", a honeycomb radiator of 25 mm (1") tubes, 250 mm (10") long took the exhaust through the tube spaces, with the air from the propeller passing through the tubes themselves. 'The recovery was very satisfactory with the tubes clean, but decreased seriously after 10 hours running. Internals had stratified with virtually no deposit at the hottest zone, then dry carbon, oily carbon, and where water condensed, slimy oily carbon. A separator box with strands of hairy wool was adopted.'" Therefore, it was a good idea regardless of the lifting gas its ship employed.

At the 3rd Open Meeting of the German Academy of Air Transport Research about four months after that letter,[319] K. Decker produced his 40 page report, *Eimzelbericht am Traggas und Ballastsystem*, in which was detailed the many trials and results of the exhaust water recovery effort. A quick glance at the new car design plan belies the wonder contained therein; little surprise, as Dick pointed out, the long delay caused by perfecting the new cars and apparatus. What had be spread over the skin or hung in the slipstream of the American zeppelins was neatly compacted into the engine gondola itself. The hot exhaust gas was at first cooled by air, making it less likely an engineman would be burned, and the heat was passed out the car. The second

[317] J. Duggan, "Commercial Opposition to Supplying Germany with Helium," *The Airpost Journal*, Aug 1997 (333)
[318] H. R. 9855, March 14, 1938
[319] *Dritte Offentliche Sitzung der Deutschen Akademie der Luftfahrtforschung*, am 17.Juni 1938. Courtesy Dieter Schulte, DaimlerBenz Aktiengesellschaft.

stage was water-cooled at the aft of the car, the location of the cooling water and oil cooling *kuhler* (radiator). More compact than the below deck fan in the ZRS ships, a belt driven *Kuhllft* drew air through the car and, directed by vanes, out the drawn-out stern of the pod. Most forward was the condenser itself, and below it the propshaft-driven *Zahnradwasserpumpe*, specifically a gear-type higher pressure water pump, to move the *ruck wonnenes ballastwasser* back up into the keel's tanks. Perhaps the greatest wonder therein has gone completely unappreciated.

Soot in the engine exhaust was corrosive and quickly clogged the water recovery plumbing on the ZRS ships. The only organization to operate airships with water recovery, the US Navy, had a ready supply of captive labor. As a "reward" for various military malfeasance, sailors found themselves cleaning the sooty plumbing. Diesel engines would have made *more* soot, and L-Z had no free laborers to order about, nor the downtime availability. The solution was a cleaning apparatus.[320] The June 1938 report optimistically pointed out that if they did not get the helium, the airship would still be safer, since the hydrogen need not be vented to maintain EQ. As the weeks passed, it was clear the helium was not coming. The *Dessau* took the empty "bottles" back to Germany where they were distributed for other purposes.

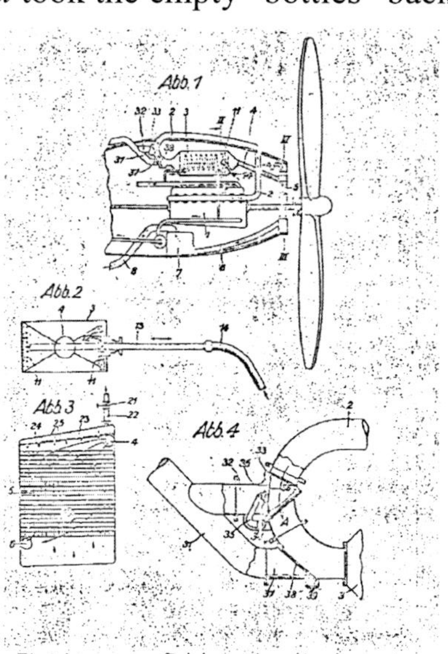

Einrichtung zum Reinigen einer Kondensanlage

If there was to be rigid airship flight, here was no choice but to inflate the new ship with hydrogen. Men bearing scars from burns suffered in the LZ-129 fire would be the first people who one would think would have a problem taking positions under giant gasbags that supposedly killed some of their shipmates. Obviously, they were instead convinced the problem had been solved, and the ship was safe to fly. Electrical gradients before launch and landing were also carefully monitored via the *Donnerwagen*. When the LZ-130 finally took to the skies, Dieckmann was aboard with sensing equipment. Tests were made on the ship's static charge during flight, and what would be discharged to ground upon mooring. There was a seemingly surprise discovery. The operation of her new water recovery equipment *reduced* the static charge to a considerable degree.

Thus equipped, and with her many other innovations, LZ-130 would have been the safest of the rigid airships. A 1939 issue of *Harper's* ran an article about transatlantic travel that season. A giant fold-out diagram (sought after by collectors today) detailed Germany's most luxurious entry, the LZ-130. Yet, civilian passengers would never be allowed on board. Waiting for some lift in the ban of carrying passengers, LZ-130 made some money carrying souvenir mail from its Frankfurt base to various cities, as far away as Eger in Czechoslovakia. Yet there would be one more military use left in the old Count's dream. Suddenly only radio detection gear graced her sumptuous athwartship passage. An old WWI spy basket was taken out of museum storage, refurbished, and used to obtain clear signal sampling during intelligence gathering missions. As if protesting the luxury liner's role as a spy, the missions gathered no useful signal intelligence because the spies were listening on the wrong frequencies. Meyer summed up what happened after the Nazis no longer found any use for the airships, excepting charging admission to see their suspended frameworks. "On March 1st, 1940, General Göring appeared there with a full retinue of high military aviation brass and civilian experts to survey the scene that he had already

[320] The 1938 report contained this patent drawing detailed installation of condenser *cleaning* equipment, also contained in the power pod. Courtesy M-B.

condemned to destruction. In vain the Zeppeliners gave their detailed commentary to convince him of the possible future significance of airships. Climbing around in the stripped structure of LZ 130, the air minister responded to their efforts to inform him with the kind of folksy chatter that had long endeared him to the German public: 'Na, that thing's for the birds! ... one match and the hold contraption will go up in flames!'" Oh, that one could have been a fly on the wall to hear the airshipmen carefully explain to the air minister how they knew what caused the LZ-129 fire, and how their new hydrogen filled airship could not suffer the same fate. They knew the old fabric held a charge; none could build up on the new skin. They knew the lacing old cord would not conduct, and the blue primer added insulation; both of those problems were solved. The old engines could emit a stream of highly charged ions; the new tractor configuration was proven to help disburse them safely away. One can only imagine the frustration of watching the Air Minister's no doubt vacuous stare, the words bouncing off his hat, his mind made up. The madness of knowing the truth, but having the Nazi goons refuse to reverse their public policy least an embarrassing mistake become common knowledge, has to rank as the most antagonizing experience of the rigid airship era. Here was the most magnificent example of the airshipman's art; fast, luxurious, safe, and unable to carry a single passenger, because doing so would harm a reputation! Not to admit the truth even *after* the war was the Nazi's final insult to our civilization. No coming forward with the truth helped retain hydrogen's fearsome reputation, so its logical adoption as a fuel was delayed for needless decades.

Von Schiller had written before the accident, "If the technique of airship navigation was retarded for years by the imprudent attitude of the Entente, this was because the victorious powers had insufficient experience in the management of the technically perfect instruments and refused to allow us to continue the development." Yet these hated Allied powers victorious in the Great War could not kill the rigid airship. Its final destruction would take only a handful of brown-shirted morons in his own country that could see military value only in the ten miles of aluminum girders therein. A common theme of airship histories states the lack of helium ended airship travel; this is to ignore what happened in England following the Great War. An extensive infrastructure of airship bases, many with hangars, were located near major cities. Trained crews were available to convert the readily available and quite capable North Sea type airship into a ten-passenger liner. With Germany's airship structure being dismantled, Great Britain could have taken a giant leap forward in luxury air travel. Yet for reasons still inexplicable even to historians in the UK, all this was simply forgotten to death.

It was all but over for the rigid airship. Following the LZ-127's 590th flight, the *Graf* had been gutted and hung up as a tourist attraction. As Jane's lamented, "From 1930 to 1936 it was in regular service on the Germany-South America route with conspicuous success. It made its 100th crossing of the South Atlantic in passenger and postal service in November, 1935, and in December of the same year flew its millionth kilometer. Up until then it had carried 10,400 passengers. It was withdrawn from service in May, 1937, after being in continuous use for eight years, during which time no special repairs were needed." Soon both *Grafs* were disassembled, the LZ-130 by Army troops; L-Z workers withstood the Nazi goons and refused the task. The Nazis ordered that nothing be kept, even for museum purposes. Their hangars were then blown up, as if to forever erase this unpleasant memory from the Reich that was supposed to last for the next thousand years.

Back at Lakehurst, the former LZ-126 had not even been much of a tourist attraction before it too was stripped, gutted and cut up for scrap. Likewise, very little of her exists today. The almighty gift of helium could not save the rigid airship – because the hydrogen wasn't the problem. "The fate of the *Hindenburg* at Lakehurst more than two years later obscured these prior disasters but in fact did no more than underscore the existing verdict against dirigibles. There could be no such thing as a weather-proof dirigible lifted by helium, itself a structurally lethal ten percent heavier than hydrogen."[321] Yet there would be further American attempts as airplane carriers. "In 1937 the Bureau of Aeronautics prepared a design for a 10,600,000 cubic foot rigid airship, 897 feet long, which would carry nine Northrop dive bombers. Moreover, in 1938, the Navy obtained Congressional authorization to build a 3,500,000 cubic foot rigid airship for training purposes, which would have carried four Vought fighters. But neither of these projects materialized in three dimensions. The same was true for the Goodyear Air-Craft[322] Corporations' post war design of 10,000,000 cubic foot commercial airship which would have embarked and disembarked its passengers in a skyhook equipped Lockheed Lodestar. But by the end of World War II, hook-on flying had left the airship behind and had passed into the jet age."[323]

Promising results in assisting landings and take-offs were enjoyed with the *Schwenkpropellers*. The official reason for not equipping LZ-130 with this enhancement was said to be their lack of solving the basic problem of managing airship "static balance," yet this car modification appears to incorporate some water-recovery apparatus as well. Eugen Hartmann, standing aft, designed and then tested the creation in flight.[324]

The colossal mistrust about the accident and hydrogen grew to legendary proportions in the following decades, with nature's simplest element became even more super powerful and mysteriously dangerous with each telling. Eckener's 'imaginings' later became Eckener's 'conclusions', which eventually became Eckener's 'insistence,' ad nausuem.

"Given the extent and speed of the fire that consumed the LZ-129 *Hindenburg*, it is surprising that there were any survivors at all."[325] This is a common sentiment; it is so? Devastating as it was, a levelheaded assessment of the accident is in order if we are to overcome our fear and get some benefit from this sacrifice made to the God of safety. Recapping, we remember Rosendahl wrote: "The *Hindenburg* arrived in perfectly normal condition, made a skillful and normal approach, and was being landed in an entirely normal way." If they had stuck with the dope they knew, if they had continued using the proven low landing, or had there not been a storm or perhaps even a gust, it would have been an accident waiting for another chance to happen. Maybe that peculiar set of circumstances would never have come again in spite of the American's insistence on high landings.

Morrison spoke for everyone subjected to the blistering heat and brilliant flames when he cried, "Everything crashed to the ground and there is not a possible chance of anyone being saved." However, there was no iron oxide on the lower covering, hence no thermite. While radiant at

---

[321] A. Gray, "'Stand By For Crash:' The Demise of the Naval Dirigible, *Air Power History*, Fall 1992 pg.6
[322] Changed from Goodyear-Zeppelin, for obvious reasons in a war with Germany.
[323] R. K. Smith, "Forty Years of Skyhooks," *Air Progress*, Fall 1962, pg. 20
[324] "*Eugen Hartmann: Konstruckter des Schwenkpropellers feierte 90*," *Zeppelin Briefe* #27, June 1995.
[325] C. Chant, THE ZEPPELIN: THE HISTORY OF GERMAN AIRSHIPS FROM 1900 TO 1937, Barnes & Noble Books, New York, 2000

over 3,000°F, the thin covering, lesser temperature silk tapestries and other lightweight airship elements were consumed quickly, reducing the exposure time – and the ability for the flames to initiate secondary ignitions, like the occupant's clothing and then their flesh. *After* Morrison recorded "...there is not a possible chance of anyone being saved," he ran out to the wreck and helped burn victim New Yorker Philip Mangone from the mess. He also found Spah. He saw others leaving the burning wreck. Returning to his microphone, he told of helping them and gasped, "It is my sincere hope you don't... you don't think... it was as bad as I made it sound there at first." But that part of his recording is *never* played. The damage was done.

Since none of the motion picture cameras got rolling until the fire was almost half over, the newsreel companies swapped their three snippets and put together a sequence of the three different lens viewpoints. Creating a single piece of film that could not actually be claimed by any one company, it would have probably faded into the obscurity enjoyed by most other newsreel film that would cost potential users $3,000 to $4,000 to broadcast. When the US Navy obtained it and another copy was donated to the US Air Force, it became "public domain," meaning no royalties need be paid. This single edit of motion picture film has emerged as the most-played historical visual record in history. "Curators have not hesitated to exploit the accident as museum entertainment. At the National Air & Space Museum, footage of the fire was shown continuously in the now-disestablished Balloon & Airship Gallery – as if the crash defined an entire technology. One cannot imagine accident footage of, say, a DC-3 or a 747 being granted such attention in the nearby galleries."[326] When Pan Am flight 103 exploded over Lockerbie, there were mangled bodies amid wreckage scattered over the Scottish countryside. Yet at thousands of dollars per minute of stock footage, you usually see only 'in-house' file clips.

Indeed, the images and a snippet of Morrison's recording were projected on a special water spray-screen program given on the Lake of Constance on July 2, 2000, as part of the 100th anniversary of the first flight. It's free to use; and it will be used, from now until the end of time. There is no denying its impact. Better than a bullfight or gladiators in Rome, it's a 'snuff' film that can be played and replayed. Airshipmen and hydrogen proponents alike must turn the insult around. We must wear it like a badge of honor, proudly and for good reason. There had been 97 souls aboard, and 35 were fatally injured. The 13 passengers would be the only *fahrgast* lost in the history of the service. There were hundreds of people on the ground below, and only one – Allen O. Hagaman[327] was lost trying to run clear. Was this a disaster? By today's standards, it's nothing short of a miracle! How did the over-reaction ever get so out of hand?

A reality check is overdue. "Public reaction to the disaster of the airship *Hindenburg* in 1937 with the loss of 36 lives, preceded by the crash of the British R101 in 1930 with 48 fatalities, was such that passenger transport by airship was halted for half a century... In the 4-day 1991 Easter holidays Spain alone, 147 people were killed on the Spanish roads – normal for the vacation period. The news barely received a mention in the local newspaper."[328] In his paper for the 3rd Int'l Airship Conference, "The Airship Accidents And Catastrophes," Dr. M. Arie wrote: "The most frequent types of the airship incidents were: the incidents on the ground, fire and damage during the landing - 30,9; 27,2 and 20,6% correspondingly from all quantity of the incidents...we considered 136 cases of accidents and catastrophes with 119 airships. In 26

---

[326] W. F. Althoff, ARCTIC MISSION, LTA Institute, New Zealand, 1999, pg. 249

[327] One published report had a relative stating an engine car fell on him, while others say Hagaman died of a heart attack trying to run clear. Coroner's report says he died as a result of 'Zeppelin Explosion,' contributory cause severe burns.

[328] F. H. Hawkins, HUMAN FACTORS IN FLIGHT, Ashgate Publishing Ltd., U.K., courtesy Greg Lumann

catastrophes concerning which we have information 415 persons were killed including 4 members of the ground crew." As stated, the Zeppelin LZ-129 was designed and built with the best thinking of the day, tolerating materials known to be less than safe in order to achieve a desired performance at a specified cost. The next-to-emptiness airship construction meant wafer-thin materials were in a most advantageous burning thickness and orientation. Still, that also meant they were consumed in seconds, before even their blistering radiant heat had time to work on secondary ignitions, such as passenger clothing. Owing to the 16°C (61°F) temperature that May night, many people in the airship were wearing several layers of clothing. In many cases the quickly burning airship materials simply did not radiate enough heat *for a sufficient period* to do any more than singe the heavier clothing. Had they not been so terrorized as to jump, many more might have survived with burns only on their exposed flesh. One elderly couple simply huddled in their cabin until Navy men shouted for them to come out. Loss amid the crew would have been greatly reduced by nothing more complicated than a sturdy handrail forward.

Just as WWI Zeps' ammunition sometimes survived their fiery falls, even *paper* survived in the LZ-129. The airships usually carried mail, and the LZ-129's 63rd flight was no exception. Many lightweight airmail letters were burned around their edges but addresses could be read.[329] Even though stored in flammable canvas sacks, the blistering heat simply did not last long enough to penetrate and destroy all the paper inside. Sought after by collectors today, there is even a problem with counterfeit "*Hindenburg* fire" covers.[330] Here again there had been historical precedent, since the displacement vessel's thin materials, while highly flammable, were gone in seconds. Thompson: "Luckily for British intelligence, one of the few things that survived the [L32] blaze was a scorched copy of the German signal book." Hydrogen burning above would of course be incapable of radiating enough heat to ignite the paper; but even the highly flammable fabrics were gone in a few seconds. Even in the horrifying LZ-129 accident, one can see that the basic safety of buoyant flight was still working in favor of the occupants. Though in his death throes, the airship slowly surrendered its altitude ("it seemed like the ground was coming up to me"). This has to include the lifting gas; the pure cooling hydrogen actually worked to *save* lives, to a degree, amid the quickly burning flammables that were the airship. The fraction that did burn released paltry heat that was headed the other way, and could not harm the passengers. Bob Rutan even found a cell log book near the remains of a LZ-129 engine car.

Since hydrogen was blamed in the fire's ignition, it was eventually blamed for the demise of the rigid airship. "Enthusiasm for airships crashed with the *Hindenburg* in 1937."[331] Even some hydrogen professionals seem to accept that hydrogen must have made the fire worse and therefore at least played a role. "But the disastrous explosion (sic) of the *Hindenburg* in Lakehurst, N. J., the advent of World War II and the rise of jet air travel turned the Zeppelin into a symbol of a bygone era."[332] A more realistic assessment says "LZ-130 was an evolutionary dead end. These problems were recognized by the Zeppelin Company. A much larger LZ-131 was designed, and construction actually begun in 1938. LZ-131 would still have to deal with the overwhelmingly unfavorable lift-to-weight ratio of helium, and it is highly debatable that this severe technical drawback would have ever been overcome by sheer size, let alone in time to displace airplanes as the winning technology of intercontinental air travel... In light of these

[329] In a National Public Radio interview, a stamp collector claimed an LZ-129 crewman can be seen tossing mailbag(s) out of the settling wreck. Author can't verify this claim, but it would seem about the ultimate dedication to getting the mail through!

[330] *The Zeppelin Collector* periodically reports on this problem.

[331] C. N. Barnard, "Blimps," *Smithsonian*, June 1998

[332] K. G. Hagenlocher, "A Zeppelin for the 21st Century," *Scientific American*, November 1999

problems, it seems unrealistic to pine for the premature death of airship travel due to lack of helium."[333] Likewise, Peter Allen would write, "Whether a slower and more painstaking development and the use of helium instead of hydrogen to provide lift would have kept the big airship alive is questionable." Dr. Monk Gibbon would be quoted as saying the rigid airship "was an aeronautical cul-de-sac."

In the effort to have Goodyear-Zeppelin offer its unused facilities to build an LZ-129 replacement ship while LZ-130 was being finished, Arnstein prepared plans with two options. The second provided an 8.25-million ft^3 (235,000 m^3) airship 880 ft (268 m) long and 156 ft (41.5 m) maximum diameter which at mission origination would be 87.5% helium and 10% hydrogen inflated. Given a 250-meter initial altitude, the design would allow a total lift of 218,000 kg. Its useful load of 98,200 kg compared favorably with the helium-only alternate of only 95,800. "The ship outlined in the attached blueprint would carry 65 passengers for operation with 100% helium and 80 passengers if operated with 10% hydrogen and *if hydrogen burning was used*.[334] Little did they know the Germans were already working on that very advantage.

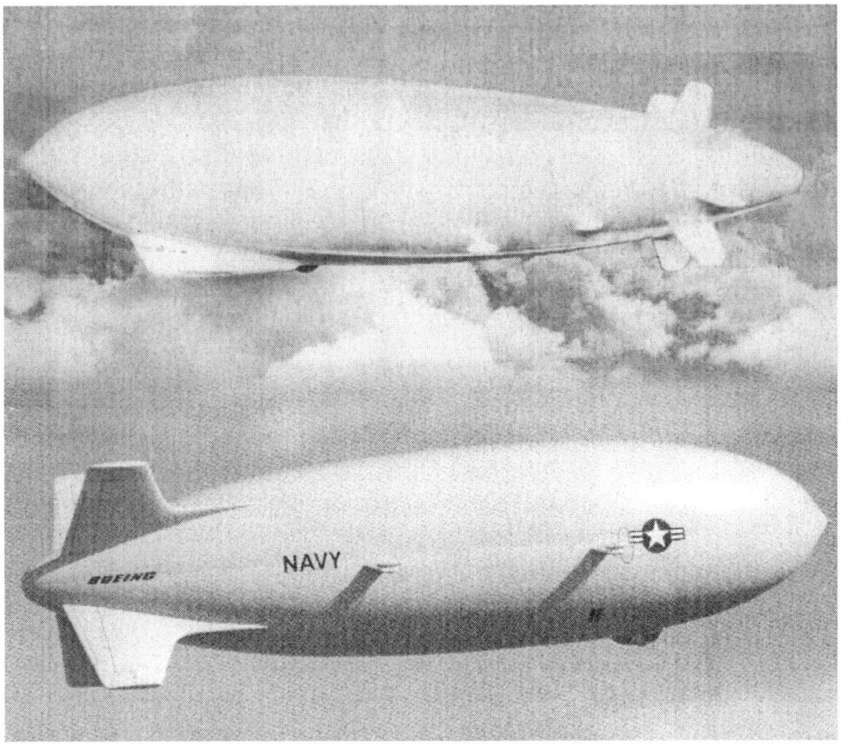

Lt. General W. E. Kepner, USAF, Ret. wrote in an August 1974 letter, "I believe the Metalclad ZMC-2 demonstrated that strong and fast airships can be built to succeed... [ZMC-2] was not indestructible, but it was the strongest airship ever built. If this could be reproduced in a larger ship it would be the answer to speed, safety and reliable operations." Upson and Pavlecka were ahead of their time; years later when Boeing explored military airships, they also passed on gasbags and designed a stressed skin monocoupe hull rigid with enclosed accessories. Author's collection; Boeing illustration; Kepner letter courtesy Robert Slate.

Clearly, the demand for large airships was not strong enough to support two companies in one country. In a series of communications, Detroit Aircraft approached Goodyear-Zeppelin about a merger of sorts. In a passionate five-page letter of January 18, 1939, W.B. Mayo spelled out how the two companies could share resources and make money giving the government what it wanted. P. W. Litchfield replied tersely, "The only type of ship which has proven satisfactory in commercial service is the rigid fabric covered airship."[335] Goodyear would be granted a patent for its own all-metal airship the following year, but the Navy wanted no part of anything that did not look like a conventional fabric-and-dope covered gasbag. Even Robinson and Keller would

[333] R. Wilcsek, "Helium and Airships: A Reality Check," *Jack Knight Air Log-PAA Notes*, Apr-Jun 1999
[334] Emphasis added. Memo dated 1 July 37 in the Arnstein Collection, University of Akron Archives.
[335] Letters found in the Arnstein Collection, University of Akron Archives, by Eric Brothers. W. E. Kepner also approached Litchfield about working with Detroit aircraft, and was also refused. The ZRN *could* have happened.

conclude, "Thus, nothing could alter the widening gap between the static performance of the big rigid airship and the spectacular technological advances of the American aviation industry." So ended the era of the rigid airship, in spite of all the world's helium being on call. Yet, God of War Mars would offer the rigid one last chance.

During the dark early days of America's involvement in World War Two, there were two voices urging utilization of proven airship technology. Lieutenant Richard L. Barkley and Lieutenant Frank. D. Buckley, US Navy, friends who served in the Pacific and had many discussions concerning LTA matters, prepared an article for the US Naval Institute.[336] It took months for the publishing process, and little notice was made of it. More than ordinary blimp pilots, both men were passionate about LTA and did not stop there. Buckley, while skipper of Airship Headquarters Squadron 3, took the initiative to fulfill the Navy's request for lightening K-ships for better performance. With a mixture of engineering expertise, politics and cum-shaw, his team modified the K-21 with B-29 bomber yoke controls, automatic cable tensioners, auto-pilot, and a host of other fly ability improvement modifications that resulted in a *savings* of 1200 pounds! As the war deepened, the two men remembered the writings of Frenchman Jean du Plessis, calling for rigids in 1921. Though by then fighting the Japanese in the Pacific, Barkley in HTA and Buckley aboard the USS *Isabel*, they wrote a commentary envelope for the translation of du Plessis by Anton Heinen. With Alice Buckley, back home in Olympia, Washington, they self-published both in "The Rigid Airship in Peace and War." Between the two works, a very powerful case is made for the return of the flying carrier. It must be remembered that neither man suspected the LZ-129 was not primarily a hydrogen accident. Space limitations forbid reprinting the entire documents, so we have attempted to paraphrase their article, commentary and notes into the following combined dissertation. We pick up following a discussion of how, in spite of warnings, the torpedo caught armored Navies by surprise in 1914:

It may be said that the use of aviation at sea, within the limits of its sea-keeping qualities, has replaced the gun as the primary offensive weapon. The provision of overwhelming offensive power at sea, therefore, seems to hinge upon the improvement of the sea-keeping qualities of aircraft. No one has ever questioned the logistic axiom that all naval forces can operate most efficiently at a fairly short distance from well-equipped bases, and that their efficiency drops off as the distance from their bases is increased. This is particularly true of aircraft, for in addition to the normal difficulties which accrue to long-range operation against an objective, the weight of the military load, or persuasive force, must be decreased to provide additional fuel to attain the necessary range. The Navy early realized that it would be necessary to provide some means of transportation, equipped with basing facilities, to bring aviation within a reasonable striking distance of its objective.

This realization led to the development of the aircraft carrier. Sea power is created to maintain sea communications - or to interfere with them. The intentions of seafaring nations are carried out at sea – they must be to ensure their continued existence. Any means of interfering with them becomes sea power. The carrier transports planes of moderate range, suitable for combat, to within striking distance of the objective, remaining well out of the immediate vicinity to avoid destruction. It is vulnerable to all the weapons of sea warfare: guns, torpedoes, mines, and aircraft. The *Glorious* was sunk by gunfire, the Illustrious so severely battered by aircraft that operation of her own planes became impossible although she did succeed in getting off a few fighters. The fact that launching or recovering of aircraft necessitates maintaining a steady course and speed for a considerable length of time makes the carrier particularly vulnerable to the submarine, which has accounted for the *Courageous* and *Ark Royal*. As a type, the carrier is

[336] "Carrier Crisis," USNI *Proceedings*, Vol. 69 #3, March 1943

expensive to construct, purely aside from cost considerations. It requires, under the best conditions, two years to complete. It absorbs 20,000 tons of material vitally needed elsewhere. These factors make the replacement of lost carriers an onerous task. For all this, only 75-100 planes are provided with a mobile base. The personnel required to operate it consists of about 2,000 well trained men plus the personnel required to operate the attending cortege of antisubmarine destroyers and perhaps a protecting cruiser or two, which are thus immobilized for other duties. This personnel loss is perhaps the most difficult to replace. At any rate, a somewhat different concept of the picture has been already indicated by the heavy emphasis recently laid on aircraft carriers. The announced program of 500,000 tons of new carrier construction, plus the conversion of other naval and merchant types to carriers, will enable us, it is hoped, to provide our primary offensive weapon, naval aviation, with crushing superiority.

To date, the numbers of aircraft taking part in action at sea have been small compared to those used in land operations. The primary reason for this is that there are relatively few aircraft carriers in existence. The very nature of the aircraft carrier makes it difficult to produce and replace. Hybrid conversions are being used - and should be - to augment the strength of our primary offensive power at sea, but many of these are bound to be disappointing from the standpoint of results obtained. The insufficiency of carriers seriously hampers the sea-keeping qualities of naval aircraft. Press releases have indicated that many cripples, which could have been polished off by aircraft - or fast battleships for that matter - escaped from the Battle of Midway. Sufficient aircraft of suitable types could have followed and annihilated these units, preventing their future use against us. Airship-based aircraft could have been gainfully employed here to pursue and annihilate these forces-without the danger of being ambushed by enemy submarines. The construction of sufficient surface craft, particularly battleships, to accomplish such a task requires a formidable investment in material, time, labor, and personnel. In addition, such means are vulnerable to all the weapons of sea warfare, whereas the airship is vulnerable only to aircraft attacks. The adaptability of aircraft to mass production assures us relatively easy attainment of decisive superiority, and the rigid airship can provide basing facilities for these aircraft. So far, results seem to indicate that the relatively short-range, carrier-based dive bombers and torpedo planes have a greater effectiveness than the long-range horizontal bomber in fulfilling the precise requirements of sea warfare.

We must concede that the naval operations to date have been on a small scale insofar as the number of aircraft employed is concerned. Future actions will undoubtedly involve numbers greater by many times-even considering only the employment of surface carriers. But we must wait until these carriers are available before we can employ them. We have time, therefore, to undertake the development of the airship carrier.

The idea of a defensive fighter airplane, dropped from a Zeppelin to ward off an attacker while the mothership more easily gained altitude (left) was tested during the Great War. The ultimate expression came with the ZRS ships. The USS *Macon* (ZRS-5) was scheduled to scout with the fleet from Hawaii when lost in 1935. Today's airships would serve as a base for a squadron of unmanned aerial vehicles (UAVs). Photo courtesy Dr. Douglas Robinson, "The Flying Carriers" painting by James Simpson.

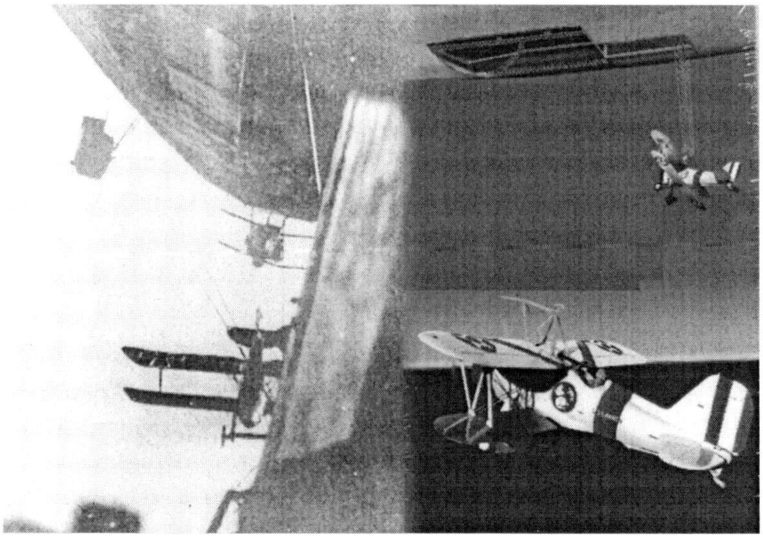

It is in the nature of things that the movement of any object is governed by the medium through which it must pass. The movements of surface ships are governed by the seas, but more important when considering the question of mobility, is the fact that their broader movements are controlled by land formations. Thus, generally speaking, the mobility of a seaborne aviation base is influenced by the sea (which regulates its speed) and by the land (which regulates its source of travel and its real speed from point to point). An airplane carrier going from New York to San Francisco, airline distance some 2,300 miles, must travel about 6,000 miles, and nearly 15,000 if the Panama Canal, our vital sea link, is ruptured. While it is possible for the aviation units to fly this distance, making necessary intermediate stops, in about two days, the sea borne base cannot possibly arrive at the destination in less than ten days and maybe months. Herein lies the gravest weakness of naval aviation today - its inability to concentrate rapidly in large formations. The basis of this weakness clearly lies in the characteristics, operational and constructional, of the chosen carrier. Until naval aviation takes to the air, bases included, it will not attain its real full strength. Such a means, holding great possibility, lies in the development of the rigid airship as an aircraft carrier.

The airship carrier idea, as well as others set down by the author [du Plessis], no doubt was considered purely visionary by many who took the trouble to read his works. Perhaps they were. But vision is the most important essential to imagination, without which there can be no progress. And progress, as it pertains to the development and use of the tools of warfare, has many times spelled the difference between victory and national disaster. The speed and range of the airship carrier, coupled with its availability in large numbers, would make the concentration of aircraft when and where they are needed a much simpler problem than it is today. It would provide mobile bases for aircraft types which have been found to be the most effective at sea (i.e., dive bombers and torpedo planes) for much less expenditure of time, labor, and material than a similar concentration effected by surface carriers. It is also probable that larger and heavier airplane types than can be operated from surface carriers could be operated from airships. The problem of training plane pilots to operate from airships is quite simple, as we have found from our own experience.

It must be taken for granted that the development of the airship as a major means of extending the striking power of naval aviation would be strenuously opposed from many quarters. The rather high loss percentage with our airships led to a demand to do away with the type rather than the causes for the failures. Quite the contrary has been true with commercial airline disasters, which were investigated and made to contribute to the excellence of our commercial air service today. The structural problems of the airship have always been regarded with an air of uncertainty. Much attention has been drawn to the difficulties of airship construction by the rather formidable groups of technicians assembled from time to time to investigate them. Most recent of these was the Durand Committee, convened shortly after the loss of the Macon to investigate the engineering difficulties and make recommendations to the Navy as to the continuance of airship construction. Treating our airship troubles as primarily of engineering origin, they arrived at an affirmative conclusion with regard to continuance, recommending a mincing step by step policy of gradual size increase.

The greatest objection to be held against such a proposal will be the alleged vulnerability of the airship. This criticism will become sharper as soon as it is thoroughly understood that the airship demands hydrogen to ensure continuous availability. There is no doubt that by far the greater part of this opposition would be strictly honest in its intentions. The roots of most of it can be found in the misconceptions which we have outlined briefly in the first section of our discussion. Public opposition is due primarily to lack of information, “artists' conceptions” of airships rent asunder by the atmosphere, editorials and the like. However, the general reputation of the rigid airship, based largely on the results of our own operation, does not lend itself to the placing of much confidence in such development. Many harmful misconceptions regarding airships have taken firm root in the public opinion. One hears occasionally the argument that we have built only three airships - how can we expect not to have failures. After all, look at the third airplane, submarine, cruiser, etc.

Although these factors have filtered somewhat into naval opinion, naval opposition is based primarily on the failure of the airship to deliver the desired results. It is quite generally recognized in military circles that all weapons are vulnerable to countermeasures. It is also recognized that the vulnerability of any weapon is a function of the application of employment technique. The airship, having three times the sustained and top speeds of the carrier, need not expose itself to the gunfire of hostile surface craft. It is not vulnerable to the torpedo or the mine. Being enveloped, together with its planes, in the same air mass in which it operates, it can choose any course for launching or recovery. A top speed of at least 80 knots can easily be attained. With the development of arresting gear capable of absorbing a speed differential of 20 knots, the employment of planes having a landing speed of up to 100 knots would be entirely possible. The employment of such planes, with the additional advantage of high wing loadings, greatly increasing their load and/or range, would enable the airship to launch its striking force further from the place of its employment than can the surface carrier. With proper handling facilities, larger and heavier planes than those utilized aboard surface carriers could be operated. In addition, retractable landing gear and its attendant weight could be eliminated, for it is of no value anyway in the event of a forced landing on the water.

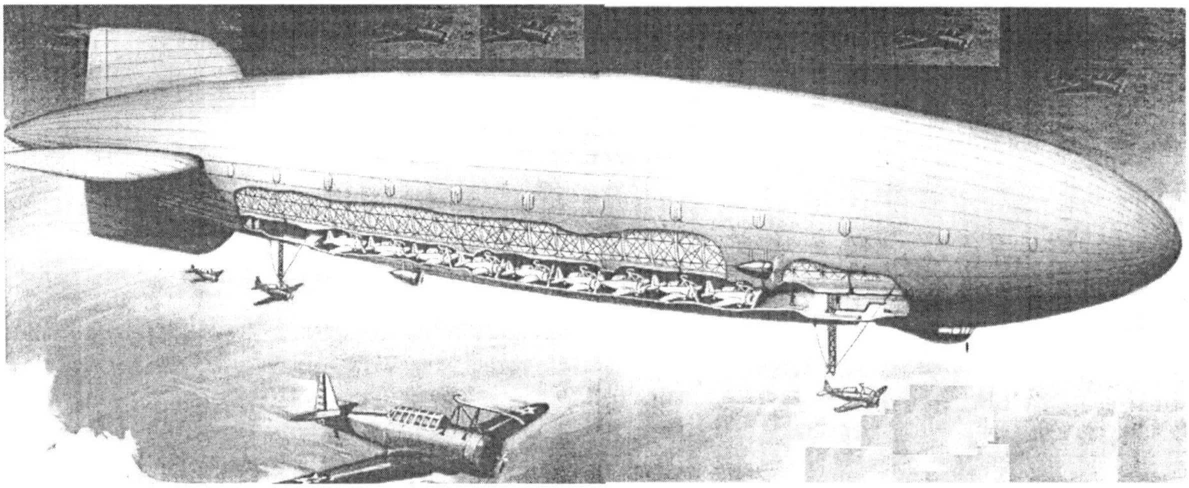

Goodyear-Zeppelin updated its ZRCV design several times during the war. The last version would have fit into the standard timber 1000-foot long hangar, of which 17 were built. G-Z concept courtesy LTAS.

It has been pointed out that the bombing Zeppelins of World War I defended themselves by making use of their great rate of climb and high operating altitude. The airship carrier will have to do practically all its operating at relatively low altitudes. Certain possibilities of airships were just beginning to be demonstrated by the *Macon* when the ship was lost. This old conception which previously governed its development must be immediately junked. It is not a negative scout or a patrol vessel. Its aircraft do not exist only to augment its ability to scout. It is merely transportation for an offensive force of aviation, which it serves as an auxiliary-with all that may imply. The airship must be considered the auxiliary - the tender which makes possible the range and striking-power increase of naval aviation.

Since it is acting as a transport medium rather than as a weapon itself, it must remain in the denser air of the lower altitudes in order to maintain its lift at the maximum possible value. Its defense lies in its ability to remain at a great horizontal distance from areas which are a source of danger to it. The high performance of its aircraft as compared to carrier-based types, and its relatively high speed combine to enable it to avoid attack more easily than can the surface carrier. It fears no reefs or rocks nor does it leave a betraying wake or oil slick hours behind. In an hour it moves 80 miles. In addition, it is vulnerable only to air attack, unless it should be brought by error within the range of anti-aircraft fire.

The personnel investment of a surface aircraft carrier is relatively tremendous, requiring not only personnel for the carrier but the immobilization of other units and their personnel to provide escort facilities. The loss of a carrier airship would involve the loss of not more than 200 men. The airship is very inexpensive with regard to the number of personnel involved to operate a given number of aircraft. The operating personnel of airships can be easily and rapidly trained.

Today the non-rigid is being talked up to the exclusion of the rigid airship, though history has conclusively proven that far greater advantages are to be gained from the use of the rigid airship. In the United States, rigid airships occupied the attention of the Navy to the exclusion of the nonrigid until 1935. After the failure of the *Akron* and *Macon*, considerable conjecture was bandied about that we had attempted too large a jump in volume. It was pointed out that they were nearly twice the volume of the *Graf Zeppelin*. This opinion was dispelled by the operation of the *Hindenburg*, which was slightly larger than either of our ships. The Germans held a similar view prior to the close of the First World War. The largest war-time airship had a volume three times greater than that of the prewar commercial types. It was felt that the technical advances made during the war would permit the construction of small commercial types which would be much easier to handle than the larger naval airships. On commissioning the post-war commercial airship *Bodensee*, of the same volume as the pre-war types, it was found to the general surprise of all concerned, that the large naval airships had been much easier to operate, with the exception of ground handling. The reason was shocking in its simplicity. Any vessel of relatively large displacement is less subject to the vagaries of the medium in which it operates, as compared to a small one. This is quite apparent to the airship student who bounces about in a small training blimp, enviously observing the relatively serene behavior of a larger patrol blimp operating in the same vicinity.

The use of the rigid airship as an aircraft carrier will require much larger volumes than have been heretofore constructed. Volumes of approximately 20,000,000 cubic feet - three times those of the *Hindenburg* and *Akron* types-are necessary to the development of efficient carriers. Such volumetric increase does not introduce any new structural or operational difficulties. It perhaps does introduce some terrifying misgivings where there is a lack of imagination. In reassurance, it is pointed out that the *Hindenburg* had a volume three times greater than that of the standard World War airship, which in turn was three times larger than the pre-war commercial types. According to the *Hindenburg's* operators, she was far easier to operate than any previous type. As to size, such an airship would require a dimensional increase of about 50 per cent as compared to the *Hindenburg*, i.e., a length of 1,200 feet (Hindenburg 803 feet) and diameter of 200 feet (Hindenburg 133 feet).

To realize the airship as a plane carrier capable of operating 50 planes, a volume of 20,000,000 cubic feet will be required. Assuming a structural dead weight of 300 tons (40% dead weight; 34% was actually attained in 1918 in ships of about 2,400,000 cubic feet), a useful load of 400 tons remains under average operating conditions (about 600 F., 30" Hg). With 150 tons of fuel to drive it 12,000 nautical miles at 60 knots, in still air, such a ship could carry 50 planes averaging 3 tons each, with 100 tons remaining for crew, provisions, extra fuel and bombs for planes. Up to 10% of the standard gross lift of an airship (750 tons in this case) can be carried dynamically after take-off, that is the ship can fly "heavy" that amount. An additional 75 tons can be loaded aboard after take-off, either by picking up planes or fueling in flight.

The development of the rigid airship would depend on hydrogen for several reasons. Contrary to press conceived belief, we do not possess sufficient helium to operate a fleet of airship carriers. According to figures released in the Army and Navy Journal of September 6, 1941, 57,000,000 cubic feet per year of helium would be required by 1944, upon completion of the 48 nonrigid airships then authorized. Authorization was being requested to bring the yearly output of helium up to 60,000,000 cubic feet. Since that time, a total of 200 blimps have been authorized by Congress. So it would not be too surprising to see even the nonrigid shift to hydrogen, assuming of course that they are delivered on time. Basing an estimate on the rather limited peace-time operation of the *Akron* and *Macon*, 60,000,000 cubic feet per year would not suffice to operate two 20,000,000-cubic foot airship carriers. It should be noted that this hypothesis does not consider the important problem of transporting helium to airship operating bases. Scarcity of ship bottoms to transport helium, when other commodities are more urgently required, would seriously hinder the operation of airships. In railroad parlance, some 100 carloads would be required to provide initial inflation for one 20,000,000 cubic footer.

The relative lift values of helium and hydrogen are frequently misunderstood - primarily because they have been frequently misrepresented. Helium, at standard conditions (100 per cent purity, 29."92 hg, 320F.) has 93 per cent the lift of hydrogen. In practical operation, because helium is never available at a purity greater than 97 per cent, and because of operational considerations which result in a further lowering of purity of helium relative to that of hydrogen, the lift difference is rarely less than 10 per cent. As the temperature of the air mass in which the airship operates increases, a further disparity is introduced which increases the lift advantage of hydrogen. To the 20,000,000-cubic foot aircraft carrier, structural weight and fuel load (range) remaining constant, substitution of helium means a reduction of the military load from 250 tons to 175 tons - a loss of 30 per cent. It is quite conceivable that the additional 75 tons carried by the hydrogen airship, disposed in additional aircraft, armament, or bombs, could easily mean the difference between accomplishment and failure of a mission, or even the difference between saving and losing the airship itself.

Although R-101 designer V. C. Richmond said "... the fabric work of the rigid airship (i.e. the gas bags and the outer cover) represents the least satisfactory part..." some Brits wished they'd had one at sea in 1939. Years later, the conduct of the Pacific theatre might have been quite different had the flying carriers been built, as speculated in 'ZRS' by Rowan Partridge, and depicted here by Cortney Skinner.

It is quite generally realized that helium lifts less than hydrogen. What is not understood is how this lift difference effects the performance of an airship. It has been frequently said in print that helium lifts 93 per cent of what hydrogen will. This is approximately true at standard conditions (32 F., 29.92" Hg, 100% purity). But this statement is always ambiguously left hanging in the air, leaving the reader with the impression that a helium-filled ship will carry 93 per cent the pay (or military) load of the same ship filled with hydrogen. Actually, at normal operating conditions, the difference of the standard gross lift is more nearly 10 per cent due to the lower purity with which helium operations are conducted. But in substituting helium for hydrogen in an airship, the structural weight of the ship is constant, service load including fuel is constant (for the same range) and the entire loss of lift must come out of the only remaining variable - the pay or military load. Let us assume an airship of 10,000,000-cubic-foot capacity of 50 percent dead weight and 30 percent service load.

	Hydrogen	Helium
Standard gross lift	375.0 tons	337.5 tons
Dead weight	187.5 tons	187.5 tons
Useful load	187.5 tons	150.0 tons
Service load (fuel, etc.)	112.5 tons	112.5 tons
Military load	75.0 tons	37.5 tons

The entire loss of lift reduces the military load by 50 percent, not 7 per cent, as we have been lead to believe. The standard gross lift of such an airship (100 per cent pure hydrogen at 320F. 29."92 hg) would be 750 short tons. Of this, 300 tons would comprise the structural dead weight of the airship (allowing 40 per cent dead-weight tonnage). A useful load of 400 tons would be available assuming normal operating conditions of 600F. 29."92 hg, with 98 per cent pure hydrogen. A fuel load of 150 tons would drive the airship some 12,000 nautical miles in still air. It could carry and operate 50 planes averaging 3 tons each, leaving 100 tons for aircraft fuel, bombs, etc., less the weight of the airship and aircraft crews (about 200 men) and their necessary equipment. The above estimate of performance presumes a static equilibrium take-off. Such an airship could carry (operating "heavy") at least 5 per cent of its standard gross lift dynamically after take-off-about 40 tons in this case, and operate in that condition indefinitely. A maximum of 10 per cent of the standard gross lift can be carried dynamically for periods of a few hours. This overloading can be done by taking extra weight aboard in flight such as by fueling from another airship or landing additional aircraft. Under nearly all operating conditions, a useful load of 400 tons would be available. A top speed of at least 80 knots could be expected from such an airship. Two airships of 20,000,000 cubic feet could deliver 50 fighter planes per week from San Francisco to Sydney.

Much remains to be done in the development of mechanical landing gear, to further reduce the personnel absorbed in the ground crew. Satisfactory solution of this problem could be achieved quickly if actively prosecuted. Even so, the utilization of a 300-man ground crew to handle several airships compares favorably with the enormous personnel required to operate carrier based aircraft. This personnel is in addition composed, with few exceptions, of untrained men. But the most efficient operation demands the development of mechanical landing gear to the highest degree of perfection. The development of equipment to moor airships at sea would greatly increase the flexibility of their operation. Interesting experiments were carried out along this line by Lieutenant Clinton S. Rounds, U. S. Navy, at Lakehurst in 1939. The system consisted of a drogue, to which the airship rode by means of a line attached to its mooring point, and a ballast bag suspended from a point beneath the airship's center of gravity. The ballast bag, fifty feet below the airship and nearly full of sea water, prevented the "light" ship from rising. The drogue or sea anchor held the ship's head into the wind. With this system the old nonrigid J-4 was used to demonstrate the feasibility of riding out high winds at sea, of changing personnel, etc. Adapted to the rigid, similar apparatus would permit refueling and taking on supplies at sea, thus greatly decreasing the dependence of the ship upon its base. To complement this apparatus. Lieutenant Rounds developed a means for pumping sea water into the airship to compensate for the weight of fuel burned. The second unit built succeeded in pumping eight gallons per minute to the ship at an altitude of 85-100 feet, sufficient to compensate for one half hour's fuel consumption at cruising speed. The complete water pick-up apparatus weighed about 150 pounds.

Perhaps the most important advantage of using the airship as a carrier is its adaptability to mass production. During World War I some 78 Zeppelin and 20 Schutte-Lanz rigids were built; the average production for the period of the war placed one ship in commission every sixteen days. The greatest obstacle in the path of the airship is the cloak of mystery which has been thrown around its construction and operation. As to construction, competition between at least two companies should be encouraged and subsidized. The competition of the smaller Schutte-Lanz airship organization was one of the major factors in the effective development of the Zeppelin airship. Contracting for the first 27 of the 48 originally authorized blimps from one company has been a second factor responsible for their slow rate of delivery. The Army, in ordering barrage balloons, contracted with at least five different companies, at greater cost perhaps, but the production has been gratifying. A blimp is, after all, nothing more than an enlarged barrage balloon provided with power and control facilities. The idea has been fostered here, however, that the construction of airships is such a difficult task that it is better to entrust it to a single company which possesses all the expert knowledge required. This idea has filtered down to the nonrigid, with the naval result that only one company is building them today. If necessary, the Navy might provide competition by entering upon the construction of rigid airships itself, as it did in the case of the Shenandoah. The airship is not such a mysterious construction problem that it requires the monopolized services of a single, highly specialized construction company. Nor does it require a monopoly of highly specialized operators possessing superskill. Press contention notwithstanding, the past history of airship operation amply confirms this fact. The construction of such airships simply involves the employment of more facilities and men to attain the same rate of production. Such volume increase presents no new technical problems in construction or operation. Assuming that we can equal the German rate of airship production, utilizing 12 erection hangars instead of 3, 90 airships per year could be produced, basing 4,500 aircraft averaging 3 tons each; or 2,700 averaging 5 tons each. Produced at this rate, the cost equivalent of the carrier airship should not exceed that of the modern destroyer, now in mass production.

As a point to illustrate the efficiency of a well-organized airship construction plant, it is appropriate to point out the case of the wartime zeppelin L59. The L57 was originally constructed to make the trip, discussed several times by du Plessis, from Jamboli, Bulgaria, to provide relief for the German forces in East Africa. It was merely a lengthened version of the standard 2,000,000-cubic-foot production airship. On its first test flight, he ship demonstrated rather peculiar aerodynamic flight characteristics and was so badly damaged on landing that it had to be destroyed. Immediately, orders were issued to enlarge the L59, which had already completed its first trial flight. The L59 was cut in two while still inflated, two additional bays were inserted, and it took the air four days later, her volume enlarged by of a 400,000 cubic feet. This fact, together with the rate of production achieved by the Germans, tends to disarm the airship of the smoke screen thrown around it by those who intend to capitalize upon the mystery of its construction. For these interests, it is essential to maintain the abracadabra attitude which exists in the matter of airship construction. It is essential to completely break this attitude if we are to have suitable airships in sufficient. quantity to satisfy the requirements of naval aviation.

Rigid carriers could and should have their construction centers located well inland, both as a defensive measure and to make use of labor which is not available to coastal shipyards. Should the airship be found only equal to the surface carrier as a base for aircraft, more attention could be diverted to the vital problem of producing cargo bottoms. It consumes some 300 tons of material as compared to 20,000 tons for a surface carrier. It is true that the material required is vitally needed in other war industries, but the development of a new and efficient aircraft carrier would most certainly establish the desirability of favoring it with a high priority classification.

The implications to be drawn from the above are quite clear: the hangar must be used only for the construction and extensive overhaul of airships. All operation must be carried out from mast bases, to assure the constant availability of the airship and the aircraft based thereon. An airship immobilized in a hangar is helpless against attack, which fact was demonstrated to the German operators during the First World War. Apparently the lesson did not penetrate. The greater problem is the development of proper tactical employment of the airship carrier. For this reason its operation must be constantly dominated by line aviation personnel. The airship carrier must be developed and operated in exactly the same manner and by the same forces as the aircraft carrier. It is the responsibility of the Navy in general and of naval aviation in particular. The development of the airship carrier will require the whole-hearted support of the naval aviation organization, of which it must become a part. The responsibility for this development is clearly that of naval aviation. The possibility exists of developing an auxiliary which can multiply many times the strength and sea-keeping qualities of our primary naval offensive weapon. We would be unwise to ignore it."

In spite of Admiral King's generally favorable view of LTA, and a very close vote in Congress, the ZRCV remains a footnote to history. What part of this snafu was due to the general perception that hydrogen would be unacceptably dangerous can only be guessed. However, Buckley & Barkley's call to stop wasting precious resources on vulnerable flattop ships has only gained in importance in the past half century. Today the sea mines move, and rocket torpedoes travel at more than 200 miles per hour. Today aircraft carriers cost billions and tie up 6,500 man crews, expending most effort protecting themselves, just to place a few airplanes on target. It is time we re-examined these prophet's ideas in a modern context of shrinking defense dollars.

In October of 2000, American balloonists were finally able to enjoy the efficiency of their European counterparts; the ban on hydrogen ballooning was officially lifted. H_2 balloons – with pilots Sullivan, Hale, Brielmann, and Herchend - took part in the Great American Gas Balloon Challenge Race at Albuquerque's International Balloon Festival. Eric Brothers photo. (right) Hydrogen Fuel Station dispensing compressed gas and liquid hydrogen built by Air Products for Ford Motor Company in Dearborn, MI. AP Inc. photo.

The first small but official step in recognition of history's error and correcting the hydrogen misconception came early in 1999. Dr. Tom Crouch, Chairman of the Aeronautics Department of the National Air & Space Museum, worked out a new placard for the movie model to be hung in the lobby. Mr. Donald Egan, NASM Director, wrote Senator Tom Harkin, who had made Dr. Bain's work part of the Congressional Record.[337] Egan attached a copy of the new label. It reads:

"This 1/28th scale model of the airship *Hindenburg* was built by Universal Studios for the feature film *HINDENBURG*...As the *Hindenburg* attempted to land at Lakehurst, New Jersey, on May 6th, 1937, it burst into flames. Static electricity apparently ignited the flammable lacquer-doped fabric covering starting the resulting conflagration. Amazingly, 62 of the 97 passengers and crew on board survived the disaster."

[337] U S Senate Congressional Record, 6 OCT 98, §11631. www.fas.org/spp/civil/congress/1998/s981006-zep.htm

Meanwhile, automobile manufactures were turning to hydrogen to overcome challenges imposed by problems dealing with the world's most-traded commodity and the massive impact this petroleum fuel usage imposes on our fragile environment. Billings Energy had first converted a car to hydrogen power in 1965, and got a contract to convert a post office delivery car in 1977. In 1999 Volkswagen AG rolled out their "3 liter" car, the special edition *Lupo*, which travels 100 km on three liters of gasoline. This was quite an achievement; that is the equivalent of 91 mpg. Yet in 2000, the company would roll out its *Bora* (*Jetta* in US) Hy-Motion, literally a *zero*-liter car. It travels 100 kilometers, or even a million kilometers, without using *any* of the substance that has polluted more ecosystems and killed more people in accidents, as well as wars fought over it, than any other commodity in the history of civilization. In 2001, BMW sent its fleet of hydrogen-powered sedans on a world tour. There is no need for concern with this industry's success. People will always purchase the more cost-effective method of transportation offered to them. They will do so through the creation of an infrastructure to support it, and do so even if there is a perception of some danger. They will have to; non-renewable fossil fuels must eventually become too expensive as they run out.

Consuming hydrogen in a high-compression turbocharged engine delivers more spontaneous and powerful response than gasoline equivalents. Admittedly, "...certain essential industrial, political and economic conditions will have to be met before the hydrogen age can begin for real," says BMW. Paving the way, LH_2 refueling stations in Hamburg, Berlin, Munich and Milan have created a "hydrogen superhighway" prototype for Europe.

Hydrogen's use in the space program is expanding. "...catalyst beds of this type would be used on Mars to produce hydrogen from an endothermic reaction between carbon dioxide and methane... Hydrogen is then available for a propellant for launching payloads from the Martian surface or as a fuel for fuel cells generating power for surface mobility or stationary electricity."[338]

Water can be split with less than one and a half volts, taking only about 4.8-kilowatt hours to electrolyze water for one cubic meter of gas. That is about the *least* efficient method. There is promising work in "thermal electrolysis" that does the splitting just with high heat alone, no other energy required. Lower temperatures are possible using bromine or iodine. We will probably be able to make hydrogen even in the process of cleaning sewage, using blue-green algae that produce feed and fertilizer.

The airship, on the other hand, is at a crossroads. Even the new well-funded, well-engineered airship building organizations will ultimately fail if based solely on disappearing helium. Acceptance that hydrogen was *not* the instigator in most of the airship's historic fiery accidents will play a pivotal role in buoyant flight – or lack of gasborne flight - in our near surface transportation corridor in the 21st century.

---

[338] "Metal-Supported Catalyst Beds for Reacting CO2 with CH4," NASA Tech Briefs, March 2001, pg. 64.

Chapter Eleven: "Executioner" as Savior

Scientists at the turn of the previous century ridiculed the dream of the flying machine, insisting it was no more than a carnival stunt to *plane* across the air. Simon Newcomb, called one of the greatest mathematicians of the age, produced a formula that proved the heavier than air machine impractical owing to the constant need for uninterrupted power. He ridiculed the Government's wasting fifty thousand (gold) dollars on the obviously foolhardy creation of Samuel Langley. Newcomb no doubt felt vindicated when the catapult fouled and the *Aerodrome* contraption fell into the Potomac River. Even after the Wright brothers catapulted into fame, he insisted the flying machine was so inherently dangerous as to be impractical. In 1906, Newcomb wrote " If an engine were necessary not only to propel a ship, but also to make her float – if, on the occasion of any accident she immediately went to the bottom with all on board – there would not, at the present day, be any such thing as steam navigation." Think about it: even the heaviest Jet Ski won't drown its pilot if the motor fails. Newcomb logically concluded "If, therefore, we are ever to have aerial navigation with our present knowledge of natural capabilities, it is to the airship floating on the air, to which we are to look." [339] Six years later Augustus Post would still warn the aeroplane was badly in need of refinement because it was "fast becoming 'a brief-lived, dangerous toy for colossal fortunes.'"

"If it is to remain in the air, an aeroplane must constantly move like a skater on thin ice. The skater must move fast enough to reach a new section of ice before he falls; the aeroplane must move fast enough to reach a new section of air before it falls."[340] Generations have set aside this concept, instead squelching the terror associated with the need to continually rush forward at high speed in order to defy gravity. Even so familiar a precaution of slowing down one's automobile, when conditions suggest a reduction in forward speed, is an unaffordable luxury in an airplane. The inability to slow down when weather demands and navaids are out would be quite useful, as the victims of the Korean flight #801 crash on Guam would attest, if they could.

Remembering the old Biblical truism about what happens to people that live by the sword, one could conclude heavier-than-air flying is an inherently dangerous business. As the Florida Aviation Historical Society pointed out, Japanese fighter ace Hiroyoshi Nishizawa shot down over 80 planes but died as a passenger in a cargo plane. Stunt pilot partners Paul Manz and Frank Tallman thrilled many a movie going audience until Paul was killed flying a bastardized contraption of a script engineer's airplane. Yet Frank was killed in his ordinary Piper Aztec in an everyday storm just miles from home. Add to the basic risk with an unproved design; 25 pilots were killed before the Navy concluded they really didn't know why the Vought F7U Cutlass kept crashing. Airplanes that rack up millions of safe miles have all the more problem overcoming their inertia for necessary redesign even in the wake of horrible accidents. In September, 2000, Boeing finally agreed to a 737 rudder redesign after several massively fatal crashes hinted something was improper.

Part of that danger comes from the petroleum fuels airplanes need to not only make progress against wind resistance but to defy gravity as well. This heavy liquid fuel can get you into

---

[339] S. Newcomb, SIDELIGHTS ON ASTRONOMY, Chapter XXI "The Outlook For The Flying Machine," Harper & Brothers Publishers, 1906. (Courtesy Mrs. E. Van Treuren, age 90.) President of the Royal Society Lord Kelvin, Herman von Helmholtz, and Joseph Guy-Lussac maitained HTA flight was *impossible*.

[340] W. Kaempfert, "Why Flying Machines Fly," 1911, included in EARLY FLIGHT, Edited by Frank Oppel, Castle, Book Sales, Secaucus, NJ, 1987

trouble even before it burns. A B-1B bomber was heavily damaged on the ground when transferring fuel upset the center of gravity. More susceptible in flight, popular singer John Denver may have been killed simply because he did not drain his fuel tanks evenly. Most dangerous is the fully fueled takeoff roll toward rotation speed, when even a simple problem can result in a 747 being transformed into a hideous mass of flaming wreckage, as the few survivors of Singapore Flight SQ006 would tell you. After the Concorde crash of 2000, we forgot wheel and tire problems on takeoff were hardly just limited to SSTs. A National DC-8-61 suffered a similar fate at Jeddah on July 11, 1991, and killed more than twice as many people – 265 in all. Oh, that those were the only two.

When, for even the silliest reason, airplanes can't move forward at high speed, they can't defy gravity, illustrated in accidents like the infamous Eastern Flight 401 crash. As any pilot will tell you, the only thing more useless than the altitude above you or the runway behind you is the fuel left back at the base. Airplane operators actually say you can't have too much of the deadly liquid unless it's on fire. It's not the falling or the sudden stop at the end that claims all the victims. It may not be the structure itself, even though airplane components are built of the same material that Solid Rocket Motors use for fuel. Most likely what kills the occupants is the inferno caused by tons of petroleum fuel bursting forth and soaking everything. An airplane must carry much more of the deadly liquid concoction, because it must provide energy both to lift *and* propel the airplane. Adding the danger of petroleum, Newcomb's argument is hard to beat.

We depend on generations of technology to overpower nature's tendency to punish the foolhardy. Periodically we find some combination of events that sidestep these safeguards, and the result is usually an instant loss of life greater than all of the airship accidents ever *put together*. We pay good money to be stuffed in to large winged aluminum tubes with the expectation we'll be transported with our luggage to some distant destination. Years ago the piston-powered airplane was touted as a cheaper alternative to the jet, owing to the 'stovepipe's' thirst for fuel. The jet and turboprop vehicles demand every available cavity be bulging with explosively dangerous petroleum fuel. More recently evidence suggests the massive fuel load can leak fumes into the airplane's passenger space and cause harm due to minute ingestion.[341] Some accidents provide spectacular, in-flight disintegrations like TWA flight 800. More often, fires on impact just leave a big black hole in the ground. Always having to rush forward is an under-appreciated liability. Countless victims of collisions with hard objects would verify this problem, like the souls aboard the 757 aircraft that hit near Cali, Columbia, if they could. In the ongoing battle between objects made of aluminum going hundreds of miles and hour and the ground going zero miles per hour, the ground has yet to lose. We shouldn't be surprised that, even with flash paper-like covering and huge quantities of gaseous hydrogen they could not purify, and the even partial use of gaseous fuel, the Zeppelin passenger service had a very respectable per *passenger mile* safety record.

Of course removing the explosively flammable components would mean there would have been no accident to the airship that night. However, if secondary, less flammable construction materials were also replaced with noncombustiles, then only the petroleum fuel, lubricant oil and the hydrogen lifting gas would be left to incinerate in an accident from some other cause. In the famous accident, the hydrogen was gone quickly, but the belching black smoke of the petroleum products rose from the wreck burning into the next morning. Those images tell us a great deal.

[341] Another study says the packed-together cramped seating contributes to a higher *postflight* death rate!

"Yet, a more objective analysis reveals that hydrogen is no more dangerous than other flammable materials used as fuels, and even has some significant safety advantages. Ironically, a case in point is the *Hindenburg* incident itself. Although it exploded [sic] hundreds of feet off the ground, 62 of the airship's 97 passengers [sic] survived, partly because the lightweight gas rose clear of the craft as it burned... Kerosene jet fuel would have clung to the wreckage and burned furiously for a long time, likely killing everyone in the vicinity."[342] Indeed, even petrol *vapor* is heavier than air. The belching black smoke rising like a funeral pyre in the morning light reminds us that even the LZ-130 would still have this major safety problem. *Graf Zeppelin* (II) was vastly improved with lessons learned from the LZ-129 accident. Yet the vessel would still have contained liquid fuel sloshing around in necessarily fragile tanks, just waiting to douse some poor soul like the victim who "... didn't have a stitch of clothes on him, all he had was his shoes and they were smoldering." The horribly burned victims in the photographs got off easy compared to the typical air crash today, where nothing is left of the passengers but a big black spot on the ground.

Gasoline has always been an enemy of airships, regardless of what lifting agent was used to displace air. A fire, created by a gasoline being used to clean the car and ignited when a technician switched on radio equipment, occurred inside a hangar. Airships lost in the fire were SSZ-23, SSZ-38, SSZ-54 and the rigid airship R.27. The Germans are thought to have suffered an equally stupid accident, when crewmen cleaning L51's aft car sparked a fire that destroyed the ship – and the L46, L47, L51, L58, and the wooden-framed SL20. That voracious gasoline-fed fire killed 10 men and injured 134. Popular thinking would assume hydrogen, a fuel, caused most wartime fires, though this accident was still a complete mystery to von Schiller when he mentioned it in his 1937 writing. No surprise, the literature seems to contain no references to fires when there was no gasoline involved, and specifically the *lack* of fire when it was not present. The L50 (build number 89) was running low on gasoline trying a high-altitude escape across the Alps. "By the time the ship approached the mountains most of the mechanics had passed out... the unattended motors carried the L-50 full speed ahead toward the peaks near Dommartin. The ship smashed into a peak, knocking off the control car and the aft gondola... Looking up, they saw their ship sailing over the Alps. Dazedly they watched the L-50 hit another mountain, bounce high into the air, and then fade into the mist."[343]

These early airships were of course made of highly flammable materials. No matter, if petroleum fuels are present, they could even destroy a metalclad. As airship design worker P. L. Teed warned in 1925, "petrol vapor is 3.5 times heavier than air, consequently in airships should any of this fuel leak from its storage tanks, or be split, etc. the vapor naturally tends owing to its weight to fall to the lowest available point whether this be the bottom of the engine car or the keel, in either of which places, owing to the heat of the engines or to sparks made accidentally by the dropping of tools, breaking of electrical leads, etc., a fire might be started."

We are vaguely familiar with the danger of gasoline fire in our daily automobile driving, we just don't think about it. "In a review of the [Federal Standard for Fuel System Integrity] in 1990, NHTSA estimated that 1,400 deaths and 9,000 injuries were occurring *annually* in fire-related crashes."[344] Of course you are far safer in an airplane. According to the National Safety Council, only 13,000 people have died in nonmilitary plane crashes since 1914. There are no figures on how many of those were fuel victims, but many accidents are telling. "The inflight breakup, the

---

342 A. Dane, "The Hydrogen Age", *Popular Mechanics*, September 1990, pg. 21
343 J. Toland, "Zeppelins Against England," SHIPS IN THE SKY, Holt, Rinehart & Winston, 1957
344 Emphasis added. Los Angeles *Times*, September 18, 2000

board determined, was initiated by a Jet A fuel/air explosion in the center wing tank… The fuel /air vapor in the ullage of the [center wing tank] was flammable at the time of the accident and the explosion would have been capable of generating sufficient internal pressure to break apart the tank… Since 1966, the FAA has issued about 40 airworthiness directives related to fuel tank safety… The NTSB believes inerting to be 'the most promising method for dramatically reducing fuel tank flammability.'"[345] Again, hydrogen's lower explosive limit in air is 18. Mapco Alaska Petroleum lists its Jet Fuel, grade A as composed of benzene, cumene, naphthalene, toluene, etc. and having a lower explosive limit of 1.0. Farmland Industries lists its Jet Fuel, grade Q, as containing only kerosene and trimethylbenzene, and having a lower explosive limit of 0.7!

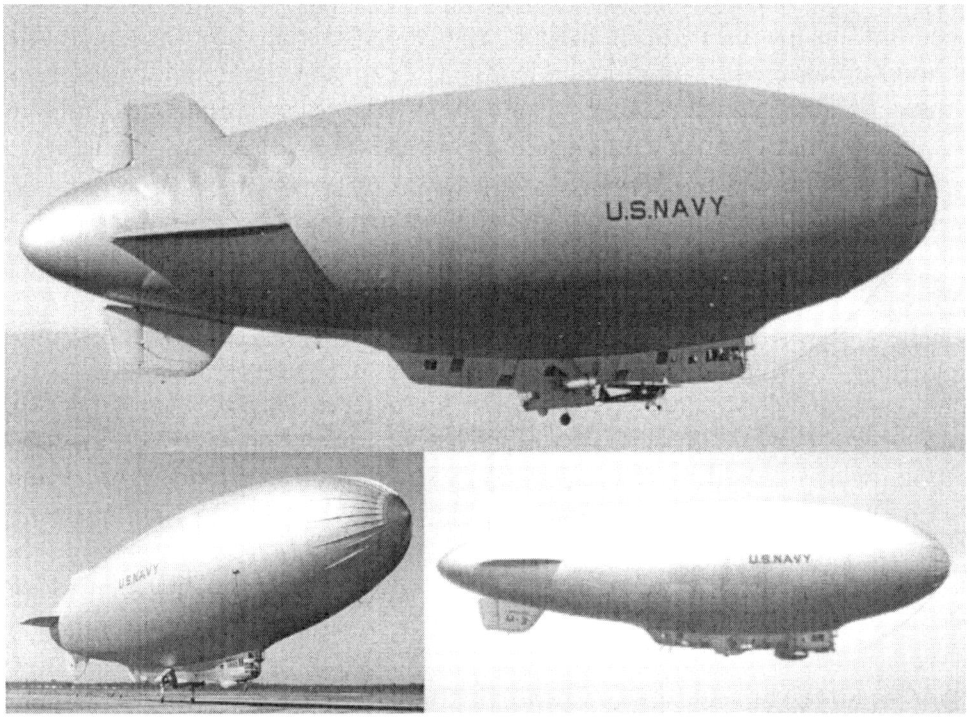

Dr. Karl Arnstein's favorite non-rigid, the mighty "Mike." M-1 (top) carried an airplane as part of standoff weapon tests; M-2 (left) pioneered tricycle landing gear; M-3 (right) proved the AWACS concept; M-4 proved nuke subs could be tracked by their heat trails. Only the four 170 + hour duration ships were built before the Navy lost interest. Courtesy Gordon Vaeth, Hepburn Walker, Jr. and James Shock.

Attempts to make one date back to the late 1800s, but there never has been an 'airship engine' as such – only adaptations of automobile, locomotive, motorcycle, motorboat or airplane powerplants. In fact the early British SS and their American equivalent, the B-ship, were simply adapted airplane fuselages, engine and all. However, there was an "airship fuel" for the LZ-127.

Diesel engines employing their "safety fuel" were on the Zeppelin wish list since the 1920's when Maybach's first effort, the GO 56, developed only 410 hp despite its 4,620 pound mass. In comparison to gasoline, diesel seemed the far lesser of the evils; yet it was no panacea either, first discovered by the British. "It was found that there was no significant difference in accident-frequency rates between gasoline- and Diesel-powered planes… In retrospect, the Beardmore seemed to have everything possible wrong with it. To begin with, it was a straight eight, a configuration that almost always results in a heavy engine. Then, it was a four-stroke, rather than the generally better supercharged two-stroke. Instead of using unit injectors (pump and injector combined), it had pumps at the end of the engine with long tubes making the fuel flow subject to pulses and irregularities. A monster in size as well as concept, it had a bore of 8.35 in. (210mm) and a stroke of 12 in. (305 mm), giving a displacement of 5132 cu in. (84L). It gave 585 hp at 900 rpm and weighted 4000 lb. (1712 kg), for a miserable power-to-weight ratio of .84 lb/hp

---

[345] F. Fiorino, "NTSB Dissects Tank Design Philosophy," *Aviation Week,* 28 AUG 2000

(3.11 kg/hp). It was started by a small gasoline engine and required the full-time attendance of a crew member to keep running. The excess weight of the Beardmores contributed to the fiery demise of the R-101 on its first voyage... Incidentally, an idea of the magnitude of this last of the Zeppelins can be gained from the fact that even with the economical LOF 6, the *Hindenburg* required a fuel tank capacity of 66 tons (60,000 kg)."[346] A lightweight Packard radial diesel was once fitted to the Goodyear *Defender*.[347]

Was all this effort on both sides of the Atlantic directed at oil-burning engines really justified? Cracking "Texas Tea" for the production of diesel fuel oil produces a complex mixture of paraffinic, olefinic, naphthenic and aromatic hydrocarbons. Unlike non-toxic hydrogen, but like gasoline, warnings of all the bad things that actually happen after exposure (as well as long term effects that are suspected of happening) go on for pages in the Material Safety Data Sheets (MSDS). In the refining process after the naphthlene, next up the line of distillates is hydrodesulfurized kerosene, and then kerosene itself. After the hydrodesulfrized middle distillates comes No. 2 diesel fuel. Seemingly much safer than gasoline, No. 2's flammable limit in air is 0.6 percent and its autoigniton temperature is 495°F. Like all oils, it won't dissolve in firefighting water. Flowing diesel can generate enough static electricity to actually *ignite itself*. Water spray intensifies the flames, and the oil floats to continue to burn on top. Nor does safety improve markedly as the fuel is consumed. Mostly empty petroleum fuel tanks are little more than bombs awaiting a suitable ignition source, so even the LZ-130 would have been susceptible to fuel tank explosions. Very high on the list of concerns of a combat pilot as he enters hostile territory is the fact he's surrounded by rapidly emptying tanks of kerosene-like jet fuel. JP4's lower explosive limit in air is 1.3%. In contrast to hydrogen's 18%, should one find out a way to keep such a concentration together, with JP4 "the invisible vapors are heavier than air and travel some distance to ignition sources and ignite/explode." The USAF Materials and Manufacturing Directorate issued, "'Presently, when an F-16 enters an area where there could be unfriendly fire, Halon 1301 is injected into the fuel tanks for one-time use to prevent fuel vapors from being explosive,' said Dr. Juan Viai, Research scientist at ML's Airbase and Environmental Technology Division." Halon 1301 is being banned since it attacks the ozone layer, so airline customers may soon be footing the bill for nitrogen – rich inerting. "Just add nitrogen. That may be the recipe for preventing another catastrophic fuel tank explosion like the one that brought down TWA Flight 800 in July 1996, killing 230 people."[348] (Other fatal fuel tank explosions include a Thai 737 and a Philippine 737.) The last US Navy officer to fly on a rigid airship noted LZ-129's last return from Lakehurst consumed 28,553 Kg (63,000 lb.) of fuel oil, only a part of the total 66 tons available. However, even ullage inerting and then filling LZ-130 with helium would have been incompatible with the heavy liquid fuel tankage, safe or not.

Paul Wilkinson[349] wrote: "...the main engines [Daimler-Benz DB602[350]] ... fuel consumption is 1,200 lb. of fuel per hour oil for all engines. All engines run on ordinary commercial fuel oil." But this was not possible, since a typical four day run to South America would find the vehicle *57.6 tons lighter* than when it started. A helium LZ-130 would be heading for the stratosphere, unable to recover the lost weight, and daring not to valve the gas which could not be replaced at

---

[346] H. Smith, A HISTORY OF AIRCRAFT PISTON ENGINES, Sunflower University Press, Manhattan Kansas, 1986. (Courtesy Charles Landers.)

[347] R. Jackson, AIRSHIPS, Doubleday, NY, 1973

[348] W. Phillips, "De-icing for summer," *Popular Science*, Feb. 2001, pg. 28

[349] P. H. Wilkinson, DIESEL AVIATION ENGINES, National Aeronautics Council, Inc. 1942

[350] DB 602 engines have "... steel cylinders have Daimler-Benz pre-combustion chambers. The connecting rods have roller-bearing big ends and the crankshaft runs on nine bearings. There are two inlet valves and two exhaust valves to each cylinder."

Santa Cruz. Water recovery couldn't help, as C. P. Burgess wrote, "Water recovery from the exhaust gas appears impracticable with diesel engines because of the low hydrogen content of the fuel..." It was a show-stopping problem. Bauer & Duggan: "One flight from Frankfurt to Lakehurst/USA required an average consumption of fifty four tonnes of diesel fuel. In weight, this corresponded to 48,000 cubic meters of hydrogen. The weight of the fuel consumed led to a constant diminution in the weight of the ship compared to what it was at the start. As the ship became constantly lighter, if had to valve off gas in order to remain at the same height. On arrival at the end of the flight, in order to replace lost gas, which was necessary in order to refuel and take on cargo, a corresponding quantity of gas had to be taken on board. The 'loss' of valving off gas had obviously always brought with it considerable disadvantages. Next to the cost, which in the case of helium would be too high to contemplate, there were other problems; and in particular, the difficulty of maneuvering of the ship with an every-decreasing buoyancy was especially noticeable at the end of the flight with the landing itself proving very complicated."

LZ-131, which was to be first of the super-Zeppelins, would have been the first airship with the designed-in lessons of those built to seven million cubic feet in size, but without the helium-hydrogen confusion. The LZ-131 gas capacity was to be nearly 8 million cubic feet, and in fact its overall hull volume was to be 8,250,000 cu ft. The earlier DB 602 engines gave way to the Daimler-Benz MB808, a hefty 18-cylinder diesel brute, of which ten were delivered. (This family of engine would evolve into a 20-cylinder model for watercraft, and example of which is on display at the *Aeronauticum* museum in Nordholz.) Fuel weight loss would have been even greater in this super-ship.

The answer was right overhead. *The Century Magazine* had predicted that by 1915, "Such freight might be carried on an air-line at sixty miles an hour, and more, by man-made cloud, which is not inconceivable when we consider that the Wetzel air-ship, now being projected in Germany, is being planned to transport thirty seven tons of passengers. Immense supplies of gas would be required; but that cost would be negligible, owing to the saving of leakage through the burning of hydrogen as fuel, which offsets the weight lost by the consumption of gasoline." This 1910 article explained this would also offset the cost of fuel. "A new process for producing hydrogen, makes the inflating of ships the size of the Zeppelin IV cost $526, as compared to $3091 a year ago. Gas drawn from the chambers and burned in the motors, to compensate for weight lost by consumption of gasoline, would cost $155, leaving $198 for the gasoline consumed."

It took more than a decade to successfully prove this theory. Shute reported, "At first it seemed expedient to design a special engine for R.100 running on hydrogen and kerosene." "...in 1926, [engine design guru H. R.] Ricardo carried out some tests on behalf of the Air Ministry to investigate the possibility of burning hydrogen which has to be valved from air ships to compensate for the loss of weight as the diesel fuel was consumed. This offered the possibility of extending the range of the air ship by conserving the diesel fuel... The hydrogen was introduced into the induction air at a choke tube fitted in the induction system and at a pressure of 0.018 psi... The valving ratio of hydrogen (1 lb. H_2 for 13 lb. oil) can be matched at about 80-psi bmep. Below this load the engine can easily consume all the hydrogen which has to be valved."[351] As Ricardo recalled in his own memoir, "...the maximum cylinder pressure would only be about 250 to 300 lb. per sq. inch as compared with about 600 lb. in the petrol, and

[351] W. M. Scott, in a General Motors research report, date unknown, courtesy James Shock. The lack of reversing ability meant one Beardmore was mounted backwards, used only to slow down during mooring!

probably nearly 1000 lb. per sq. inch in the diesel engine; thus we could safely employ much lighter scantlings and working parts. Again, the whole cycle temperature would be so much lower that we need hardly worry about thermal stresses, and in short the design problem resembled much more nearly that of a steam than an internal combustion engine." Shute: "After a year's work it became evident that this engine would not be fully developed before it was required for installation in the ship." R.100 wound up with Rolls Royce 650 hp *Condor* V12 IIIBs, mounted in tandem, two per car. "In England, considerable single-cylinder development work was done at the Royal Aircraft Factory, and subsequently, Beardmore produced an 8-cylinder, in-line, water cooled engine known as the *Tornado*-III. It was rated at 550 hp at 900 rpm and weighed 4,550 lb., or 7.8 lb. per hp."[352] There was some work on testing the 1500 bhp Ettore Lanzerotti Spina Stromboli engine for R.101. The six massive 12-inch bore, 16-inch stroke cylinders were designed to deliver peak torque at 900 rpm, allowing a 20 foot diameter prop to be driven direct from the crankshaft. These were evidently considered to replace the Beardmore engines.[353] There was still hope for at least engine modifications for airship use.

After the R.38's gasoline deaths, a lesson had been learned. P. R. Stokes wrote, "The 1924 requirement had specified, with regard to power plant, that they were, 'to be operated on fuel which could safely be carried and used in sub-tropical or tropical climates.' This presupposing oil engines instead of conventional petrol engines, although 'Diesel' engines had only recently been introduced in transportation and none had flown... Beardmore's of Inchinnan, Scotland, were a major engineering conglomerate who as well as building airship, aero and marine oil engines, were pioneering compression ignition rail traction under the leadership of Alan E.L. Chorlton... In 1922 Beardmore built six cylinder diesel units for rail cars operating to 500 hp at 1200 rpm in service with Canadian National Railways and then Britain... In July [1930, there] was serious discussion of fitment of gas bag tankage to allow engine operation on 10% oil and 90% fuel gas, in the manner of the Ricardo proposal, to save 12 tons weight in future flights to India. At R.101's departure Wing Cdr Cave-Brown-Cave was to remain at Cardington in pursuit of this work." Tests of the R.101 burn-off-buoyancy scheme showed the optimal mix was 1 pound of hydrogen per 13 pounds of diesel fuel. R.101's loss ended that promise.

After the ZRS ships had offered operational data for large helium filled gasoline powered airships, NACA performed a series of tests to see if hydrogen could be consumed in diesels along with the oil, so as to match the lift lost to weight being consumed. The tests showed the combination of hydrogen and fuel oil burned more efficiently than oil alone, that more lift could be consumed than was required, and that more water weight was in the exhaust than that of the fuel oil itself.[354]

The Germans realized the same thing and tackled the problem head-on. Bauer & Duggan: "Experiments were conducted on all possible gasses and gas mixtures to establish their suitability for operating a fast running internal combustion engine, their resistance to cold temperatures as well as their effect on cellulose and their corrosive effect on metal with which they came in contact... all investigations had shown that carburetted[355] hydrogen was suitable for

---

[352] L H Morrison, DIESEL ENGINES – A COMPLETE HOME STUDY COURSE, Revised and Edited by C F Foell, Diesel Publications, New York, 1942. courtesy the late Francis Scales. The course shows a Beardmore diesel, a 12 cylinder, horizontal opposed-cylinder rated at 500 hp at 1750 rpm, and weighted 1450 lb. for 2.9 lb. per hp.
[353] Air Commodore F. R. (Rob) Banks, I KEPT NO DIARY, Airlife Publications, 1978, courtesy F. Scales.
[354] H. Gerrish & H. Foster, "Hydrogen As An Auxiliary Fuel in Compression-Ignition Engines," NACA Report No. 535, courtesy Eric Brothers.
[355] Webster's Dictionary defines this as "to mix or charge (a gas or air) with volatile compounds of carbon in order to increase the potential heat energy"

airship operation…" as if foreseeing a modern product like Hythane©.[356] So, the answer came by making a diesel fuel with hydrogen *built in*. Bauer & Duggan: "With the use of German synthetic gas oil from Ruhrchemie, the hydrogen content was 15% so that in this case the water was 135% of the weight of fuel…The weight of the recovered water can exceed that of the burnt fuel since the weight of the recovered particles of oxygen would be taken from the outer air and not from the fuel." Dick wrote the performance with this light oil, 'Kogasin 2,' was helped by its lack of fouling sulfur, thereby extending the radiator overhaul period to that of the engines, 1000 hours. This was another masterstroke. Hydrogen delivers even more, downright surprising results when used as a combustion additive. The generation of carbon monoxide is a problem for internal combustion engines. Increasing gasoline's hydrogen content by just five percent can cut an auto's emissions by 30 to 40 percent. One approach toward their reduction readily lends itself toward hydrogen, whose wide flammability range and gaseous nature allow fuel-air premixing and reduced fuel-air ratios. When injected with a timed mechanism, a considerable improvement in efficiency will result; this too has airship applications. C. P. Burgess explored the idea of using hydrogen injection for suggested diesel engines, then being designed by Allison, to enhance water recovery for the USS *Macon*. "The specific fuel consumption of a Diesel engine should be about .37 lb./hp./hr., and the oil consumption of .02 lb./hp./hr., instead of .48 lb./hp./hr. with the Maybach engines. With the same initial fuel load, this means a gain of 23% in endurance."[357] This would not only be logical for military airships where gasoline is a no-no, but also because several very efficient aero-diesels are soon to be available. The Dutch Rigid Airship is, at this writing, slated for diesel powerplants.

Using hydrogen, a safer airship could have been built by a means no more complicated than *reducing the amount of dangerous petroleum fuel carried*. Making the airship even less of a fire hazard is no more complicated than eliminating the most dangerous energy storage left in the design. Consider what a revolution it would be to construct a vehicle without this basic safety defect, that of petroleum fuel that propels, and with the wing lifts, but also has a kill ratio per passenger mile.[358]

In his argument against helium, P. L. Teed assumed some gaseous hydrogen would be burned in the ship's engines. "In a hydrogen filled airship the propelling machinery can be run partly on liquid and partly on gaseous fuel, (.32 pounds of kerosene and .02 pounds of hydrogen per horse power hour), while in a helium filled vessel the engine must be operated entirely by liquid fuel (.45 pounds of petrol per horse power hour), consequently this greater consumption for a given length of flight, combined with the lesser carrying power of the helium filled ship, very greatly reduces its effective range…" Teed provided figures that showed the maximum full speed length of flight for such a ship would be 40.6% greater than a helium/gasoline airship. A sad second best, putting hydrogen in the fuel at least made water recovery possible. The next logical step, to balance the energy carried with its own weight, would be to store the additive in gaseous form and mix it on board. Thereby, the additive's lift would "pay" for its weight by carrying it, and produce ballast water in the process.

---

[356] Trademark of Hydrogen Components, Inc.

[357] C.P. Burgess, Design Memorandum No. 146, Maybach Engines and Water recovery versus Diesel Engines and Hydrogen Ballonets in U.S.S. *Macon*," October, 1933.

[358] Ironically just as the author was writing this sentence the news flashed that a Concorde supersonic airliner had crashed with almost 20,000 gallons of fuel on board, 25 July, 2000. Air France flight 4590 could have been carrying 31,650 gallons (212,000 pounds) of fuel had it needed to fly its full range of 3,550 nautical miles. Narrowly missing a hotel and impacting a field, the resulting fireball killed only 4 people on the ground. Of course there was very little left of the 109 hapless souls inside the ill-fated craft.

Following that, one would make the logical progression to an engine that performs some of the airship's work and burn lift in the process. Burgess had already dealt with that as well: "...an alternative arrangement is to burn the hydrogen in a special engine, giving the exhaust gas rich in water vapor, from which water ballast may be recovered with much less apparatus than is required by a gasoline engine." In the Burgess memo #146 discussion, it was suggested that besides the 35 pounds of water ballast recovered, "Adding this to the 71 lb. of lift lost by the consumption of the hydrogen gives a total loss of 106 lb. of net lift per 1000 cu. ft. of hydrogen." This ratio of disposable lift and recoverable weight would cover a wide range of operations.

The Germans put the idea into practice. Bauer & Duggan: "Additionally, experiments in relation to the water recovery system were carried out. Modifications to a BMW engine which burnt hydrogen and recovered water ballast, thus balancing the weight of the ship. Experiments with this fifth engine on the early flights of the LZ-130 until the end of 1938." That fifth engine would pretty much need to run all the time. Wilkinson wrote, "Two 50 hp Mercedes-Benz diesel-electric power plants on board supply the electricity for lighting, heating and cooking..." This might have been the answer, but if they had considered such an idea, it was thrown over the side in an effort to save weight for helium use. So LZ-130 was right back to the stone age, more helium to lift more fuel; not enough lift available so it had to start superheated. This is particularly ugly because a lift casualty would have the ship coming down in a pool of its own petroleum fuel. The heavy weight would be the final thing dumped in a last-ditch effort to remain aloft. If hydrogen fuel had no other advantage, it would be worth having the potential flammables *heading the opposite direction* of a plummeting airship, dumping fuel.

Thankfully, we have a way of eliminating that entire petroleum hazard (except lube oil and grease) if we can grasp that hydrogen causes fewer potential problems than any of the energy carriers (oil, natural gas, coal, benzene, propane, etc.) we are presently using. It's water. All we have to do is separate its hydrogen and oxygen before we consume them, and then get water back after the process. There is but one great stumbling block in our march toward a more efficient future. "This famous crash has traditionally been blamed for the decline of the airship industry, and there is little chance the public would accept a hydrogen filled airship with the image of burning wreckage firmly planted in their minds. In fact, by making use of the technological advances which have occurred between the 1930s and the present, hydrogen could be used with greater safety than gasoline for automobiles or jet fuel for jumbo airliners. Nonetheless, unless scientists and engineers can break the fear syndrome surrounding the use of hydrogen, airships of the future will be designed with the use of helium in mind."[359] That's okay, as long as there *is* helium to use. We don't have to *inflate* the main gas cell(s) with hydrogen. Just use it as *fuel.*

Since 1937, hydrogen gas seemed the greatest evil. Yet before the accident made the very mention of the word seem horrific, hydrogen *fuel* for airships was sought even before petroleum fuels had much of a toehold in this country. "Since the early 1890's Myers was looking for a suitable power plant for the Skycycle and in correspondence to Octave Chanute was looking at hydrogen as a fuel in 1894."[360] (Ironically, Sir James Dewar first produced liquid hydrogen in 1898.) Myers had the vision we lack: Hydrogen fuel would offer unmatched airship efficiency. Likewise, the Captain of the most successful airship did not entirely agree with changes that came after the LZ-127. "The LZ-129 and LZ-130 are propelled by diesel engines. It is impossible to obtain powerful enough gasoline engines for this type, and it was also desirable to

---

[359] W. J. White, AIRSHIPS FOR THE FUTURE, Sterling Publishing, NY 1976

[360] Researcher Maureen Lynch found an 1896 Myres letter to Chanute that said "... I presume your new motor is a turbine, either steam...or a new form of engine."

eliminate completely the easily inflammable gasoline...Does this mean that gaseous fuel had not proved satisfactory? It does not. Properly, gas is still the ideal fuel from the technical point of view..."

Graf's **dual keel arrangement (left) allowed management of the hydrogen cells above (being inflated at right) and the fuel gas bags below. LZ-127 was so efficient the new larger 129 could not outrange her. Ernest J. King wrote in 1934 that the use of a fuel having a density nearly equal to that of air is the most promising method of maintaining the airship's equilibrium in flight. Courtesy NAA.**

Once we actually believe what the safety experts have been saying for years, we can then utilize hydrogen fuel's special advantages for our buoyant vehicles. The US Navy's ZPG-2 airship, of 975,000 cu. ft. of helium in capacity, was the most successful of the postwar designs. The ZPG-2 could carry 12,940 pounds of aviation gasoline, which under CAPT M. H. Eppes, established a 200-hour flight endurance record. The only powered flying vehicle that ever broke this incredible record was another ZPG-2. CDR Jack Hunt and his crew stayed aloft without refueling for an incredible 264.2 hours. Those Navy ZPG-2s theoretically got 17,780 Btu for every pound of the highly volatile avgas they carried. The avgas caused its own share of problems.[361] By contrast, liquid hydrogen could deliver 51,600 Btu per pound carried. With no other improvements, just swapping the gasoline with hydrogen fuel, the ZPG-2s endurance could have been greatly increased. So would its survivability, since just removing the petroleum greatly decreases the chance of accident or injury. Industrial safety has shown why. In judging the degree of hazard, three other criteria are examined besides the fire danger: Corrosivity, Reactivity, and Toxicity. Gasoline, jet fuel and diesel oil suffer *all four* problems; hydrogen, only the ignitability. So what about it? The MSDS uses capital letters to emphasize an important point: "The amount of hydrogen gas necessary to produce oxygen-deficient atmospheres is well within the flammability range." Indeed; if you have a dangerous amount of fuel leak into your room, don't expect to breathe normally. Airshipmen would have had a second safety warning, something like "When the level of air contaminating your lifting gas allows it to become flammable, be aware you are no longer defying gravity." Obviously, keeping the hydrogen in the tank away from air is more than just good safety policy, it's an airshipman's reason for being. It is a natural partnership long ignored in safety discussions. Hydrogen can't burn; the more pure it is, the greater load it lifts. There is no conflict; what's good for safety is also perfect for performance. In contrast to gasoline and diesel fuels, hydrogen does not wash off or thin the engine's vital lubrication oil. There is no possibility of fouling a spark plug or dirtying the combustion chamber or developing valve deposits. BMW first demonstrated a practical hydrogen powered sedan in 1990. Ten years

[361] Aboard ZPG-2 Bu No 126716, during a 24-hour training flight on 18 November 1954, a small gasoline leak developed and, while attempting to repair it, the fuel tank fitting broke off the tank. Gasoline flooded the bottom of the control car. Thankfully the internal engines were well sealed-off from the interior air of the car, and a fire did not occur. On landing, with one engine and no take-off power for a wave-off, the tower reported a fire aboard. The "fire" turned out to be the vaporized fuel streaming from the car.

later, the fifth generation car's 37-gallon liquid hydrogen tank is refilled in less than five minutes, giving the car a range of 250 miles. One would hardly notice any differences under the hood containing the standard-looking twelve-cylinder BMW engine, seemingly identical until one measures performance. By the late 1990s, the BMW 7-series H_2 cars were capable of 400 km per robotic, three-minute fill-up. Their 5.4 liter 12 cylinder turbo engines can get the big car up to 100 km/h in less than 10 seconds.

Also important to spinning a big propeller is *where* on the torque curve the power is available. "The optimal efficiency of a hydrogen fueled motor is achieved earlier in the load range than a gasoline engine..." Furthermore, "Compared to gasoline [hydrogen's] actual thermodynamic process more closely approximates the equal-volume process, which serves as a comparative process for the Otto-cycle engine. A requirement of this equal volume process is the total combustion of the cylinder charge at top dead center where the flame speed *must approach infinity.* The considerably higher flame speed of hydrogen compared to gasoline under the same conditions is therefore the cause for improved conformity to the equal-volume process which in this respect is superior to any other process with slow combustion." Hydrogen's very small kinematic viscosity and the strong turbulence associated with it "...provides improved idling with multi-cylinder engines as a result of the uniform distribution of fuel to the cylinders...fuel enhancement is not necessary for hydrogen engines at cold start or during acceleration." Virtually knock-proof fuel efficiency is welcome anywhere. Critics bemoan hydrogen's lower heating value per unit volume as a major disadvantage, but that negative is actually advantageous to an airship. "In the case of a heat engine, the use of hydrogen permits a gain of efficiency of 20-30%; for instance, for a consumption of six liters per one hundred kilometers, thirty liters of petrol would be needed to travel five hundred kilometers compared with 6.5 kilograms of hydrogen."[362] Small wonder why Shell Chairman Mark Moody-Stuart created Shell Hydrogen. Others, including Texaco and BP Amoco, are readying for the hydrogen fuel business.

Yet there is much more synergy using hydrogen fuel in buoyant vehicles than just increased safety and improved engine performance. At STP, the partial pressure of the exhaust's water vapor will cause water to condense out if cooled below 120° to 140°F. This causes liberation of the heat of vaporization for any water condensed. At the height of experience using USS *Akron's* refined Mark IV watery recovery apparatus, slightly better than a pound of water was recovered for each pound of hydrogen-rich gasoline burned. Hydrogen fuel by itself returns a theoretical 8.94 pounds of water per pound of fuel burned, nearly *nine times* as much. As a diesel additive envisioned by Burgess, "Combustion of 1000 cu. ft. of hydrogen yields 47 lb. of water vapor, of which 35 lb. may be counted as recoverable." The weak point in all the water recovery equipment was its increasing degradation as soot coated its guts, and cleaning was awarded as punishment. Dropping hydrocarbons eliminates the soot. Stainless steel mufflers, likely added for crew comfort in a modern long-endurance airship, would probably recover enough water simply by welding on cooling fins and plumbing in a set of drains, and cooling them with the airship's helium. Nor would cold temperatures be a problem. In the ZRS sized ship, 12,000 pounds of anti-freeze could be dumped. Jerome Hunsaker answered Eckener's request for water recovery details on December 23, 1930, writing, "the effect of freezing temperature is a nuisance which seems to be practically eliminated by fuel gas."

Hydrogen's low octane rating can be bumped up by mixture cooling, such as water injection – and as we've seen, there will be plenty of water. The Zeppelin Company experimented with

[362] B. Duret and A. Saudin, "Microspheres for on-board Hydrogen Storage," *Int. J. Hydrogen Energy*, Vol. 19, No. 9, 1994, pp. 757-764.

water recovery just from the air itself using silica gel sand, using engine heat to extract the water and recycle the gel. C. P. Burgess noted in his memo #160 the apparatus weighed 1200 kg but would gather 500 kg. of water an hour if the outside humidity was 55% at about 60 degrees F. It would make pure, useable water instead of the filthy ballast produced by cooling exhaust gases. Not certain the apparatus could be made practical, the US Navy never built one; perhaps a modern version would offer benefits. With progress made in the area of hydrogen-generating electrolyzers, one might assume they could be carried aboard an airship and make hydrogen additive right out of the ballast water recovered from the exhaust. The claim has been made that the extra hydrogen in the fuel mix would more than pay for the energy needed to create it. If this is too optimistic, it might become practical as efficiency of elcctrolysis continues to evolve.[363]

Put it together: 100 Standard Cubic Feet of hydrogen weighs 0.521 lb. (0.236 KG); in a balloon, it lifts itself and about 6 pounds of cell material, airship and payload. Consuming the 100 SCF delivers between 28,317 KJ and 33,486 KJ (6,768 Kcal and 80004 Kcal) of energy; the six pounds of lift disappears, and about four pounds of weight could be *gained* in the process.

Hydrogen can also deliver useable torque if harnessed in a different kind combustion chamber. Its quick burn time yields much more *kinetic* energy, quite useful in spinning a propeller or alternator via a *rotary* engine. With no valves to float, the rotor doesn't mind hydrogen's six-times-faster-than-gas burn rate. It just spins faster. A rotary's combustion chambers are naturally cooler than a reciprocating engine, and its separate intake chambers naturally solve hydrogen's inherent tendencies to backfire in unmodified piston engines. Mazda demonstrated its first hydrogen powered rotary engine car in 1991.[364] The HR-X concept car's 1308cc engine, running a stock 9.7 compression ratio, produces 100 hp, almost the same as the original gasoline version. Recently Moller has shown a combined rotary engine-generator occupying only one cubic foot but producing 15 horsepower; one can only imagine the possibilities with hydrogen fuel. The rotary engine's major problems all but disappear when petroleum fuels are removed. By 1994 Mazda had squeezed a 500-lb. metal hydride hydrogen fuel "tank" into a 120 hp *Miata*. "Today, the company has a total of 11 hydrogen-fueled cars in test stages. One is operated by a steel company in Japan, where the by-product of steel making – hydrogen – is abundantly available as a fuel for the vehicle."[365] Rotary engine fans claim hydrogen's low ignition temperature is not a problem with the Wankel. It was possible to run one without a throttle just by varying the hydrogen-in-air ratio around stoichiometric. Its low weight per unit power makes it attractive for airship use, perhaps overcoming its lesser fuel efficiency. Water injection finds the combustion utilizing steam for extra kick. Think the Germans were wise to unload the Wankel on the Japanese? NASA's Rotary Engine Technology Enablement Program had the goals of 2000-hour ops between overhauls, 28,000 foot altitude, and 5 hp per cubic inch. With hydrogen, Wankels have greater possibilities, possibly overcoming their low-compression inefficiency with hydrogen, which of course needs more air in the process.

BMW sees small *fuel cells* replacing batteries in hydrogen internal combustion engine cars, calling them "electrical APUs." In their fifth generation hydrogen cars, a lead-acid battery sized PEM fuel cell puts out 5 kw, more than enough to run the car's electrical accessories. These new

---

[363] Those unfamiliar with the laws of thermodynamics believe suppressed inventors long ago perfected an "engine that runs on water," meaning the electrolysis was so efficient the generation of the hydrogen fuel took only a fraction of the energy released when it is burned in the combustion chamber. Nothing mystical about it; the concept needs only the right catalyst or perhaps even a resonant electrolysis chamber.
[364] "Hydrogen Rotary Engine," *Popular Science*, February 1992
[365] J. Dunne, "Hydrogen-Powered Miata," *Popular Mechanics*, October 1994

smaller units can cool themselves with their own water output, needing only a vent fan for air.[366] It is to this combination of separate motive power and electrical energy generation that would seem to be most appealing to a modern airship designer. We could power electronics and cool them, blow in ballonet air, even run megawatt radar, while the motive engines idle.

If money is less an object and turbines are used, hydrogen has been called the perfect fuel. More efficient because of increased turbine inlet temperatures, a hydrogen-powered turbine could be smaller, lighter, and would require less maintenance. Airships can also take advantage of regenerative plumbing, too bulky and aerodynamically unsuitable for airplanes. Likewise, a combination fuel cell/gas turbine power plant is possible by teaming the high-temperature fuel cell with a gas turbine, air compressor, combustor, and metallic heat exchanger. Compressed air and fuel would pass through a metallic gas-to-gas heat exchanger to recover heat from the combustion product gases leaving the gas turbine. Thus heated, the fuel and air streams would pass into the anode and cathode fuel cell compartments respectively, where the electrochemical reactions would take place. Fuel cell exhaust gases would be mixed and burned, raising the turbine inlet temperature while replacing the conventional combustion section of the gas turbine system. Expansion of the fuel cell exhaust gases through the gas turbine would provide an inexpensive means for recovery of the fuel cell waste heat. Synergistic effects of the combined fuel cell/gas turbine could lead to electrical conversion efficiencies of 72-74 percent for systems under 10 MW capacity.

In case you have not yet read chapters one through eight, and still think hydrogen was the weak point in the Zeppelin scheme, consider a report prepared for the Department of Energy. Factory Mutual Research Corporation of Norwood, Massachusetts, compiled a data bank of hydrogen accident reports in order to analyze the information and anticipate common hazards associated with hydrogen use. The study shows a multitude of hydrogen uses dating back to 1965, and the safety record is a good one. With the use of hydrogen among a wide range of applications, experience with handling hydrogen in its different forms is long standing. This report, along with many others, has helped form the conclusion that hydrogen, handled properly, will be a safe alternative to fossil fuels. Not only the knowledge, but also the rules, are already in place. The Department of Transportation regulates the distribution of hydrogen over the nation's roads. Gasoline tankers involved in traffic accidents routinely incinerate their drivers and nearby victims. In an accident where a eighteen-wheeler carrying 14,000 gallons of liquid hydrogen was broadsided by a car, the auto's gas tank burst and set everything else on fire – without any casualty to the hydrogen. There are at this writing about 30,000 shipments of liquid hydrogen tankers to hundreds of locations in the US annually. Fact is, in more than 40 years of working with hydrogen, NASA seems to have no record of anyone being hurt by it. That includes the extra danger of super cold liquid hydrogen. Seemingly benign nitrogen killed two NASA contract workers in 1981; life-giving oxygen alone has killed five astronauts. In August of 1986, in Cameroon, some 1,700 people were killed in a single night by carbon dioxide. Again, *hydrogen does not appear to have actually hurt anyone in forty years of US use.* The safety record of hydrogen speaks for itself; we should start enjoying its advantages.

Using hydrogen as an energy carrier means it is very easy to move energy; sending electricity a long distance is four times as expensive as shipping the equivalent hydrogen by pipeline. This might be well for ground based systems, haven't these people seen the movies of the *Hindenburg* burning? While airshipmen have long been horrified by just the mention of filling a balloon with

[366] D. Simanaitis, "Is there an H_2 APU in your future? BMW sidesteps the fuel-cell bandwagon," *Road & Track*, November 1999, pg. 114

it, other users of hydrogen are as varied as today's global marketplace. Hydrogen is a key component in the manufacture of chemicals such as ammonia and methanol. It is used in large quantities in refineries for manufacturing gasoline and heating oil. It is used to make fertilizers, glass, plastics, resins, refined metals, vitamins, cosmetics, semiconductor circuits, soaps, lubricants, cleaners, margarine, even peanut butter. It is needed to synthesize nylon and polyurethane. Tens of millions of pounds of hydrogen are used daily in production plants across the country and around the world; more than 50 million pounds *daily* in the US *alone*.

Again, are these people fools? Don't they understand it's mysterious, the experts were never able to pinpoint the way hydrogen 'blew up' the *Hindenburg*? How can they handle such a spooky-dangerous commodity? Actually, it's routine. People involved in the manufacture of equipment know the standards for the design of hydrogen systems, and individuals who work with these hydrogen systems are familiar with safe handling procedures. This knowledge is developed through training in the operation of the complete system, understanding the properties of hydrogen and observing the necessary safety precautions involved in hydrogen production, handling and use. From the hydrogen rich 'town' gas experience of the distant past, to the hydrogen gas experience of today, a solid foundation of knowledge has been established on how to handle hydrogen safely. Hydrogen is an energy carrier; to be a fuel it must have combustible properties. These warrant the same caution any fuel should be given, and some cautions that are unique to hydrogen. Do we have the experience to handle this seemingly exotic fuel safely? In a word, yes. It is distributed by over-the-road trailers and rail and barge, and by pipeline. Approximately 700 miles of hydrogen pipeline exist in the U.S., Germany and England. This is small compared to natural gas systems, but it is important to note that there are hydrogen pipelines in place today that deliver gas to users successfully. Air Products and Chemicals, Inc. has delivered more than 600 million gallons of LH_2 to NASA alone.

As we stated in chapter two, hydrogen generation and handling has had an excellent historical safety record, and it continues today. Associations like the American National Standards Institute and the American Society of Mechanical Engineers publish standards for components used in hydrogen equipment. While gasbag men have long believed it impossible, the Material Safety Data Sheet states, "The fire and explosion hazards can be controlled by appropriate design and operating procedures that prevent the formation of fuel-oxidant mixtures and by removing or otherwise inerting potential sources of ignition in the area where the hydrogen will be used."

The Compressed Gas Association sets standards for many gases including hydrogen and their published data cover gas production, handling and use. The standards set by these organizations are high. Typical tube trailers hold up to 126,000 cubic feet of hydrogen under up to 2640 psi pressure. Tube modules carry up to 150,000 cu ft at 2400 psi. Both are manufactured and maintained under DOT-3A/AA specifications that pressurize the cylinders at least three times the service pressure. The excellent safety record continuing with ever increasing production and use is no accident. The nearly unblemished record is due to the use of proper safety precautions, familiarization with hydrogen properties, and strict adherence to rules, regulations and standards established through years of experience and research. It is industry's long standing commitment to these high standards that enables hydrogen to be safely handled and used today.

We can also consider that keeping hydrogen away from oxygen is the perfect way of preventing undesired reaction, such as accidental ignition. As Marc de Piolenc wrote, "Recombination catalysts developed for large storage batteries allow H_2 and O to re-combine non-explosively at concentrations well below the explosion threshold. Running the [hydrogen fuel] cell contents through a bed of this stuff would allow hydrogen to combine with any oxygen present, producing water, which is drained out." An airship also has a great place to keep it; a helium cell would represent about the safest place one could store the fuel. This was not a new idea when Eberhard Lempertz of L-Z was granted a patent that called for "the usual fuel gas is stored in an admixture of an inert gas which is not flammable… The compression of the motor causes good ignitability."[367] Rocketdyne patented *Tridyne*, which uses helium to hold hydrogen "harmless" until a catalyst extracts the fuel for energy production. Their process is proprietary, but separating the two gasses would be no more complicated than adding a palladium filter to the standard, transportable membrane helium purification process. (If we are going to be using helium for cooling electronics and engines, we'll have to be moving it, and in-flight purification is an added possibility.)

If the worst did happen, an oxy-hydrogen fire can be snuffed out with most any commercially available fire extinguisher. Water is not as effective, but neither will it spread the fire wildly as in the case of burning gasoline, lube oil or diesel fuel. The accident that destroyed the LZ-129 was caused by an unforeseen vulnerability in what was thought to be the pinnacle of the airshipman's art. Hydrogen was abandoned, but industry has long been developing counter measures to battle more dangerous and frequent *petroleum* fires. The new 'Cold Fire' extinguishing agents from Firefrezee Worldwide, Inc. works to remove heat *and* hydrocarbons from the fire triangle. Hardly a week passes without a child being horribly burned by gasoline having to then be taken to one of the Shriner's Burn Hospitals. Firefrezee's burn treatment, made from the same environmentally friendly organics, shows great promise in treatment of even gasoline injuries.

Taking the extra step of making the lifting gas container fire resistant greatly increases an airship's survivability, regardless of what gas is inside. It is easy to imagine an airship constructed of fire-resistant materials. Contrast the silk-decorated, nitrate-doped fabric walls of the LZ-129 passenger compartments with today's 3M Nextel 312. Thin as paper and weighing only 2.3 ounces per square yard, Nextel 312 survives the FAA's 15 minute 2,000 degree firewall test without melting, burning or shrinking. Should a fire occur in a hydrogen fuel system by some as yet unimaginable scenario, new passive technologies are available to prevent the spread of fire. The US Navy is adopting a new intmescent paint that expands at a given flame temperature to reduce both flame spread and heat transmission. Coupled with hydrogen's low radiance, NoFire Technologies' A-18NV coating would greatly enhance airship survivability. Firefrezee Worldwide, Inc. also makes 'Fire Block,' which is employed to treat almost all Class A materials to prevent flames from spreading.

Giant new envelopes and fuel tanks can roll out of the factory will greater integrity when new leak detection technology is utilized. The *Physics News Update*[368] reported on a new "photoacoustic" technique that can detect a leak so small it might take a week to fill a thimble. The enclosure is filled with an inert tracer gas and then scanned with a CO_2 laser. Heated when hit by the laser, the leaking gas makes a detectable sound pulse. Materials safe against nature's electrical trickery are also coming. BF Goodrich has introduced Estane thermoplastic polyurethane, which is non-halogen, flame retardant, a low smoke generator and provides a good flame resistance in profiles, tubes and both molded and/or extruded parts. Their Stat-Rite S-250, a static dissipative polypropylene polymer alloy system, provides two orders of magnitude lower surface resistance (10^8 ohms) than existing polymers. The new 'Sorbothane' line of products absorbs shock, isolate vibration, and muffle sound.

---

[367] US Patent 1,990,993 granted 12 Feb 35. Arnstein mentioned such an effort at G-Z in 1930.

[368] Newsletter of the American Institute of Physics, June 15, 2000 issue

New technology awaiting applications in LTA are electric actuators and all-axis-thrust propellers. Courtesy ARFL Horizons, and Roy Gibbens, AIAA.

Rigid airship enthusiasts have long wondered what might be accomplished with construction materials not available in their era, the Zeppelins having come and gone before even the blind rivet was invented. Boron fibers, made by chemical vapor deposition using hydrogen reduction of boron trichloride gas on to tungsten, first offered high strength, compression and temperature resistance. *Cargolifter's* 118,846 cubic meter 'vectran' test balloon, of greater displacement than the LZ-127, withstood *eight* inches of water gage pressure. Montreal's Nexia Biotechnologies announced their BioSteel© fiber that claims better strength in a product some 25% lighter than even the newer aramid fibers. (Previously unthinkable farming of spider silk was made possible by their inserting spider genes into goat mammary glands, each animal producing the silk of 10,000 spiders.) Even these new developments may be outpaced by chemical-resistant carbon nanotube high-strength fiber technology. Composite constructions too are evolving; structures would benefit from newly developed polyphenylene sulfide glass fabric composites. Nickel coated carbon fiber composites shield delicate electronics.

Likewise, airshipmen have long dreamed of applying modern manufacturing techniques, such as auto ultrasonic cutting. Overcoming resistance to stop using their big-investment riveting machines, manufacturers are slowly catching on to the versatile and better friction stir welding technique. Electroforming and other new technologies are more suited to affordable small production runs. Since no airship model has or ever will enjoy a production run of more than a few hundred, DTM Corporation's Sinterstation laser prototyping system is an example of welcome technology. Feeding in only CAD (drawing) data, a company can develop prototype parts from a great variety of materials that can withstand high temperatures and harsh conditions. This was recently extended to include metal composites by Metal Matrix Composites, Inc. From CAD data, their "3-D printing" quickly produces a porous ceramic preform that is repeatedly "printed" with an organic binder and sealed in what they call a "Tool-less Mold."

Lightness and strength are no longer exclusive in materials; safety and efficiency can and often do go together as well. The inherent efficiency of buoyant flight is represented by the unrefueled powered flight endurance record. It is still held today by a Navy ZPG-2 non-rigid airship, Bu No 141561. Designed back when *steam locomotives* where still being built, its car and fins given to a museum before the Navy had given up on hi-octane gasoline, long since judged as being too dangerous for sustained shipboard use. Far safer than petroleum fuels, hydrogen has 52 times greater buoyancy and 12.2 times greater diffusion coefficient than gasoline, so it will disperse much more quickly than gasoline or even natural gas. Similarly, hydrogen's lower flammability limit is *four times greater* than that of gasoline.

If airships seem to have little military value today, one can only consider the ridicule one would endure by suggesting hydrogen fuel for Naval airships destined to go into harm's way. Ford Motor Company contrasted its electric car's hydrogen storage system to ordinary feeble gasoline tanks: "Each tank is tested at 1.5 times its rated operating pressure, and samples from each lot are pressure tested to failure. Each tank design must be qualified at 2.25 times normal operating pressure. Each class of tank is also subjected to gunfire and must not explode but leak only through the bullet-hole." BMW's tanks have survived safety valve-blocked smashings.

Similarly, hydrogen's need for a more robust tank extends its safety when hydrides are used for storage. The book HYDROGEN: FUEL OF THE FUTURE reports, "In another test, Billings Energy Research Corp. shot a hole in a iron-titanium hydride tank charged to full capacity [with hydrogen]. The armor-piercing incendiary bullet ripped through the casing of the tank, but there was no explosion; only burning gas, escaping from the hole, was detected... For comparison, a gasoline tank was shot with the same ammunition. The explosion threw flaming fuel over a wide area. This illustrates a safety advantage of all gaseous fuels; they dissipate more rapidly and don't soak into porous materials." Again, once burning, the ascending gas doesn't initiate secondary ignitions to any great degree. "Surprisingly, perhaps, hydrogen burns with almost no heat radiation. It is anticipated that the aluminum fuselage of a modern airliner would be enough to withstand a fuel fire and protect the passengers within." [369]

Furthermore, "In a collision in open spaces, a safety-engineered hydrogen [Fuel Cell Vehicle] should have less potential hazard than either a natural gas vehicle or a gasoline vehicle due to four factors. First, carbon fiber wrapped composite storage tanks (the leading high pressure storage tank material due to its low weight) are able to withstand greater impacts than the vehicle itself without rupture, thereby minimizing the risks of a large release of hydrogen as a result of a collision. Second, hydrogen, if released, disperses much faster than gasoline due to much greater buoyancy, reducing the risks of a post-collision fire. Third, the FCV will carry 60% less total energy than a gasoline or natural gas vehicle, resulting in less potential hazard should it ignite. Finally, the design recommended here includes an inertia-activated switch in each FCV that, in the event of a collision, will simultaneously shut off the flow of hydrogen via a solenoid valve or valves, and will cut electrical power from the battery."[370] Liquid hydrogen storage is making slow advances. The vacuum jacket bottle and lines may soon evolve into thinner jackets with water vapor traces in the vacuum, steaming away heat to a different part of the container.

Check out what the US Army is proposing for its combat soldier's uniform in shooting wars to come: "Microturbines fueled by liquid hydrogen would provide a steady two watts of power that can surge to 20 watts if needed. Ten ounces of fuel contained in lightweight, plug-in cartridges would last six days."[371] Probably before the military adopts such a new technology, youngsters will be toting wearable computers powered by tiny hydrogen fuel cells. The internet buzzed, "Speculation about what "it" is has run wild, prompting massive coverage by national and international media... 'Ginger' has to do with a ground-breaking, scooter-type vehicle that can balance on two wheels. But the real revelation is the power behind it - hydrogen, which runs basically emission-free... The ramifications of a 'hydrogen economy' would be profound on everything from the environment to the energy business to global politics." As we shall see, whatever revolution comes, it will be of greatest benefit to buoyant vehicles.

---

[369] T. Newton, "Rising Power," *Flight International*, 5-11 FEB 97, pg. 37.

[370] "Direct-Hydrogen-Fueled Proton-Exchange-Membrane Fuel Cell System for Transportation Applications: Hydrogen Vehicle Safety Report" by Ford Motor Company, May 1997, page xi

[371] F. Vizard, "Future Soldier," *Popular Mechanics*, July 2000, pg. 45

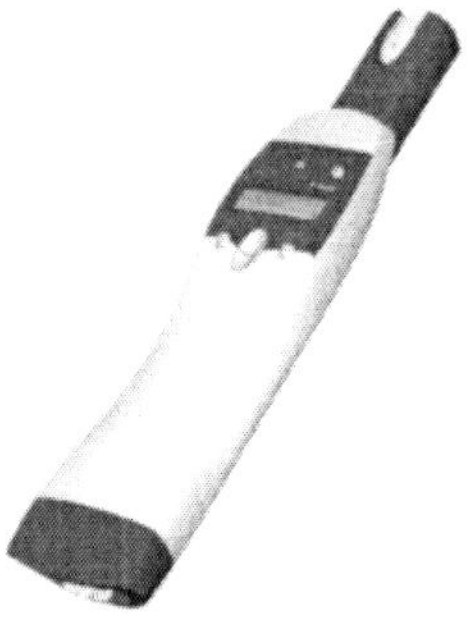

The double-walled steel liquid hydrogen tank in the BMW 7-series holds 140 liters, adding 350 kilometers to the standard gasoline range. DCH Technologies' new hand held hydrogen detector (right) is accurate down to 10 parts per million. Courtesy BMW and DCH.

Slight leaks would of course be as much a concern as with other gaseous fuels, owing to the difficulty in detecting the actual leak site. Adding the odorant as the Zeppeliners did would be cheap, and offer a useful backup. Odorant would be less effective with a pressure airship because it does not indicate the concentration nor pinpoint the origin. Unlike the pioneers, we now have thermal conductivity meters available for use. Twenty years ago Rocketdyne perfected a palladium-based tape that absorbs the gas and changes the color of its thermochromic paint. A 1975 study suggested "For continuous hydrogen leakage monitoring of the atmosphere during filling, launching, landing, and refilling operations, a catalytic combustion detector with multiple remote-sensing heads could be employed."[372] Miniaturization and mass production have made it possible for a network of inexpensive sensors to be placed in grids to instantly warn of a leak and pinpoint its location. The DCH Corporation has licensed the Robust Hydrogen Sensor technology and expects to be competitive in what is likely a $2 billion annual business.

Hazards to property are manageable; the human side has also been considered. OSHA has no Permissible Exposure Limit (PEL) for hydrogen and NIOSH has no Recommended Exposure Limit (REL) for it either. Again airship history offers a baseline in assessing the fuel's effect on the human body. Abbott: "Although hydrogen is both colorless and tasteless and itself not poisonous, it has unpleasant effects when inhaled; in quantity it can cause asphyxiation. One man was overcome on No.9 [R.9] while checking for leaks... another fell into the gasbag of SSZ-55 and slashed his way out through both ballonet and envelope and only just in time to avoid suffocation." In our modern safety-conscious world, the MSDS states, "Hydrogen is nontoxic and classified as a simple asphyxiant."

Of obvious concern for system design is the element's interaction with materials it would contact. The MSDS states, "Since hydrogen is non-corrosive, special construction materials are usually not required." In plant design, recommendations about the proximity of nearby structures show that in system sizes greater the 150,000 cu ft, heavy timber or ordinary construction buildings can be as close as 25 feet; sidewalks as close as 15 feet; fire-resistive construction, zero feet. Just as the masters who fireproofed the LZ-130 the best way possible at the time, we have learned the lesson, *don't put your hydrogen in something that will burn.*

There are compelling reasons to adopt hydrogen fuel for airships beyond the desire to get rid of dangerous petroleum fuels; it's not just some exotic new technology that has hi-tech rocket

[372] C. J. Huang and C. Dalton, NASA Contractor Report CR-2636, CARGO TRANSPORTATION BYAIRSHIPS: A SYSTEMS STUDY, Univ. of Houston

appeal. Amid new technologies also available today are high capacity, long lasting and environmentally friendly nickel-metal-hydride batteries. These units feature a nickel compound anode and a hydrogen occluded alloy cathode, and have twice the energy capacity of standard Ni-Cad batteries. Japan's NEC Corporation is developing batteries that work more like capacitors. "When the battery is fully charged, hydrogen atoms, which each have one electron and one proton, are bound within the cathode material. When an electrical circuit is opened, the electrons split from the atoms and flow through the circuit toward an anode, giving off electricity."[373] Yet we might not need any of them. Taking a cue from the Space Shuttle, which contains no power batteries at all, today's ultra-reliable power backup systems rely on *fuel cells*. Just as the airship's lifting gas made its lift independent on its motive engines, its separate generators allowed full electrical loads even with engines stopped. Today, we have fuel cells placed in back-up power systems that require high reliability – 99.999%, or no more than a few seconds per year down time. Fuel cells also date back to the 19th century, but have enjoyed a rapid evolution thanks to the space program.

All three of a Space Shuttle's mid-1970's International Fuel Cells (a South Windsor, Connecticut company, a subsidiary of United Technologies) need oxygen and *combined* produce 21 kw. Twenty years later the company and its competitors have produced greatly improved models, like the Proton Exchange Membrane (PEM) fuel cell system producing 259 kw fit into a the Mercedes NECAR II van prototype, fed from 45,000 liters of compressed hydrogen stored in seven roof mounted tanks.[374] We need not carry tricky cryogenic oxygen as in Space Shuttle fuel cells, because at the airship's low altitudes we can get all the oxygen we need from the surrounding air. Collecting heavy oxygen molecules from the air to make water in the process means our airship could be gaining disposable weight even without running its engines for water recovery – or at all. Even cleaner than an IC engine, the only 'exhaust' is all the pure water a crew can drink. Promising amid the various fuel cell types is the PEM type, which rely on a platinum electrolyte and an acid-impregnated polymer membrane. In the early 1990s a method of reducing the required platinum by a factor of 40 made the technology more cost competitive. Early in 2000, Canada's Ballard Corporation rolled out its series 900 fuel cell, capable of being manufactured in a drive train for about $60 per kilowatt. That's about $4500 for a 100 hp drivetrain, compared with about $3000 for a piston engine equivalent.[375] Motorola has demonstrated a prototype of a ceramic-based microfluidic fuel delivery system for a miniature direct methanol fuel cell (DMFC). The goal is five times the energy density of conventional Li-Ion rechargeable batteries – and that's *including* the reformer to break out the hydrogen. Manufacturing technology is also finding improved fuel cell part making methods. Apex Plastic Technologies and Premix/Quantum have collaborated to create a new thermoset injection process to replace the more time-consuming cell plate-making compression molding process.

Once again airship applications offer unique opportunities. Joanna Marchant wrote, "Today's fuel cells are heavy blocks as big as a car engine. The idea for an energy-generating skin was hatched by Lockheed Martin engineer Laurence Williams... he could solve the problem by marrying fuel-cell technology with advanced flexible polymers making a skin for the airship that would also generate power... the membrane is an acidic polymer layer that acts as the electrolyte. At the anode, on the inside of the skin, hydrogen gas loses its electrons to form hydrogen ions which travel through the polymer layer. When they reach the cathode on the outside of the polymer sandwich they react with oxygen from the air to form water vapour. The

---

[373] D. Normile, "Proton Power," *Popular Science*, November 2000, pg. 48
[374] G. Kobe, "Batteries Not Included," *Automotive Industries*, June 1996, pg. 61
[375] Dan McCosh, "Hydrogen on Wheels", *Popular Science* May 2000, pg. 61

energy liberated pulls electrons into the electrode, driving current round the circuit. Thin layers of platinum catalyse the reaction, while a fine wire mesh covering each surface acts as the electrodes... In an airship, the hydrogen would not only serve as fuel, but also provide extra lift."

The first airships using hydrogen fuel will be powered by solar cells in concert with fuel cells. "AeroVironment has already shown that Helio's predecessors can fly at record-setting altitudes in the daytime. The giant step to be taken is to make solar-powered aircraft also fly through the night... the company is pinning its hopes on efficient fuel cells that will store excess solar energy during the day by electrolyzing water into hydrogen and oxygen, then converting the gases back into water and electricity during the night."[376] Sounds heavy for an airship, since nearly 5 mwh of electricity are required to make 1000 cubic meters of hydrogen and 500 cubic meters of oxygen. When passed through a fuel cell, this hydrogen will only yield 1 mwh as electricity and 1 mwh as heat. This yields an (electrical) efficiency of "only" 20%. Since the sun's power is 'free,' however, this bears closer examination. These unmanned vehicles will be operated as communication relays stations, well above the clouds, utilizing solar power, then switching back to fuel cells at night. Lockheed-Martin and a host of smaller competitors are racing to perfect these up-to-five million cubic foot capacity unmanned airships. Some may employ new lightweight gallium-arsenide on germanium solar cells, likely 18% efficient, to split fuel-cell generated water back into hydrogen and oxygen in a closed loop system that could run indefinitely. "A new type of semiconductor material can split water into hydrogen and oxygen directly from sunlight. Developed by the National Renewable Energy Laboratory (NREL) in Golden, Colorado, it produces hydrogen twice as efficiently as electrolysis, which uses electricity to split the molecules. The cell [is] made of platinum, gallium arsenide (GaAs) and gallium indium phosphide (GalnP)..."[377] Proton Energy Systems are developing their Unigen® regenerative fuel cell system to do the same thing on the ground using any combination of wind, solar, and hydroelectric to split water and then turn the hydrogen into electricity. Matsushita has demonstrated a co-generative system that generates 1 kw from its one- liter enclosure of what they call their polymer electrolyte membrane fuel cell.

Critics point out fuel cells are capable of producing only direct current, much less efficient for spinning motors than alternating current, and that conversion drops efficiency and adds cost. But here again research is offering solutions to compete with even those rotary-engine alternators. "Oak Ridge National Laboratory just demonstrated a smaller, lighter more manufacturable, more reliable AC inverter for electric vehicles. The so-called resonant snubber inverter (RSI) is so efficient, it more than doubles the Partnership for a New Generation of Vehicles goal of 5 kw/kg. The 9-in x 12-in x 6-in inverter weighs an incredible 20 lb.!"[378] The hydrogen fueled, electric-powered airship enjoys the efficiency of short transmissions lines.

Electric powered motorcars were popular before IC engine cars overcame them, way back when homespun hydrogen gasbags were thrilling airship crowds. We should remember the very first airship to safely return to its starting point was even before that, the all-electric *La France* of 1884. Owing to some very interesting leverage advantages associated with large-diameter ring motors, all-electric propulsion will be part of manned airships in the future. Modern materials will likely allow new records to be set. As Dr. Kirchner wrote, "A U.S. Navy project toward realizing this [High Altitude Station Keeping] concept was the High Altitude Superpressure Powered Aerostat (HASPA). The mission of HASPA was to develop a surveillance vehicle

---

[376] M. A. Dornheim, "Special Fuel Cells Key to Months-Long Flight", *Aviation Week & ST*, Feb. 28, 2000

[377] "Hydrogen From Solar Cells," Popular Mechanics, Sept. 1998

[378] "Newsline," *Automotive Industries*, August, 1996 pg. 13

capable of flying at an altitude of 70,000 feet for periods up to 30 days. On-board power was derived from several different sources; solar voltaic cells, fuel cells and batteries. Two electric motors produced 5 hp for driving the stern propeller." Electric airship propulsion is certainly coming back now, perhaps with the HALE (High Altitude Long Endurance) unmanned aerostatic craft and likely its competitors (Sky-Net and Sky Station International, to name two) as well. LTA can benefit from very large programs under way to produce a zero-emission automobile.

The California Fuel Cell Partnership was formed in April 1999 to pave the way for demonstrating fuel cell powered vehicles. Private companies and public agencies formed the partnership to help advance the technology that promises practical, affordable, and environmentally friendly transportation solutions for California and the world. While some automakers simply pieced some available components into an existing model to jump on the bandwagon, others optimized for fuel cell operation. Volkswagen's entry uses a fuel cell system whose most efficient operation is realized at mid throttle, which is the throttle range used the most during driving. Departing from the off-the-shelf pressure systems, the *Bora HyMotion* tank has a capacity of 50 liters of liquid hydrogen stored at -253° C. This is an energy equal to 12 liters of petrol, delivering a range of about 350 kilometers. The asynchronous electric motor has a power output of 75 kw and with 240 Nm of torque in a large revving range, the *Bora* accelerates from 0 to 100 km/h in 12.6 sec reaching a top speed of 140 km/h.[379] General Motors has demonstrated the HyrdoGen1 minivan whose fuel cell weighs but 149 pounds and fits under the hood. A 75 liter LH_2 tank allows 250 mile range, and the vehicle can hit 90 mph via is 74-hp three phase electric motor. Electric motor driven, hydrogen fueled vehicles are running about as you read this.[380] Their simpler power trains offer greater longevity and cost effectiveness, and "...a fuel cell car running on hydrogen scores 2.8 times the gasoline car's performance."

By 1999, Physical Sciences, Inc., announced power densities on the order of 740 watts per liter in its *ambient* temperature and pressure system, claimed to be 88% better than standard PEM fuel cells.[381] B. Duet and A. Saudi wrote: "In the case of a fuel cell (the trend today) where efficiency can attain 55%, the same requirements [6 liters per 100 km, then 500 km =] would correspond to just over 4 kg of hydrogen." Although the fuel equipment cost is currently higher, the cost-per-mile is already lower because fuel cell drive trains are so much more efficient. Motorola has demonstrated a prototype of a ceramic-based microfluidic fuel delivery system for a miniature direct methanol fuel cell (DMFC). The goal is five times the energy density of conventional Li-Ion rechargeable batteries – and that's *including* the reformer to break out the hydrogen. Indeed, low impact-of-loading might just be the magic bullet to solve the overwhelming problem with this most efficient way of motivating an airship. The month the LZ-129 was lost, C. P. Burgess would reach the conclusion in design memo 237. "The maximum propulsive efficiency can undoubtedly be attained by using a single, slow turning, large diameter propeller at the tip of a well drawn out stern... the propeller efficiency is increased by the forward wake velocity." The idea showed promise in wind tunnel tests and Goodyear had drawn up a design ready to build and mount on a tri-tail ZS2G-1. The decrease in car interior noise alone would have added to crew endurance. The program was canceled before it was given a chance, but submariners went on to perfect and employ the idea.[382] As it had in Goodyear's full scale "Silent Joe" test

[379] The *Bora* is known as the *Jetta* in the US. Info from www.VW.com.

[380] D. McCosh, "Hydrogen on Wheels," POPULAR SCIENCE, May 2000, pg.52

[381] *The Hydrogen And Fuel Cell Letter*, Vol. XIV No. 2, February 1999

[382] Designers of the USS *Albacore* used airships and their data to perfect the revolutionary hull. Submarine officers actually took control yokes out of a large blimp and adapted them for use on the sub, since despite the overwhelming

program, the long sought stern propulsion dream was also demonstrated by Raven Industries in miniature form. "Microblimp's power came from a stern-mounted 4 hp Wankel engine driving an eight-foot diameter, molded polyurethane, three bladed propeller. Directional control was achieved by gimbaling the engine-propeller assembly."[383]

GAC Wake Propeller was tested on a modified ZPG-3W one-twentieth scale model in the NASA-Langley wind tunnel in 1961. Submarines harnessed the concept instead. Courtesy Jeff Cook.

With all this heat and water about, there is a long-forgotten technology that might find a modern application. Several historic airships employed steam-cooled engines. T. J. Goodey writes, "We propose using steam as lift gas, both for free balloons and for airships. In the airship case, we propose also using a steam engine for propulsion… To remain gaseous at sea level ISA, steam (the vapour phase of H_2O) must be maintained at a minimum temperature of 373°K. The lift provided is 6.26 N/m^3. This is about 60% of the lift provided by helium and more than twice the lift provided by hot air. Steam is non-corrosive, non-poisonous, cheap, and odor-free. It cannot ignite and can be easily produced anywhere…" A steam engine is certainly a competitive alternative to the use of an 'explosion engine' for powering a steam airship…Using a steam engine for airship propulsion yields several advantages… We think that there is better potential in using a reciprocating steam engine or a vane motor. These can be constructed with very respectable power to weight ratios, have low maintenance requirements, and operate at high torque and low rotational speed so that they can be directly coupled to large slowly rotating propellers, which excel in thrust efficiency, especially at the low airspeeds typical of airships. Furthermore, the typically quiet operation of a steam engine eliminates the requirement for any silencer… the two-cylinder reciprocating steam engine fitted to the Stanley Steamer automobile set a land speed record of 127 mph in 1906. This engine weighed about 85 kg and was reputed to develop 250 hp. Modern practice could improve substantially on these figures. A typical rotational speed for such an engine operating non-expansively at high pressure is 1100 rpm. This type of engine can run for long periods between services and is very straightforward to maintain, and its inherent reliability is much greater than that of an internal combustion engine, especially

3 fold difference in density of medium, the sub and blimp have a great deal in common. Information courtesy Robert Largess.

[383] Dr. E. J. Kirschner, AEROSPACE BALLOONS, Aero/Tab Books, 1985

a spark ignition engine."[384] We must not overlook the possibilities of another method of hydrogen production. Powernova Technologies hopes to overcome the high temperatures required for steam reforming with a hydrocarbon catalytic conversion process running less than 200°C – less than typical exhaust gas temperatures.

However, it is not only motivation, but also *control,* that hydrogen-fueled electrical airships will really shine. Electro-reology, a British Ministry of Defense secret until announced in 1984,[385] has evolved rapidly. Replacing hydraulic systems in aerospace vehicles, electrical control surface actuators, now already supersonic, are a natural for fly-by-light control systems. Adoption of aero-surface control independent of engine power adds another layer of survivability and would eliminate the need for bulky, toxic hydrazine-powered pump-based traditional hydraulic systems. The new Airbus A380 "...is the first system to replace some of the hydraulics with electric power; the next generation of aircraft will likely do away with all of the big hydraulics altogether."[386] *Air Force Technology Horizons* magazine pointed out, "Power-By-Wire technology payoffs include elimination of the central hydraulic system, reduction of logistics support, increased survivability, significant reliability improvement (increased mean time between failure), more efficient use of secondary power, and improved ability to turn off or reconfigure a damaged or inoperative flight control surface."

The British have wasted little time in applying the technology where it would be most immediately needed, in rotary-wing aircraft. "The replacement of hydraulically powered systems by electrically powered equivalents will permit rationalization of the helicopter power supply arrangement, and potential elimination of conventional hydraulic systems."[387] It is in this direction the US Navy's surface fleet is headed. The new DD-21 class warships will replace hydraulics and pneumatics with electrical systems, not only more reliable and battle hardened but also offering more design flexibility along with their reduced bulk and weight. Likewise encouraging to this trend, safer electrical circuitry is coming. "A new device being developed by Eaton Corp. could protect electrical wiring from arcing, which is more likely to occur with aging aircraft. Although circuit breakers protect aircraft wiring against thermal damage from current overloads, they do not detect arcing, which can start fires without tripping the breakers."[388] What we can do with all that easily manufactured electrical power will cause us to rethink major airship design elements. Extra power could fight the effects of weather. BF Goodrich is working to develop Dartmouth College Professor Victor Pretence's technology to use electric currents against ice formation. "...electrodes attached directly to surfaces can break down ice and water through electrolysis. When applied to aircraft surfaces, gas bubbles of hydrogen and oxygen... generate pressure on the aircraft surface that enables it to shed ice."[389]

Critics point out that fuel cells are bulky and aren't particularly efficient, with more than half the potential energy of the hydrogen being converted to heat. A problem for other vehicles, perhaps, but not an airship, which has plenty of room and can use the heat to enhance lift. Heating the lifting gas is the ace-in-the-hole that is available after all other efforts have been exhausted. When we can't add any more gas because we need the ballonets at a given volume to reach desired altitude, we can always add lift by falling back on hot air balloon principles. Volker

---

384 T. J. Goodey, "Steam Balloons And Steam Airships," 3rd Annual AA Conference
385 D. Scott, "Amazing Hardening Fluid," *Popular Science*, April 1984
386 B. Sweetman, "Double Decker," *Popular Science,* October 2000, pg. 56
387 Alan Bean of FHL quoted by S. Birch in *Aerospace Engineering*, October 2000, pg. 6
388 "Circuit Breakers Are Not Enough," *Aviation Week and Space Technology*, 21 AUG 2000 pg. 72
389 F. Bokulich, "Aircraft structure and ice protection," Aerospace Engineering, SEP 2000 pg 13

Steiner presented "How to Improve the Performance of an LTA Vehicle" to the 1997 AIAA LTA Systems Conference. Therein he presented benefits of heating helium and hydrogen, offering a rule of thumb. "A deviation of the buoyant-gas-temperature compared to ambient atmospheric conditions by 2°C (2°K) increases or reduces the lift by 1%." This is very important in helping overcome what we will see is the major problem at airship launch, minimum lift at maximum weight. That problem is worsened by the fact it is usually warmer on the deck, as well as more humid on deck. Higher air temperature and humidity rob lift.

An airship's heat sources could be used effectively for getting the fully loaded airship started on its mission. The "Standard Atmosphere" according to NACA Report No. 218 shows temperature of the atmosphere drops about 3.5 degrees for every thousand feet gained in altitude. That chart shows that at sea level when the local pressure is 29.92 and the temperature is 59 degF., then at 5000 feet it's 24.89 and 41.2 deg. At 10,000 ft the pressure is down to 20.58 and it's 23.3 deg., and toward the upper limit of airship operations at 15,000, the pressure is only 16.88 while it's a chilly 5.5 F. The warmer airship reacts two ways. If warmer, it is always loosing some heat through direct conduction via the airflow and it also radiates heat away. Both losses reduce lift, but lift is gained through heat absorption, or *insolation.* New higher-pressure fabrics should allow a ship's heat sources to deliver lift faster than it is lost, which could at the least be used for minor adjustments in static condition. In his paper for the 3rd International Airship Conference, Alan Sherwood wrote, "It was concluded that heating and cooling is the most practical concept. A system applicable to a typical 7000m^3 airship was schemed, using: fuel burning with a heat exchanger heating, evaporative water spray for cooling, and zeolite adsorption or gas cycle cooling to remove water. The estimated performance is 20-25% lift change in 1 minute, for a 250 kg weight penalty." Indeed, compartmented hot air airships today can be erected and operated against winds exceeding five knots. More heat resistant fabrics are coming that will turn these pressure airships into real competitors to the helium variety. Temperature control is seen as fine-tuning static condition even on balloons designed for missions on Mars.[390]

Old news to the Zeppeliners faced with using the less efficient helium in an airship designed for hydrogen. "The *Hindenburg*, inflated with hydrogen, carried fifty passengers and ten tons of mail and freight... The [LZ-130] airship, operating in the worst conditions of weather and temperature, will thus be able to carry forty passengers...and from five to six tons of mail and freight. By heating the gas 3° Centigrade over the existing atmospheric temperature, then allowing it to cool as the ship leaves the ground, she attains the required flying height without loss of volume." [391] Not depending on the difference in a hangar-warm ship being rushed out into the cold, the Zeppeliners had planned to add to helium's weaker lift by adding electric heaters to LZ-130's cells, dropping the heaters off and carrying the extra payload dynamically when the gas cooled in motion over the first few hours of fuel consumption.

Unlike LZ-130, a modern airship's fuel cells would provide onboard electrical power that could be used to heat the gas. One published airship design utilizes fuel cells to generate heating-element amperage for temporary, mission-originating lift, a timely application of unused capacity before mission electrical power is needed. Winter icing conditions lead to ZPG-2s being equipped with electrically heated propellers and center ballonet air-inlet dampers.[392] With fuel

[390] J. A. Jones, "Two Techniques for Controlling Altitudes of Lifting Balloons," *NASA Tech Briefs*, Mar 99

[391] WONDERS OF WORLD AVIATION, Ed. By C. Winchester, Waverly Book Co. London, 1938. Courtesy the late Francis Scales.

[392] Also fitted were control cable ruddervator horn pulleys, protective coatings on the top surfaces of the lower fins, and protective blankets inside the envelope in the plane of the props. See W. Lewis and P .J. Perkins, NACA

cell power we could possibly weave electric elements into the envelope. Not only could we use early-mission excess generating capacity to warm elements to make heat, we could also fight envelope icing. Themselves generating heat, smaller fuel cell cooling equipment could be utilized if the very gas we want to heat, the lifting helium, was used as a more efficient fuel cell cooling medium.

Engineers have known for decades that a great deal of energy produced by internal combustion engines is wasted as heat, and still more is wasted cooling vital parts. Back in the early 80's, racing mechanic Smokey Unick developed his own version of an "adiabatic" engine, or one that makes some use of its waste heat. His two cylinder, 150 hp unit fit under the hood of a VW Rabbit, was very clean, and thanks to a high-torque lower end and some tall gearing, got 70 mpg on Daytona's track.[393] This writer touched the engine and interviewed Unick, who explained his patented process recovered heat from the water jacket, lube oil and exhaust manifold to make the mixture more volatile. Some would suggest his patent actually beats vapor lock, but however it worked, no one has seen fit to mass-produce the engine. Owing to hydrogen fuel's low radiant heat, the benefits of building an "adiabatic" engine for its use can only stimulate the imagination. Neither must we forget the great promise of the Sterling engine, whose proponents claim, with hydrogen, is the closest thing to perpetual motion as a mechanical device can achieve.

Scientists at the Los Alamos National Laboratory produced a fuel cell that recovers some of its own heat for more efficient operation. This "adiabatic" fuel cell uses evaporative cooling to recover water for a simpler yet superior design to the conventional cooling systems. Hydrogen Burner Technology has proposed the Advanced Energy Module, a fuel cell designed to both power a home and heat its water and air. Airships have long been able to harness otherwise wasted heat energy. Helium filled airships always had the synergy of cooling the engine exhaust to recovery water; using the helium gas for cooling goes beyond adding disposable lift. "If helium is used instead of air for cooling the radiators and water recovery apparatus for airships, about 13% less radiating surface is required for the same temperature range, and the power absorbed in circulating the helium is only about one-sixth as much as required for air cooling."[394] In this way, some of the weight of the water recovery radiators could be paid off. It also represents lift that is to some degree controllable.

A hydrogen-fueled airship would be "green" even in the manner its fuel is produced from renewable solar power, wind power, or biomass like peanut shells – any combination of these offer zero emissions for the entire fuel cycle. There is hardly any place on Earth where one or more of these methods are not possible, or a combination cannot be made effective. Manufacturing the gas does consume more energy than delivered. There is no adverse effect, however, when some of that energy can be solar, hydro or wind power. We have not considered using the "green" method of fuel production via solar cell electrolysis, whereby some of that by-product water can be turned back into fuel. To do all this in flight *seems like* a perpetual motion scheme doomed to failure owing to the inefficiencies built into the process. Also non-polluting would be emergency fuel dumping. Discharging a ton of gasoline or heavy diesel might finish an airship company legally, but dumping an entire load of hydrogen over the most fragile ecosystem anywhere wouldn't harm a snail darter. Windmills have always been earthbound for the simply

---

Technical Note 4220, "A Flight Evaluation and Analysis of the Effect of Icing Conditions on the ZPG-2 Airship," Lewis Flight Propulsion Lab, Cleveland, Ohio, April 1958.

[393] E. F. Lindsley, "Smokey's amazing 150-hp, 2-cylinder engine," *Popular Science*, April 1983

[394] C.P. Burgess, "Helium vs. Air as a Cooling Medium For Radiators," Bu Air Design Memo #177, Aer-D-13-MW. Courtesy Marc de Piolenc, ABAC

understood thermodynamic laws. Here again, airship history reveals a forgotten trick: leverage. Major Gregory Gotlieb made a passionate case for preserving LTA technological history with his talk to the second Airship Association Conference, in which he discussed "air boosters." These little devices consisted of a larger diameter circulating-type fan mounted on a common shaft with a smaller turbine-type fan. As the wind stream spun the big prop like a windmill, the smaller interior turbo unit rammed air into a scoop to pressurize an aerostat's ballonet. The foolproof device uses no power, and uses mechanical advantage to deliver higher pressure air than is otherwise in the slipstream. Like an aerostat, an airship will also have solar and some "wind" power, so co-generative technology is worthy of our study.

Even the most vocal critics begrudgingly agree that most of hydrogen fuel's problems are not technological, but rather financial. Effort expended toward hydrogen fuel in the rest of the transportation world is constantly finding new answers to the main problem, fuel *storage*. The National Fire Protection Association publishes guidelines for hydrogen storage systems, paramount to the safety of the overall system. The problem of renegade electrical energy and petroleum fuel tanks aloft continued to kill people long after rigid airships were history. Lightweight aluminum tanks, seemingly safe until a break exposes raw untreated aluminum to react violently with their contents, are often the culprit. In 1999 a hot-air balloon drifted across power lines and managed to complete a circuit between two conductors. While the two on-board steel fuel tanks were not affected, the aluminum propane tank exploded with fatal violence. Like other safety rules written in blood, aluminum propane bottles are no longer permitted in balloons.

Composite hydrogen tanks are a good possibility, in spite of the well publicized de-bonding of the X-33's composite LH_2 tank and subsequent re-adoption of aluminum, standard in most hydrogen storage that flies. Scaled Composites of Mojave, California, is ready to make composite tanks per customer order; they see containing hydrogen as no great problem. The hydrogen tank for the *Helios* aircraft, built by Kaiser Compositek, encompasses 32.8 cu. ft. with a burst pressure of 600 psid. yet weighs only 28 pounds.

Pressure storage initially seems out of the question for airships due to weight, yet industry keeps trying to change that. IMPCO Technologies, Lawrence Livermore National Laboratories and Thiokol Propulsion have tested a high performance prototype hydrogen storage cylinder designed for fuel cell vehicle applications. The cylinder achieved a mass performance record of 11.3 percent hydrogen storage by weight. These wrapped tanks typically withstand 2.25 times their design pressure without rupture. Burgess pointed out that the weight of the fuel container was proportional to its BTU; gaseous fuel mixes would have larger, but lighter, containers.

Another promising technology employs a 'hydrogen sponge.' At first very heavy, the iron-titanium hydride system mounted in a 1978 Dodge Omni weighed a hefty 400 pounds. The increased weight did not nullify hydrogen's greater efficiency in these early test cars. Compared to 30 mpg in the stock gas cars, the first ten hydrogen-powered versions got the equivalent of 44 mpg. Later, "Basically, we've figured out a way to store hydrogen in solid form... the amount of solid hydrogen in our plates is two percent of the total weight. When we get to six to seven percent, we will match the gas tank of a typical car in weight and range."[395] It is assumed hydride storage would be too heavy for airship use, but Energy Conversion Devices, using a high percentage of magnesium with several other metals in the hydride powders, has achieved seven grams of hydrogen per 100 grams of hydride. Three grams was the limit for their hydride. It's

[395]Bob Stempel of Energy Conversion Devices in *Automotive Industries*, September, 1997.

gas, not liquid, that is consumed in the combustion chamber or fuel cell bed. (Even the Space Shuttle main engine actually burns gaseous hydrogen, and the ignition is started by nothing more than a sophisticated spark plug.) The two existing methods of gaseous hydrogen fuel storage may still be too heavy for airship use. However, the robust tanks would be harder to hurt, just as high-pressure tanks represent a fortress in comparison with feeble gasoline tanks. "...liquid hydrogen tanks for cars were subjected to experiments which included a fire test and the impact of a big concrete weight. These tests as well confirmed the high level of safety of hydrogen technology."[396] Conformable tanks would allow fuel storage in oddly shaped places. Seen as an achievable goal for these lightweight, astonishing 5000-PSI tanks, is a finished weight of less than 10% of the fuel weight. Decades before composite tanks were even science fiction, C. P. Burgess stated insulated fuel tanks should weigh no more than 25% of the fuel they carried, and "...containers with 6 inches of insulation around them should be considerably safer against fire than ordinary gasoline tanks." Hydrogen is compatible with many materials, as the 1940 manual observes. "Hydrogen is considered an inert gas because it enters into direct combination with only a few elements, and at ordinary temperatures materials used for containers are not acted upon by pure hydrogen." (Even highly reactive *oxygen* can be stored in lightweight composite tanks when equipped with compatible metallic liners.[397]) In the early 90's tiny glass balls, or microspheres, demonstrated storage similar to hydrides but with less weight. While the "loading" of hydrogen fuel into this storage medium involves pressurization, this could be done at the mast. Releasing the fuel requires only heat, like hydrides, and sources will be readily available aboard.

Also promising are experiments with carbon nanotubes (CNTs) that contain microscopic pores that absorb high densities of hydrogen via capillary action – at room temperature. By 2000 at least eight research institutes made no secret of their work in CNT technology. A team of physicists in Singapore reported achieving 20% hydrogen density with lithium-doped CNTs, meaning the prototype system could store the energy of 6 kwh per liter. The Hyperion Company expects to manufacture 600,000 pounds of the material per year. What might be seen as a three-dimensional version, also a possibility as of this writing, is called a Fullerene. Breakthroughs in nanotube production, promising storage that was recently only lab curiosities, were shown in 2000 by Dr. Michael Heben.

BMW 12-cylinder 750 hL cars produces 204 bhp and can reach 140 mph. A very efficient five kw fuel cell, replacing the battery, runs accessories including air conditioning without the engine. Courtesy BMW.

---

[396] A. Bain and U. Schmidtchen, "Afterglow of a Myth", DWV Statement No. 4, 18 Jan 2000

[397] "Lightweight Composite-Material Tanks for Cryogenic Liquids," NASA Tech Briefs, Jan. 2001. pg. 58

We might be left with the cryoadsorption form as the only option presently practical, if somewhat expensive. Given hydrogen's weak energy by volume, cooling it down to 20°K makes it more practical to transport. "Research indicates that liquid hydrogen may be a viable substitute for conventional jet fuel," says a typical article[398] discussing the pros and cons of the subject as applied to airplanes. Required foam insulation might be helium-impregnated instead of trapping air, thereby paying some of its weight penalty, since tanks could be enclosed in a helium flooded space. C. P. Burgess stated long ago, "The idea of liquid hydrogen fuel is old, but... its large scale production in the near future does not seem nearly as impracticable as the production of helium gas must have appeared 20 years ago." Indeed, what would Burgess think of a Space Shuttle Main Engine turbo pump turning at 37,000 rpm and developing 100 hp per pound, which could squirt a column of liquid hydrogen 36 miles in the air?[399] That technology is 20 years old and does not seem very fuel efficient; the SSME consumes 1000 pounds of fuel every *second* sucking its 53,500 cubic foot tank dry in eight minutes. Of course, it develops more power than seven Hoover dams in the process. It is a rather difficult concept to grasp, matching today's technology that safely pushes fuel at -423° F into a combustion chamber that reaches +6000° F (which evaporates iron) - with vehicles that faded from existence before the invention of the blind rivet. Yet visionaries have long since done so. C. P. Burgess 'did the math' on powering the USS *Macon* on LH_2. In Design Memo No. 150, Burgess gives the figures of lower LH_2 heating value as 53,000 Btu per lb, and gasoline as 20,000 Btu per lb. Allowing for the considerable boiloff of the storage of the day, he figured "...the rate of vaporization will be...316 pounds per hour...[which is] equivalent to 850 pounds of gasoline per hour, which is sufficient for cruising at about 55 knots speed, and permits an endurance of 316 hours, or more than 13 days, giving a range of 17,400 nautical miles!" Metalclad advocate Pavleka saw liquid hydrogen as the fuel answer as well – with the tanks in the helium cell.

We discussed using helium's excellent cooling properties to cool our engines, and condense their exhaust to recover water weight while adding heat to the lifting gas. Hydrogen is even an electrical insulator. This makes hydrogen the cooling medium of choice for virtually all large electrical generators today. Would we dare use the fuel for cooling the engine, while using that heat to get more lift from our gas? We should not pass up this opportunity because of imagined danger, in the face of a 100% (perfect) safety record. As Marc de Piolenc put it, "...it is clear that millions upon millions of kilowatt-hours are generated daily with high power electrical machinery bathing in the 'deadly' gas credited with downing the *Hindenburg*. When did any reader last hear of a commercial power generating station 'blowing up,' or suffering an explosion in its generators?"

Liquid hydrogen offers phenomenal cooling capacity, as demonstrated by its employment protecting rocket engines' exhaust cones. No airship has ever explored the possibilities allowed by using LH_2 in the role of electrical conductor refrigerant. "The benefits are clear for the system American Superconductor is building for Detroit Edison's Frisbee substation: 18,000 pounds of 1930s-vintage copper cable will be replaced with 250 pounds of superconductor in three ducts... Other notable power applications are superconducting magnetic-energy storage systems... and, further away, lightweight motors and transformers."[400] As if every ounce saved giving another ounce delivered wasn't enough, LH_2 as superconducting refrigerant is opening up new possibilities in the electronic world. The frequency-specific filters, high-powered lasers and other

[398] R.O. Price, "Liquid Hydrogen – an alternative to aviation fuel?" *Aerospace Engineering*, February 1991

[399] The Boeing-Rocketdyne fact sheet points out that just its two pumps develop the power of 39 locomotives – even though the whole engine weighs 7,000 lb, only 1/7th as much as a single locomotive.

[400] B. Schechter, "No Resistance," *Scientific American*, August 2000, pg. 32

promising technologies in the airship-ready field of electronic commerce – and defensive warfare. If LH_2 is not cold enough for some yet undefined airship borne experiment or equipment, there is an expired spacecraft-cooling patent awaiting an application. "I quickly found that the only other possible material that might work is solid hydrogen. Its principal advantage is a heat of vaporization around 20 times higher than helium… the vapor pressure is still a substantial 1.93 mm of Hg (257 N.m^2 = .00254 atm) at 10 K… we set about seeing how low a temperature we could achieve by just pumping on hydrogen (vacuum is free in space). The triple point is at 13.8 K and .069 atm = 6991 N/ m^2 =9.2 mm Hg, so freezing it solid is easy.[401]

The masters had foreseen much of this long ago. Karl Arnstein wrote, ""Fuel gas, burned in our regular Maybach engines, makes for much smoother running, with a corresponding increase in engine life… Another possibility is the development of [fuel] gas lighter than air… which may be economically be used in airship engines. The storage of such a fuel gas would be facilitated due to its self ventilating possibilities." The C. P. Burgess design memo had suggested that if USS *Macon* had been powered by liquid hydrogen, her engines could run continuously just using the boiloff. With our use of LH_2 as a coolant, we would use the boiloff to continuously run the fuel cells, since electric power would be easily consumable (turn on a light, run a fan) if not in constant demand, regardless of the ship's condition. Space vehicle tanks are equipped with heaters to insure enough gas is available on demand, but recent NASA work allows near perfect prediction of boiloff pressurization in cryogenic tanks.[402] Burgess had suggested putting LH_2 tanks in the *Macon's* helium cells, and in 1940, Major August Post wrote "...hydrogen... can be used as fuel, enclosing the hydrogen containers in the helium gas." In doing so, we have added a unique safety measure: The 'dangerous' gas is completely enveloped in smothering helium. Constructed of fireproof materials, it would be about the most fire-resistant scheme imaginable. Whatever storage method we choose, the adoption of hydrogen fuel has made our airship not only more operationally flexible, but also more survivable as well. The airship could already defy gravity without forward speed or engine power. Now it can also have *full* electrical power after *total engine failure.*

There was undoubtedly a US Navy cadre that resisted the transition of warships from solid coal to black oil, still being present to say "I told you so" when film of shipwrecked victims showed them coated with the vessel's fuel oil. There were likewise those who opposed the next evolution to distillate fuel, and they were still around to gloat when six of the author's shipmates were burned to death on the USS *Kitty Hawk* (CV-63) during a 1972 engine room fire. We can't expect the movement towards pure, gaseous fuel to go through unopposed in spite of safety lessons learned. Eliminating the huge petroleum storage and its explosive ullage will vastly improve vehicle safety. That would leave only small amounts of lubricating oils and greases to spread fire, but modern efforts are driving toward a day when metals are pretreated with built-in anti-friction agents. Those developments, plus refinements in frictionless magnetic and superconducting 'levitation' bearings may eliminate such radiant-heat producers altogether.

As *The Economist* put it, "The whole history of mankind's use of energy is a history of the decarbonisation of fuels. As societies have grown wealthier, they have been shifting from dirty solid fuels with a high carbon content to liquid hydrocarbon fuels with a lower carbon content, and ultimately to clean-burning gases. This is not part of any green revolution: people the world over have simply used whatever fuels have been readily available, cheap and easy to use, moving

[401] D. Sonnabend, "Hydrogen Popsicles," *Launchspace*, Oct-Dec 2000, pg. 8
[402] Hochstein, Ji, and Aydelott, "Prediction of Self-Pressurization Rate of Cryogenic Propellant Tankage," J. Propulsion, Vol. 6, No. 1, Jan-Feb 1990, pg. 11

from wood and hay to coal and then oil. As energy became easier to use, it paved the way for human development, and made the world a more comfortable place for people to live in." Airship history shows this very evolution; the first dirigible used solid fuel in an external combustion engine. You don't have to live in a tree to realize this natural process will lead to gaseous fuels when the market demands them. Factoring in aerostatic law will hasten this evolution, and within our lifetime the process of lifting a gasoline can with a helium balloon will seem little far removed from Henri Giffard's coal-stoked steamer. Will a hydrogen fueled airship be totally safe all the time? No. Mother Nature will still have some unforeseen tricks up her sleeve. A story going around home inspectors tells that, as a vacant home was being reactivated, a malfunctioning water heater had been performing electrolysis. The abnormal ullage in the tank was aided by an unusual plumbing arrangement and the inspector's bad luck of turning on, you guessed it, a malfunctioning dishwasher first before running some water. Offering containment and an ignition source, the resulting explosion damaged the kitchen. If this was really a hydrogen accident, and not just a simple steam explosion, perhaps we should study hot water heaters as potential energy sources. If a legitimate accident, this safety lesson should not be interpreted as an edict outlawing water heaters and dishwashers. They are worth the risk.

THE MYSTERIOUS ISLAND predicted, "I believe that one day hydrogen and oxygen, which together form water, will be used either alone or together as an inexhaustible source of heat and light." Jules Verne had that vision in 1874; today, the country of Iceland is moving steadily toward that reality, and other governments are conducting serious experiments. The future of the airship, however, is assured only if airshipmen learn from the past masters.

In 2001, the author can make this prediction: I believe that one day we build a hydrogen-fueled airship that has complete control of its static condition.

The Italian built *Norge*, the only airship to have flown across the top of the world. *Norge* was one of two record-breaking semi-rigid airships amid a host designed by an Italian master, General Umberto Nobile. The *Norge* expedition must be actually seen as Rome to Alaska, with minimal stops and servicing along the way. From *Il PIU' LEGGERO DELL'ARIA*, 1993, courtesy Gertruide Nobile.

Chapter Twelve: Solving the Paradox

It is strangely ironic, the US government's position that helium was so vital it could not allow the gas abroad, lest a charge of the gas be harnessed for war purposes, but it was not vital enough for America to build another rigid. Would the Nazis have used helium in combat when it was acknowledged in US manuals that a war emergency would likely demand a return to hydrogen for American needs? Buckley and Barkley wrote, "This desireability, so much emphasized in the United States, greatly loses its importance when other military aspects of the airship operation are considered ...it is very likely that Germany would choose hydrogen over helium for military airship operations, even if both were available to her. If an airship were attacked by aircraft, it would make little difference whether it were filled with helium or hydrogen." It is safe to assume if the German government had paid for a shipload of helium and taken possession before invading Austria, the gas would have been at least partially supplemented by hydrogen by the time a practical military mission was conceived. Some may have thought themselves super-men, but they bungled use of LZ-130 for spy flights. They continued, "It is of interest to note here that the refusal of helium to Germany in 1938, while it resulted in the abandonment of commercial airship service, did not prevent the operation of LZ-130 (*Graf Zeppelin* II) for the training of new airship crews. Prior to the German attack on Russia, LZ-130 had been persistently reported in operation, carrying war materials from Russia into Germany. Even this relatively small commercial airship (same volume as *Hindenburg* - 7,000,000 cubic feet) with range reduced to 2,000 miles, could carry better than 75 tons of payload." When their plunder of Norway bogged down, one group of Nazi officers came up with a plan to load up LZ-130 and sacrifice him in that battle, L-59 style. Arriving at Frankfurt ready to start the mission, one Nazi officer found the LZ-130 was already being cut up by army troops from another Nazi goon who did not talk to the first group. Clearly, the Nazis would not have found in helium some battle salvation.

Apparently, the Americans did not even believe it themselves. No rigid airships were ever started again even as the war began and the British lamented not having a flying carrier to locate German surface raiders. Many would argue the US Navy never openly embraced LTA technology, only begrudgingly accepted a limited airship inventory when peer pressure or wartime emergency demanded it. "Airships seemed the most economical answer, so the Navy set about establishing an LTA program. It is unlikely there ever has been a more unhappily managed military operation in our history."[403] Navy airshipmen nonetheless squeezed out some notable achievements along the way. The contributions to air progress even pre-date LTA's creation of airplane structural duraluminum, developed to support ZRS construction.[404] The US Navy policy of acquiring about one new pressure airship *per year* was barely followed from 1935 through the start of the war. Two years running they simply purchased one stock Goodyear blimp each, becoming the G-1 and L-1. When master designer Karl Arnstein, long since a US citizen, offered a new pressure airship with a two-part articulating car, it was never even published. By the time the u-Boats were sinking ships designed to fight submarines, the Navy had only just then increased their order for copies of the 1935-based K-2 to the timid number forty-eight authorized in 1940's "10,000 plane" program. When Arnstein was finally employed to design a modern pressure airship and produced his masterpiece, the M-ship, the Navy only ordered twenty-two at

---

[403] P. Andrews, "Lighter-Than-Air, American Heritage *Science and Technology*, Vol.9 No.1, Summer 1993

[404] Alcoa was charged with duplicating the Zeppelin "airship metal" in 1917, and developed duraluminum to build the ZR-1. Airplanes might still be made of wood if a Naval Aircraft Factory innovator had not been able to suggest using "airship metal" to fashion flying boat hulls to alleviate water weight absorbed into the wooden hulls.

a time when tens of thousands of more complicated, expensive airplanes were being turned out. In the darkest hour, June of 1942, Congress authorized up to 200 airships, even allowing rigids again. Only eighty-six more K-ships were completed. By the time the M-ship was ready, the Navy had again lost interest and only four were built. The exemplary service rendered by the LTA wings is largely unappreciated, the capable Navy men achieving greatness in spite of uninspired support.

Whether or not the Germans appreciated what *Luftshiffe* were doing to their u-Boats is a hotly contested debate even today, since many American historians belittle blimp's offensive capabilities. Interviewed in the 4 August 1942 *Voelkischer Beobachter*, Admiral Karl Dönitz bragged about convoys offering massed targets, his "operating close inshore along the USA" and how airplanes could only harass his boats. Asked if US "dwarf dirigibles" could "rate as particularly effective in defense?" he surprisingly answered - for the German audience - "It cannot be denied that even the 'blimps' have a certain effectiveness, in defense..." From Rosendahl's unpublished book, chapter "Enemy Opinion:" during interrogation, Field Marshal Herman Göring volunteered that the Germany Navy was much concerned over blimps. Assuming their mass production and wide deployment "were hampered by high [US] Navy circles" he "spoke particularly of the reconnaissance value of blimps…[German command] was relieved that we failed to do more along airship lines than we did." In a 1964 letter to Gordon Vaeth, Doenitz states: "But even if the commander, upon seeing the 'blimp', surmised a convoy, that would only be a slight disadvantage to using the "blimp" since it would hinder the U-boat's approach, at the very least severely." Doenitz said: "…the American 'blimps' were very disturbing to German U-boat activity." Indeed, evidence may eventually prove the submariners took the airship threat very seriously, possibly to the point of developing what might have been an anti-airship missile. When rumors of a sub-launched tactical missile proved to be a real anti-aircraft missile, one wonders how they expected to hit a fast-moving airplane with this crude, early device.

Appreciating the wartime contributions and the high mission availability of the WWII K-ship would help the LTA cause, as would postwar developments. Helium's ascentive force overcame bulk, weight, and aerodynamic lift unsuitability for countless prototype systems. Evaluating fleet 'darken ship' or our submarine's stealth was routine long past; blimps helped perfect *guided missiles* by radar tracking and providing a stable camera platform when airplanes couldn't keep up. More than forty years ago a US Navy airship flew the Arctic[405] dropping mail to scientists a few hundred miles from the Pole. The US Navy's last airship missions[406] were research and development. On the other hand the safety record of even the scratch-built hydrogen-filled airship compares favorably with HTA craft from prototype testing through operational service. Thanks to their more forgiving nature, modern airships should have at least as good a chance. Trouble is, they aren't being given the chance.

Lighter-Than-Air is rarely considered as an option even in proven applications, partly because of the combination uncontrollable/dangerous perception. The basic reliability of nature's only alternative to gravity is reflected in the Zeppelin service record that no passenger flight was ever

---

[405] W. Althoff, ARCTIC MISSION, LTA Institute, Auckland 1999, ISBN 0-9583693-2-1

[406] Navy airships carried bulky prototype infrared sensing equipment and provided otherwise unobtainable low-speed aerodynamic data on STOL airplane designs. For the former see: C.F. Bieber and H.L. Clark, "The 100-inch Airborne Radiometer," NAVAL RESEARCH REVIEWS, June 1959, pp. 6-13. (Courtesy the late Don Woodward.) For the latter see Henry E. Payne III, Frank R. Carter, and Wayne J. Harrison, "The Navy's Flying Wind Tunnel," AEROSPACE ENGINEERING, March 1961, 18-19, 50-61. (Courtesy Hepburn Walker, Jr.)

turned back by weather, and the LZ-129 never turned back even for mechanical reasons including a bout of multi-engine troubles; repairs were carried on in flight. "Airships have played a passing role in aviation for close to a century. Their designers have tried hard to establish them as practical forms of intercontinental transport or to make them capable of performing special tasks. But for a variety of reasons that include their public safety image, airships have never fully met the criteria that would establish them as a real success."[407]

As Buckley and Barkely noted, this is largely caused by bad press. In the darkest days of after America's entry into the war, Congress in so many words removed the restrictions against airships having a structure. In the aforementioned dissertations by the two outspoken Naval officers, additional basic airship information was included. It is worthwhile to offer a paraphrased version here to remind one of the basic survivability of the airship:

"No means of transportation has been quite so generally misunderstood as the airship. On the face of the record alone, it may logically be argued that the airship has hardly produced sufficient results to encourage anyone to try to understand it. Public fascination by the unusual has always been played up to and capitalized upon by opportunists. Upon this fascination have thrived the patent medicine quacks, fortune-tellers of various descriptions, certain pulp magazines, comic strips and to a slightly lesser extent, the daily press-et cetera. Now who can find a more suitable tool with which to mystify the public than the beautiful cigar-shaped, silver-hued airship? From its very beginning, the airship was regarded as something highly mysterious. The press is further responsible for the widely accepted belief that an airship can easily be broken in the air in the same manner as a light stick of kindling wood is broken over the knee. This vulnerability to weather is as fictitious as the 'artist's conceptions' which depict the poor helpless airship viciously torn apart by the angry wind gods. How strange, in this day and age, that the fascinating mythology of ancient time retains such a strong hold over the masses so graciously enlightened by their scientists and engineers!" Because the airship is completely submerged in the fluid in which it operates, it can be said, and frequently has been without impression in many cases, that *for the airship there is no wind.* An airship, stopped, becomes a part of the atmosphere surrounding it; no strain is imposed upon it whatsoever. But when it pushes ahead through the air, the structure of the airship opposes the forces set up by the passage of the air body around the hull of the ship. This ability of the airship to stop and remain motionless, completely surrounded by the medium in which it operates, removes all dynamic strain from its structure. It is a safety feature which is always present, the importance of which cannot be overemphasized. It is the only air vessel so favored, other than the free balloon.

The *Shenandoah,* in breaking away from the British-type tower mast, gave an excellent demonstration of the airworthiness of the well handled airship. In January, 1924, the *Shenandoah* was moored out during a severe storm, and suffered a casualty to its upper fin. The damaged fin steered the ship out of the wind, with the result that the next gust, with a recorded velocity of 64 knots, struck the ship on an angle along one side, wrenching it from the mast. Structural damage deflated the two forward gas cells. The damaged ship, with its bow ripped open and two forward cells deflated, was free ballooned for several hours until repairs could be effected. From an aerodynamic consideration, the *Shenandoah*'s situation was similar to that of an airship which had been making full speed, and which had suddenly stopped her engines. Although she was immediately carried backwards over the ground, she still had forward decelerating air speed. All available emergency ballast in the bow of the ship was dropped. Up elevator was applied, and the still considerable air speed was sufficient to carry the ship up to a safe operating altitude. The engines were started to assure their good operating order and immediately secured. No attempt was made to ram the damaged bow through the

[407] S. Birch, "HALE a new airship'" *Aerospace Engineering*, August 2000 pg. 11

atmosphere. Having gained sufficient altitude, the *Shenandoah* was then operated as a free balloon, her air speed having decelerated to zero. No strain was being imposed upon her structure by the air mass of which she was now a part, though she was being carried over the ground in a northeasterly direction at nearly 60 knots. Repairs were made to the damaged structure, and after a careful inspection, the engines were started and the ship moved slowly ahead through the air. A constant watch was kept on the repaired damage, and air speed gradually increased to full speed. Wartime test pilot Anton Heinen, engaged at the time to instruct the *Shenandoah's* crew in the operation of the new ship, brought her back safely without further incident. Nine hours after leaving Lakehurst, the *Shenandoah* was safely docked. So concerned were those who worried at Lakehurst that an official announcement, giving up the ship as lost, was made after she had reported that conditions aboard were normal.

The above incident is particularly worthy of note in that the *Shenandoah*, structurally fragile by comparison with the beefed-up bridge-like trusses of the *Akron* and *Macon*, had been in operation for a period of less than four months. An improved copy of the 1916 Zeppelin L-49, which had been forced down intact in France, her structure was fabricated at the Naval Aircraft Factory at Philadelphia, erected at Lakehurst. A similar case occurred when the British airship R-33, also a copy of a 1916 Zeppelin, was carried away from the same type of mooring. Similar procedure was followed, Lieutenant Booth himself inspecting the damaged structure as the speed of the airship was slowly increased. R-33 carried the mooring gear from the top of the mast with her. Flight Lieutenant Booth brought the R-33 back to her base safely, after the ship had been carried out to sea a considerable distance.

To attempt to lay down precise casualty procedure for airships is futile, as each case presents different circumstances for which the operator must prescribe the correct course of action. There is one cardinal rule, however, which must never be violated: In the event of any reported damage to structure or to the outer cover, stop immediately. This is as mandatory in all respects as the submarine rule: Never dive with a hatch open. It has been violated with fatal results. Purely secondary to the above, of course, is the fact that any plane or airship which finds itself in such a position can blame only the operator who put it there. The storm cannot chase the airship or the plane, either must be deliberately driven to its center. This fact is lost sight of frequently. It is of course not intended to deny the folly of attempting to drive an airship or any aircraft through a thunderhead or hail cloud. Airmen in general recognize both the desirability and the responsibility of remaining out of them. The basic principles which were so successful in the *Los Angeles, Graf Zeppelin* and *Hindenburg* were well established then. So were the principles of operation in flight. It was the violation of basic principles, structural and operational, which caused the destruction of American airships."

The obvious solution to the crying need briefly surfaced in the last decade. During the Reagan/Bush years, a prototype airship made from new and existing parts was cobbled together for testing. The NASA Ames simulator was reconfigured and programmed to simulate the operation of a very large pressure airship, the concept YEZ-2A. *Sentinel 1000* had completed a 24-hour endurance test before the program's end. The maritime services should be first in line for a return to buoyant flight.[408]

In the meantime, inanimate cargo can return to buoyant conveyance to prove the concept to nervous insurance underwriters. "Currently the cost of flying a tonne of airfreight one kilometer – available-air-tonne-kilometer (ATK), a standard measure in the airline industry – stands at around 20p – 27p… a vast proportion of the world's commerce still moves at a fraction of the price of airfreight… A long-range widebody aircraft must burn a whole kilogram of fuel to lift a kilogram of payload… if you double the dimensions of an airship, its volume increases by a

[408] R.G. Van Treuren, "Displacement Vessels For Our Atmospheric Ocean," USNI *Proceedings*, Nov. 2000, pg. 74.

factor of eight, but the surface area and power requirement (and therefore the main costs of construction and operation) only increase by a factor of four…"[409] Shaving off just a few *pennies* of each ATK would translate into *billions* of dollar savings, more than justifying vehicle development costs alone. Large private and semi-private cargo airship projects are already underway overseas. First in the air will likely be the aptly named *Cargolifter*, which will offer service for indivisible objects previously considered too large to be moved. In the United States, if the past is any guide, one might guess the military would lead the way.

In discussions about building airships with modern composite materials, computers and techniques, we have the lesson of that *Sentinel 1000* to show modern safety and other requirements yielded a vehicle that in some ways was *less* efficient than the roughly equivalent helium TC-13 of half-century before. Helium maxes out at about 60 pounds per cubic foot at STP, and heating advantages tend to disappear with speed. Evolution for Naval airships has meant ever-increasing size; the production *Sentinel 5000* design eventually grew to two and a half million cubic feet.

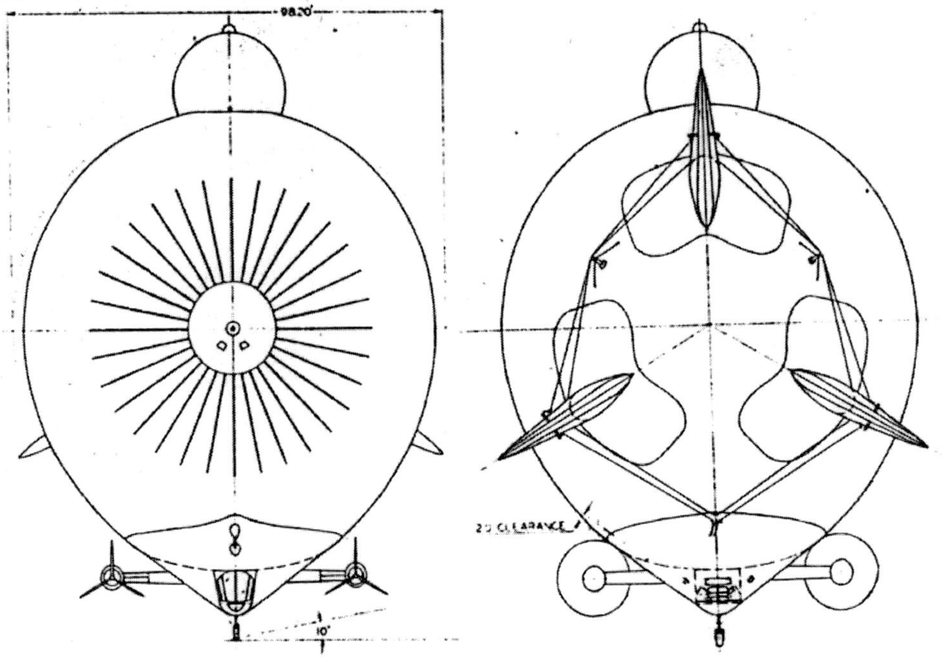

Late designs coming out of Goodyear showed cantilevered fins on two million ft^3 radar-carrying airships. Record holder CDR Jack Hunt publicly agreed with Barkley and Buckley's lament that airship design was limited to a single manufacturer, or that there was super-skill involved in their creation. Our Defense Department has repeatedly failed to create a common-airframe platform adaptable for a variety of uses; they need look no further than the airship. Courtesy Jeff Cook.

The tendency is to make the gasbag bigger not only to make up for the less efficient helium and its operational demands, but also to include ever-larger airbags to allow higher altitudes. A comparison of the ZPG-2W(M) (also referred to as the 2½ W) and the ZPG-3W find similarities in car size if not arrangement. The extra 500,000 cubic feet of enclosure from 2W(M) to 3W not only allowed more lift, but extra ballonet capacity, the air being increased from 247,300 to 383,000. This supposedly gave the 3W an everyday operating altitude of a mile, and a theoretical pressure height of 10,000 feet. An ASW variant, the ZPG-3, was offered. When it became obvious that the nuke boat was a paradigm shift that the airship could not counter the traditional "fight the battery" way, it was necessary to go back to the LTA drawing board as well. The Navy

---

409 C. Pocock, "Lighter-Than-Air, but Heavy with Promise," *Air International*, Vol. 59 no. 5. Nov. 2000, pg. 301.

did not elect to do so. Equipment lofted by the last Navy airships outclassed anything in the sky. By the 1960s, the most experienced designers of Naval airships had drawn up two million cubic foot giants with fin propulsion, and pressurized cars rivaling today's E-6 *Mercury* airplane. External suspension offered many hundred-thousand cubic feet of electronic-soothing helium in which to mount the most sophisticated sensors imaginable, extending reach well beyond that of probable attackers. Massive ballonets allowed altitudes where pressurized cars were necessary. The Admiralty was favoring carriers at the time, so what did not fit on flat-tops was folded up and put away. The designs were never published.

The modern Naval airship could take a nod from these late designs, resisting the urge to build ever larger, more expensive, and more difficult to ground-handle airships simply because others are building much larger vehicles for other purposes. Bigger is not always better, even in airships, when one considers the jobs the vehicles are being built to accomplish. If in-air refueling extends an airplanes' flight into the 24-hour realm, imagine how crews would dread serving in an air vehicle that would be making voyages of a couple weeks duration.

One of the lessons of World War II was pointed out by author Ladislas Farago in his book THE TENTH FLEET. "There was nobody in Washington in decisive authority to promote the ships we actually needed – escort craft of seagoing type, destroyers, destroyer escorts and cutters." The Navy's preoccupation with giant battleships and floating airfields found them ill equipped to fight u-Boats. "On the eve of World War II the United States had a formidable navy as far as sheer tonnage was concerned (over 1,300,000 tons all told). But it was a lopsided force afloat because well over half of this strength (853,000 tons) was in capital ships and cruisers – vessels that could make no contribution to the efficient prosecution of an anti-submarine campaign." Few took anti-submarine airships very seriously then, nor later when the book was published. Likewise, as desperately as an ASW airship is needed today, there are no prototypes in test. Just as an M-type airship could have been mass-produced and been ready for the threat of the new German "electro-boats" of 1944, so a modern airship could be lethal to subs encountered today.

The submarine's only natural enemy is faster than a nuke boat, undetectable by its sonar and invulnerable to homing torpedoes. High mission availability of the WWII ASW airship is lost in journalistic assessments that deal only with the strongly discouraged surface combat - for which they had no standoff weapon - and even that official score is wrong. Overlooked but vital was the blimp's *raison d' ˆtre*, victories against *underwater* adversaries, never credited even when using acoustic homing torpedoes.[410] Said one 1950s diesel submariner: "Planes, destroyers and helicopters are just the normal occupational hazards of our profession. But once a blimp pins you down, he can stay with you until the end of time." The last operation where airships were available, 'Whole Gale' of Feb/Mar 1960, six ZPG-2s logged a respectable 1,644 hours in some of the worst Northeast gales and snowstorms ever recorded. Yet Lundi Moore's ZPG-2 crew established today's record of 93 continuous hours on ASW station – more than 40 years ago.

---

[410] K-72 crewmen claimed use of their homing torpedo against a U-boat on 18 APR 45; the U-857 is listed as Missing In Action off the American coast in April of 1945, and a wreck was found in their area during the search for the USS *Scorpion*. Though VADM Charles Rosendahl's book was never published, it is known he located proof that the U-94 was damaged by the K-6 on 12 APR 42. Historical Action Team members proved the K-34 attacked and drove off the U-107 off Georgia 29 AUG 43. The official history of USN LTA in WWII credits only one assisted sinking and three attacks resulting in damage, in spite of *ample* evidence the Navy never attempted to verify a victory and even covered up the unsuccessful combat of the K-14 on 3 July 44.

"Let us recall also how some of our convoys, navigating in the Ionian Sea, were detoured in good time by airship. The enemy had laid a mine field in its route. The airship spotted it and reported its position. The convoy was thus saved from almost certain destruction," wrote Frenchman du Plessis. A defensive vehicle removed from the opposing weapon's medium was as priceless an asset then in 1917 as it was when ZP-14 worked mine hunting and killing in WWII. "Minesweepers, used to waiting for explosions to announce the presence of mines, liked having the blimps tell them mines lay ahead. They liked it even better when blimps told them there were no mines ahead."[411] The prototype N-1, which sustained a peaking 9,500-pound winch load and towed for 24 hours, was used in mine sweeping tests almost 50 years ago. A British study pegged a minesweeping airship supporting wire-guided mini-subs at one-third the cost of a conventional minesweeper, this coming a decade before the vessels were becoming the most costly-per-ton ships in the fleet. Today our Navy's biggest losses are from "weapons that wait."

What of anti-missile defense, a much greater problem than when a German plane first successfully attacked a loaded troop ship with an air-launched missile? Airplanes, gulping fuel constantly having to rush forward, are ill-suited to 24hr/day watchstanding. During the concept-proving Project 'Lincoln' "...airships manned an AEW station off the New England Coast continuously for ten days. Weather was the area's worst in years with combinations and variations of ice, snow, rain, and 60 knot winds."[412] The last airship delivered to the US Navy was a radar picket ship, the ZPG-3W, later designated the EZ-1A *Vigilant.* One of the four built was inadvertently flown 21,000 pounds heavier than its rated 22,366-pound useful lift. With its 40 foot radar antenna and typical 36 hour mission, the last *Vigilant* was suggested for re-inflation to serve as a prototype (Y)EZ-2A in 1987. The need to detect small, stealthy missiles against sea clutter has only become more acute since an EZ-1A stood in for Washington DC ground radar traffic control. The USS *Stark* was heavily damaged by a French missile even though she possessed anti-missile radar and guns; the ship's defensive visibility was only 25 nautical miles. Even if seen, a mach 1.5 missile would require only 100 seconds from becoming visible until impact. Impractical for airplanes, the required longwave 1,000 square foot antenna lofted to 10,000 feet would extend a fleet's vision of small, even stealth targets out to more than 120 nm. Phase differences from antennas placed bow and stern would enable interferometric processing, yielding 3D data on a single pass. Airship response need not be limited to target illumination, chaff and flare decoys; airships launched standoff weapons more than 80 years ago.

The passionate pleas for flying aircraft carriers made in the dark days of winter, 1942, were voices in the wilderness. Today, when flattop ships carrying airplanes are so expensive but one country dares even operate more than a couple, the need if even more acute. As they did in Lebanon, the Persian Gulf, and other locations, local warlords realize the massive carrier battle groups most prominent function is to defend itself. Gargantuan sums of money pay for tens of thousands of individuals and expensive equipment for the miniscule result of placing a few airplanes on the job and/or ordnance on target. An updated ZRCV, carrying Unmanned Air Vehicles, would be a cost-saving relief of major fleet assets. True, UAVs are in still underdeveloped, with cameras not working well in foggy clouds and infrared imagers not being effective in heavy moisture or indistinguishable temperature differentials. The Air Force Research Laboratory is blending a wide spectrum of emerging technologies to produce a UAV that provides revolutionary intelligence, surveillance, and reconnaissance capabilities. Armed UAVs have destroyed test targets with sub-munitions. Larger antennas on an airship with line-of-sight aiming would mean fewer communication link losses. Emission deceptive remote sensing

---

[411] A. S. Lott, MOST DANGEROUS SEA, U S Naval Institute, Annapolis, 1959
[412] D. Kasperick, "Blimps Begin Their Final Bow," *All Hands*, January 1961

would be supported by basing UAVs on the carrier dirigible. The mechanical details are not mystical, but natural, since both vehicles move in the same medium, proven 65 years ago.

"Waco rides with the MACON," proclaimed the company's newsletter, "...she carries along two Waco F-2s." Along with the 'skyhook,' Navy designation XJW-1 was also modified with a removable combing that allowed outsized cargo – even stretchers – to be loaded for an early version of *Star Trek's* 'transporter' device. It would have been the American 'running boat' when the Germans perfected their hook-on service. Basing modern, semi-autonomous Unmanned Air Vehicles aboard a modern flying carrier would offer performance unmatched even by billion-dollar flattop boats. Newsletter courtesy Eric Brothers; XJW-1 from Tailhook Photo Service; Predator courtesy General Atomics.

The Navy is not alone in its general lack of consideration of the buoyant alternative. More than 1,000 people owe their lives to blimp rescue,[413] a tertiary self-perfected mission of the WWII ASW airship. Numerous studies naming several other missions have shown the maritime patrol airship would be faster than a surface cutter and cheaper than a C-130 while it emulates both.[414] More than fifty years ago, Coast Guardsmen assisted a study that proved an airship could land on and operate from the water's surface. An airship was first to cross the top of the world more than seventy years ago and non-rigids crossed both the Atlantic and the Equator. There is no environment too harsh to monitor fishing grounds or track polluters, no place airships cannot act as command posts for oil spill or maritime disaster management and recovery. Yet in the past fifty years the Coast Guard has done little more than charter a Goodyear blimp for some tests, whose success was generally ignored as easily as postwar experiments that showed a float-equipped airship could operate from the water's surface. Airships were never given the kind of opportunities HTA enjoyed. Mainly owing their early existence to "crackpots," wartime emergency and postwar envy caused their boom time and to some degree only bureaucratic

---

[413] Conservative estimate of James R. Shock, author of numerous books and papers on all aspects of military LTA.

[414] "Analysis of Coast Guard Missions for a Maritime Patrol Airship," by Harold K. Rappoport, "Coast Guard Missions for Lighter-Than-Air Vehicles," by K. E. Williams and J. T. Milton, are typical.

momentum kept them alive thereafter. Early airships suffered from typical pioneering problems before the Great War thrust pressure ships into being, basically, wingless airplanes. There never has been an airship engine specifically; only modifications of motorcycle, automobile, boat and airplane types. While the Kelly Johnsons of the HTA world were allowed to design, build, crash, and fix even bad designs any number of times, the Arnsteins and Upsons have very few shots at getting it right. For the NACA (technical note number 194) C. P. Burgess developed "A Method For Determining The Dimensions And Horsepower Of An Airship For Any Given Performance," but like other tools it has not been applied to any great extent.

No matter the record of stunt demonstrations, operational flights or even the *Graf Zeppelin*; gasbags are unfairly seen as uncontrollable – even dangerous. The outstanding safety record and proven technique of balloon logging, for example, is ignored when stockholder's money can be squandered on a readily available if somewhat dangerous helicopter. "A study of 83 accidents with helicopters used in U. S. based logging operations revealed the vast majority, 73, took place during logging-related maneuvering and not other phases of flight… 23 were fatal and a total of 29 crewmembers were killed."[415] Safer, perhaps, opponents argue, but with airships, the lift is always there, whether it's needed or not. Someone not familiar with aerostatic law would reach the conclusion even hydrogen ships were limited by the one-two punch of carrying enough disposable ballast in relation to the expendable lifting gas. "Shortcoming of the Zeppelins was reliance on explosive and unpredictable hydrogen gas for lift. Furthermore, the huge ships had to depend on water ballast. Between the two lay a delicate balance, necessary to keep the airship aloft." [416] That statement is oversimplified, but illustrative of the problem; an airship's lift is a love-hate relationship. Indeed, it is this lack of control that hastened the rush to the newfangled aero-planes as a cheaper, easier means of getting aloft. "The cost of hydrogen-gas and of the silk containing-bags is very great, the work of proper inflation considerable, and the deterioration ruinous. A silken balloon is made worthless by fifty successive inflations with hydrogen-gas. Control is made exceedingly difficult and the length of flights are definitely limited by expansion, contraction, and escape of gas under varying atmospheric conditions."[417] If only the lift could be neutralized on command, the airship would not seem nearly so impractical. No one has demonstrated a system that can do so.

As time went on, the airship's *pressure height* was another show-stopper. Records held by airships fell to airplanes after WWI; much higher altitude records held by free balloons eventually fell to high performance airplanes and rocket craft. Airships were left wallowing in the turbulent lowest levels. When the US Navy finally started looking at airships again during the Reagan administration, an angry letter was published saying airships were an old, bad idea that keeps sitting up in the coffin. "Ask them how much lift the blimp has above 10,000 feet!" he scorned. Worse, it's not only the altitude, it's the barometric conditions. K-type airship crews operating in the tropics sometimes found pressure height *at sea level.* With helium expanding in the soggy equatorial heat, precious lift was valving away. Replacement gas had to be k-bottle shipped by surface vessel all the way from Texas. Equatorial summers found crews desperately spraying the bags with fire hoses or keeping them in daylight flight just to cool the helium down. Even the most successful hydrogen airship, the *Graf Zeppelin*, was almost done in on account of an ordinary temperature inversion. Since time began, the novice has wanted to know why a compressor can't be used to ease these problems, offering a wider range of operating options. In chapter eleven of his AIRSHIP DESIGN, C. P. Burgess calmly explains "A very common idea

---

[415] "Low, Slow and Dangerous", Industry Outlook, *Aviation Week & Space Technology,* 10 July 2000
[416] J. L. Cambias, "L-59: The Africa Ship," *Airpower*, May 1992
[417] Pitt P. Hand, "Western Men Who Would Fly'" 1910

with inventors is to reduce the lift of an airship by compressing some of the gas from the cells into high pressure containers, or else to obtain the same effect by compressing air from the atmosphere into flasks, thereby increasing the weight of the ship. Both of these schemes split on the same rock as the vacuum airship, the pressure forces entering into the problem are far too great to be controlled with the permissible limits of weight... Inventors ...are...unaware that as long as an airship is below pressure height...the buoyancy is independent of the altitude." Same even with a very efficient liquidation plant. (Worse, a cubic foot of liquid helium only makes 100 ft^3 of gas.)

Why is that, since the weight of the air displaced is decreasing as altitude increases? Thc answer is a cornerstone of aerostatic law: the gasbag will expand about $1/30^{th}$ of its volume for every 1000 feet of altitude gained. Yet lift remains the same, since the air it displaces decreases in weight exactly the same amount, as altitude increases. A rule of thumb states that at mean sea level atmospheric pressure is about 30 inches of water, and that pressure decreases one inch for every 1,000 feet of altitude gained. Therefore, an airship must be designed to loose about $1/30^{th}$ of its volume for every 1,000 feet of altitude in which it is to be operated. No problem in a hydrogen ship, but even cheap gas is not always readily available. So it was obvious long ago that a bag of air could be carried inside to allow the hydrogen to "breathe." In France, the committee for Public Security created the "Corps of Military Aerostats" that distinguished itself at Fleurus and other battles. Farsighted General Meusnier wrote down the theory of a non-rigid airship with air ballonets, and then went out and got himself killed at the siege of Cassel in 1793. Renard resurrected the idea in 1880. Six years before that, Jules Vern wrote, "Water is the coal of the future. The energy of tomorrow is water broken down into hydrogen and oxygen using electricity. These elements will secure the earth's power supply for an indefinite period."

In the case of the airbag ballonets being pushed flat as the helium expands, and you have no more air to let out, you have reached pressure height and must stop ascending. Sadly, ever since the Navy's C-7 briefly experimented with helium, we have the show-stopper problem: the *least* lift is available when it's needed *most.* At the same time, while we are now fulfilling Jules Verne's prediction of unlimited clean energy, *we need only put the two old ideas together to solve the airship's main problem.* Thus, we come to the main purpose of this book.

There will be a number of "helium heads" that will ignore or skim this book; others will study it and still remain convinced hydrogen was a vital part of the event chain that caused the LZ-129's outer cover to burn. Likewise, hydrogen fuel advocates could also blow off these results; why bother to demonstrate hydrogen safety in some archaic mode of transportation? Fine, for the helium heads, first on the list of things to do is to *put the hydrogen in something that won't burn*, put its vents on top, and get back into the air. For the latter, remember: *We're at the limit for airplane size.* To carry more, air vehicles must return to at least partial buoyancy. It's better we accept hydrogen fuel *now*, before we have no alternative. Why?

Repeating, most helium is found within a 250-mile radius of Amarillo, Texas; the Cliffside site will likely be empty in 20 years. More uses for this unique noble gas are being found faster than are ways to do without it, and more distribution means more chances of loss along with greater consumption. Soon after discovering how to isolate, compress, transport, and store helium, came the ability to lose it in great quantities. In February of 1924 the Lakehurst storage tank lid jammed and failed to rise, blowing 457,000 cubic feet of the allotted gas out the bottom. The author witnessed the world's once-largest building, the VAB, evacuated as a series of miss-steps caused a like amount of helium to escape into the high bays from a 2-inch high pressure line.

These inevitable accidents have will unfortunately continue as long as helium is in use. A new high-pressure helium line will save distribution costs on the space coast. "Other users of pressurized helium currently rely on high-pressure gas railcars (in support of Titan), compressed gas trailers, (Atlas Centaur and Delta II/III), and K-bottles (institutional shops and laboratories)... The KSC helium facility... has a capacity of 1,800 standard cubic feet per minute at 6,000 psi... a new high-pressure helium pipeline is being constructed to service launch needs at Complex 37 at Cape Canaveral Air Force Station."[418] (Since it is very difficult to liquefy helium, let alone keep it so near to absolute zero, transport will always involve pressure tanks and lines yearning to leak.) "High-pressure helium is used to flush launch vehicles, like the space shuttle, of unwanted or toxic gasses. It is also helpful for ensuring the proper pressure in the vehicles."[419]

Recent finds are promising, but only delay the inevitable; the planet Earth will probably be helium-free by the end of the 21st century. Airships won't just be an anachronism; they will be an impossibility, unless we adopt perfectly logical enhancements right now. Nor do we have to wait until helium is gone to start enjoying the advantages of hydrogen fuel. As Firoz Rasul of Ballard Power wrote, "The stone age did not end for lack of stone." Verily, it is in airships where hydrogen fuel would really show its main advantages as an energy carrier. The reason is simple: hydrogen contains nearly *three times* the energy of gasoline, by weight. A fuel weight savings factor of 2.8 is an overwhelming incentive where "bounce for the ounce" is raison d'être. We need hydrogen fuel because airships need hydrogen fuel, and *we need airships*. Airplanes are already as big as they are likely to get the next few decades. Edward Warner of MIT said long ago, "The lighter-than-air craft has inherent advantages, when enormous sizes are in view, which can hardly be counterbalanced by any technical skill that may be applied to the improvement of the heavier-than-air types. The most evident of these advantages is that the lift of an airship, depending on the volume of the gas contained in the envelope, goes up in proportion to the cube of the dimensions, while the lift of the airplane, dependant on the area of the wings, varies only in proportion to the square of the dimensions."[420] Where airships have failed in the past, we find not fire, but their inability to control their buoyancy as a primary cause of accidents. Superpressure balloons and air vehicles have taken advantage of modern high-strength materials to stretch the envelope (no pun) but the rules can only be bent, not broken. Hopefully, no young airship designer will become enamored with some new wonder fabric, least he imitate Melvin Vanniman and assume nature will compress the lifting gas as he gains altitude indefinitely.

The cost of insurance is likewise not based on inordinate occurrences of conflagration, but rather the awkwardness of airship handling. Even the most experienced crews can't always handle a buoyant airship in suddenly changing conditions, circumstances that befell Goodyear's *Star and Stripes*, causing her to be laid up with injuries and miss the 2000 football season. The history of buoyant flight is dotted with attempts to address the control of static condition, but no one has demonstrated a solution to the problem that makes airships seemingly impractical for most operations. You have little *control* over the lift; you can't "throttle back" the lift. It's like a car with no brakes but in the vertical direction; you have to drag it to a stop, then weight it down. Likewise, just as your ears must pop as you ascend, the airship would explode as it passes into ever decreasing atmospheric pressures. Therefore at mission origination, we have the greatest

---

[418] Spaceport News, Vol. 39, No. 18, September 8, 2000
[419] K. Young, *Florida Today*, 3 NOV 2000
[420] E. P. Warner, "Giant Airplanes," *The Christian Science Monitor*, July 17, 1922.

payload and the least lift, meaning pressure airship design demands carrying along a reserve capacity of air to release as you gain altitude. Of course all that air gives no lift at all, even though at launch you are at maximum weight and *need the most lift.* At high altitude, where we need the least displacement to allow expansion, we have the most.

The dictionary defines *paradox* as 'something with seemingly contradictory qualities.' We might call this dilemma the *paradox of helium aerostatics*. It is this show-stopping problem that has prevented airships from achieving a more prominent niche in our atmospheric ocean. Airships neither displace winged vehicles planing atop the air, nor missiles punching holes in the air. They do not compete with airplanes any more than rockets compete with helicopters. Yet displacement vessels have been relegated to flying billboard and camera platform duty in our country because their advantages don't outweigh their disadvantages. Aerostatic law and scarce helium are a one-two punch keeping buoyant vehicles down.

In hydrogen ships, one could simply vent gas to compensate for consumed fuel weight or to overcome pressure height. Eighty years ago Zeppelins were capable of making a vertical liftoff in their own length carrying a full fuel and military payload. Starting with full hydrogen cells for maximum air displacement, Zeppelins would be continuously vented to gain altitude beyond 20,000 feet, adjust lift for burned fuel, bomb dropping or other mission expendable weight. Regaining static equilibrium, they could settle to a heavy landing. This was of course limited by the quantities of the two consumables, and had the great disadvantage that the hydrogen would have to be replaced before the next liftoff. This was possible in part because hydrogen was generated right at the station. (In a modern version, excess shipboard electrical capacity applied to an inexhaustible supply of salty water would allow at-sea gas production through the electrolysis of brine.) Yet even cheap hydrogen was not free or abundant in nature, and the masters of the rigid airship realized this greatly limited their operation.

Von Schiller put it this way: "As the liquid fuel (gasoline, crude oil, etc.) is used up hour by hour, the airship becomes lighter in weight. Hence, if it is to come to the ground after a voyage, it must be restored to equilibrium, that is, it must be made again to hang in the air when its engines are shut off. If it were really "lighter than air," it would shoot skyward when stopped, and the ground crew, instead of holding it and managing its movements, would be lifted with it. It is accordingly, at first glance, necessary, when a Zeppelin uses a liquid fuel, to blow off one cubic meter of hydrogen for every kilogram of fuel burned. Not a very ideal state of affairs." Helium made even this awkward and expensive necessity quite impossible. Nor could the gas be vented to increase pressure height. As noted, "A further loss in performance from the use of helium follows from the necessity of starting flight with only partial inflation in order to avoid valving the costly gas as the altitude is increased." In other words, Vaniman's 1910 *America* had much more control over its static condition than the Goodyear *America* fifty years later. It boils down to gasoline vs. lift. As Burgess explained in his memo #185, "If gasoline or other liquid fuel having a lower heating value of 20,000 Btu/lb. is supported by helium lifting .062 lb./ft^3, the heating value per cu. ft. of air displaced is 20,000 X .062 = 1240 Btu/lb. This is only 77% of the heat available with lighter-than air fuel gases; and there is the additional disadvantage of having to provide water ballast recovery or pickup devices capable of providing ballast at a rate equal to the entire consumption of fuel. There is also the disadvantage of having to start a flight with the inflation limited by the ceiling to be attained..." C. P. Burgess might have agreed that if we can't stomach unhindered hydrogen lift, we could at least try to solve this fuel-vs.-lift conflict.

In answer to half the problem, coming down, handling a light post-mission airship without complicated equipment and an expensive ground crew, *hybrid* designs have evolved. Such vehicles, from the Piasecki Heli-Stat to the Zeppelin NT, depend on helicopter-like thrusters. Generally operated in heavy condition, they instantly adapt to changing loads simply by throttle adjustment. There is certainly a niche for such vehicles, even if C. P. Burgess told the youthful (Later RADM) Carl Seiberlich: "Young man, what you would have there, is a *very bad airplane*."[421] He spelled out why in his 1927 book. Perhaps the master would not have voiced an objection to some newer designs. Former airplane designers, having reached maximum parameters with the Airbus 380 design (25.2 centimeters wider than the 747-400), are adopting the lessons of the rigid airship to break the size barrier imposed by today's metals and composites. The *Skycat* and Lockheed's *Aerocraft* will be the ultimate expression of this design philosophy begun and actually built in miniature decades ago.[422] The gargantuan lifting body shaped structures will be supported by helium cells, overcoming restrictions placed by strengths of today's materials. It will be the first air vehicle that will have such bulk interior capacity that it will *gross* out before it *maxes* out. "*Aerocraft* could dramatically improve cargo transportation. It is designed to serve the market that requires transportation faster than ocean freight but cheaper than air freight."[423] The vehicles will revolutionize air cargo, now limited to airplanes that are capable of lifting lead and uranium, but quickly fill up with typical bulky, relatively light cargo. Still, even these advanced hybrid craft will have to obey the laws of aerostatics, and could benefit greatly by adopting hydrogen as fuel.

The Zeppelin New Technologies 07 is an entirely new idea, a pressure airship hull reinforced by an internal structure. The pilot's side stick input is translated into vectoring thrusters for precise control. Soon to be available will be wearable bands that marry an electromyogram with control software that simulates actually moving a stick. L-Z photos.

Make no mistake, the idea of using hydrogen part for lift and then harnessing its talents to change static condition is pretty old – and obvious. T. A. Knowles proposed an ingenious plan to help solve the paradox. The hydrogen was mixed in the helium, removed through an oxidizing agent whose by-product was water. A one-two punch, the scheme eliminated some lift while adding some weight. Copper oxide would reduce the hydrogen in an exothermic process that could add heat to the lifting gas, or be rejected, in a condensing apparatus. In a seven million cubic foot ship, with a ten percent hydrogen fill, a 70 hour mission would require 100,000 cu ft

---

421 Author's interview at 1993 NAA Reunion, Scottsdale, AZ

422 D. Fulghum & R. Wall, "Mammoth Half-Blimp Seen As Future Airlifter," *Aviation Week and Space Technology*, 22 February 1999

423 NASA press release, 30 August 1999

of the gas to be processed per hour. In Design Memo 266, C. P. Burgess dismissed the processing as being beyond practical for an airship. He also pointed out the water, containing traces of copper, would be highly corrosive to aluminum. He suggested, "A possible solution is the insertion of a ballonet diaphragm in each cell to separate the mixed gases from the pure helium, but it would add greatly to the weight and complication of the cells."

Suppose the ship's *fuel* could be used for similar results? Recalling what C. P. Burgess detailed about performance decreases associated with adopting helium, he went on to say operation with the disposable gas increased airship options. "With hydrogen, it is customary to start a long voyage fully inflated, and gas is valved as fuel is consumed and the ship gains altitude." Impossible with rare helium, Burgess offered an interesting solution, with a nod toward Haenlein of 1872. "This disadvantage of helium may be overcome through the use of coal-gas in ballonets filling the waste air space, and used as fuel in flight." That certainly was an improvement over Stevens' 1902 hot air ballonet.

The *Graf Zeppelin* took a larger step in solving the aerostatic paradox. Geiling and Sauter: "Consumption of the power gas did not make the airship lighter: ballast discharge by releasing lifting gas, or water recovery, were not necessary." The brilliance of balancing a fuel's energy content with the acentive force needed to carry it was the main secret of the most successful airship ever built, LZ-127. "In a departure from all previous Zeppelins, a series of 12 fuel gas cells occupied 26,000 cbm of the enclosed volume. This gas "blaugas" was approximately the same weight as air; using it gave the craft an endurance of over 100 hours as compared with 67 hours if hydrogen had been substituted [in the fuel gas cells] and petrol been used as fuel." [424] Atmosphere replaced the void and weight left when consuming "Blau" gas, a proprietary fuel mix named for its inventor, from the fuel bags. The last US Navy officer to ride a rigid airship recorded the mixture was 66% propane, 34% hydrogen.[425] Von Schiller explained, "It was found that the most practical gas mixture consisted of propane and hydrogen in such proportions that the specific gravity of the mixture was approximately equal to that of air. The consumption of such a gas in the ships motors does not change the weight of the ship." Compared with air, the fuel mix had a density of 1.05; with its high caloric value, the fuel gas yielded low specific consumption which one source lists as 0.168 kg/hp/hr (0.37 lbs/hp/hr).

Here was the *Graf's* real trick: Instead of being made larger to carry more heavy fuel, which required more fuel to push the big mass into the wind, calling for still more lift, ad infinitum, LZ-127 struck a balance impossible to duplicate with liquid petroleum fuel. Bauer & Duggan put it another way: "The problem of fuel being consumed and the consequent need to valve off gas was solved through the use of gaseous fuel which had the same specific gravity as air... Additionally the consumption of this gaseous fuel would be less than that of liquid fuel (10g/HPh compared to about 220gHPh)." Had Goodyear-Zeppelin been able to persuade its Navy customer to look beyond gasoline cans tied to helium balloons, the history of the American airship might have been quite different. A Goodyear-Zeppelin draft press release circa 1926 mentioned the British effort: "The ships fueled with heavy oil will have to valve hydrogen, but as this fuel is less expensive than gasoline, it will probably offset the cost of hydrogen. Here in America our Akron engineers have been working on the problem from still a different angle and plan to use natural illuminating gas alternately with gasoline, in the ship's motors, which offers still further

[424] Gasbag Journal #44, June 2000. The extra 33 hours of endurance saved the ship when it had to wait three days in the air as a revolution was fought below, for a total of more than 118 hours.

[425] Courtesy Hepburn Walker, Jr. Propane contains a bit more than five Kilocalories per gram than gasoline, making it a more efficient airship fuel; some see a mix of gasses offering the greatest advantage.

advantages for air navigation... As the natural gas is lighter than air and gasoline is heavier than air, the ship can be kept in balance and controlled by alternate use of the two fuels so that any necessity of valving the valuable helium can be obviated and still without the condensers which add weight and retard the speed by increasing air resistance. The natural gas used for fuel could be carried in gas cells, these cells to be surrounded by helium cells for greater safety." In June of 1928, Dr. Arnstein wrote to Dr. W. K. Lewis of MIT to ask what he thought of using a Navy gas mix they had received. It was 30% methane, 45% ethane, with 15% propane and 10% ethylene.

Eckener wrote Arnstein in December, 1927, that they had "...signed with the Deutsche Blaugas-G.m.b.h., of Augsburg, a license agreement which secures for us the use of a method for the production of oil gas worked out by that firm. The gas is made out of crude oil by way of distillation; it carries a specific gravity of 1.05, a low heat value of 16000 calories, and a condensation temperature of –30 deg. C. The production of 1 cubic meter of gas requires: 3.5 kg of gas oil, 0.8 kg of coke and coal, 40 lt. of water, 0.7 kw of electric current. We have ordered from [them] the erection of a gas producing plant for an hourly capacity of 120 cbm of gas ready for use." The letter suggests Goodyear-Zeppelin obtain a license, since they had been unable to verify a so-called Blaugas Company of America had really signed an agreement with the inventor, Mr. Herrmann Blau "...in the years before the war." Eckener also suggested Goodyear-Zeppelin should try to influence the Navy to build such a gas-producing plant, and/or that G-Z also build one, since the gas could be used for welding and other purposes. "The costs for the erection of our plant... about ($50,000)." Frank Howard of Standard Oil was answering Litchfield's questions in a letter of October 18, 1928, saying the Navy had asked Standard Oil to look into the matter. He told how any of their plants could supply a very similar mix in unlimited quantities: About half methane, a quarter ethane, and 20% propane with trace remainder gasses would yield a mix whose specific gravity was 0.95 and delivered about 1600 BTU. Howard stated it could easily be piped into a balloon at a cost of about $1.50 per thousand cubic feet. While the Americans did nothing, the LZ-127 was completed. A Goodyear-Zeppelin draft release then said, "Dr. Eckener's suggestion of using non-buoyant gas as a fuel will result in no variation of lift due to consumption of fuel and thereby simplify altitude navigation of the ship. Besides, heavy gaseous fuel is very much less explosive than hydrogen and in fact brings the fire hazard *even below that of gasoline*." (Emphasis added.)

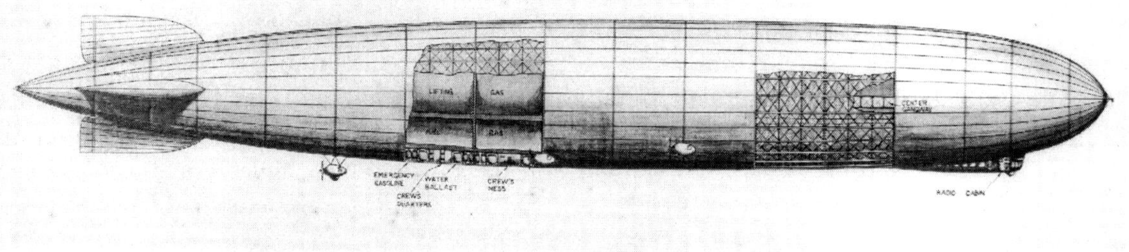

LZ-127's genius was foreseen in a Goodyear-Zeppelin draft press release. "The Germans are proposing to use a gas that is about the weight of atmosphere, planning in this way to avoid the necessity of valving out lifting gas to compensate for the loss of load which resulted from the consumption of gasoline as fuel." Arnstein added, "We do not feel that the use of heavy fuel gas... adds materially to the fire hazard already existing there." Courtesy University of Akron Archives and LTAS.

A dual fuel capability was retained, allowing heavy gasoline to be burned off whenever some excess weight needed to be consumed. In the general hysteria that fears all gaseous fuels, a second strong falsehood has become entrenched in this mindset, that liquid gasoline was carried for safety (!) purposes. The *Graf's* Captain explains its real purpose: " But a more efficient way of compensating for the initial loss of buoyancy is the use of a corresponding amount of liquid fuel, a quantity of which is always carried on board in addition to the gaseous fuel mixture already mentioned. There is also a reserve of liquid fuel equivalent in weight to the lifting power of the largest of the hydrogen cells, so that if one of these should through any circumstance be injured or gas be lost, the ship can always be stabilized again by using up the liquid fuel." He had explained how the rise to cruising altitude had blown off some lift through overpressure valves, so of course the corresponding weight of gasoline would be burned off on the way back down to landing. By comparison, the LZ-129 had to lift up to 66 tons of fuel, then valve off hydrogen as fuel tonnage was consumed. The much smaller *Graf* could actually go the *same distance* as the larger LZ-129, on far *less energy*. (Since *Hindenburg* carried a much greater payload, it is of course an apples-vs.-oranges comparison.) The concept is easier to understand when one considers that if *Graf's* lifting gas cells had been made larger to allow lifting liquid fuel, she could not have carried the payload, had the range, or been as economical.

Hugh Allen was at New York's Roosevelt Hotel on April 17, 1929 when he wired back to check an unfavorable comparison. Surely, Goodyear-Zeppelin's new helium ship designs would out-range of the smaller *Graf Zeppelin*...? Arnstein wired back that they would not! He explained his twice-as-strong design would carry twice the load and go faster, but they could not outrange the efficient *Graf*. Arnstein closed saying, "If gaseous fuels are used on new ship [,] range could be increased fifteen percent."[426] Remember, the USS *Macon* could have criss-crossed the Atlantic in 1935 in spite of her dangerous and heavy gasoline.

A Goodyear-Zeppelin memo stated, "The requirements for the fuel gas are that the specific weight shall be about 1.03 to 1.08 relatively to air, the lower heating value not less than 1700 Btu per cu. ft. at normal temperature and pressure, the condensation point below –20 degrees F at atmospheric pressure, and it must be a clean, dry gas that will leave no liquid or solid deposits in engines or containers." It listed Blaugas as composed about one quarter each Ethane and Methane, Propylene and Ethylene about sixteen and twelve percent respectively, and only 5 percent hydrogen. No accident, there was some experimentation to find the optimal engine/fuel combination for LZ-127. The Maybach marine motors had once been running a 6:1 compression ratio and an 80/20 mixture of benzene and gasoline. One source says the ratio was upped to 7:1 and the fuel was changed to a coal gas spiked with 5 to 30 percent hydrogen, for a 1928 test flight over the Mediterranean. "...[A] 14% improvement in fuel economy was achieved." Although it has not been documented, it seems only logical the experimenters had also bumped up the hydrogen content to a percentage where the mix would be increasingly lighter than air; it would have been irresistible to have the fuel lift a little more than itself. If they did, these steps were duplicated by later testers who found a normally aspirated engine could not get enough air to consume the greater quantity of hydrogen. Therefore, the final mix, with the same specific gravity of air, also happened to be about the perfect mix for an engine without a blower!

Graf Zeppelin's concept of carrying fuel energy without wasting any lift could be re-examined for improved design ideas. Since their lower flammability limits are nearly identical to hydrogen, propane, methane and natural gas are certainly not more dangerous. With these fuels, we pass the advantage of a fuel additive bearing part of the fuel's weight, and actually move into the realm of the fuel carrying its own *and* some other disposable weight as well. One thousand cubic feet of natural gas would lift 27 pounds at moderate cost. The same amount of methane would lift 34

[426] Arnstein Collection, Univ. of Akron Archives courtesy Eric Brothers

pounds while costing only slightly more. The gasses offer a specific heat of more than half a Btu per pound, which is weak in comparison to hydrogen's 3.43 Btu/lb. but, at typical prices, is actually cost competitive on an energy basis alone.

The *Graf* remains the most obvious argument against the Henny-Pennys who ignore the historical record to bemoan hydrogen lift and/or gaseous fuel. In its million miles of flight, the LZ-127 was presented with many seeming disasters in the making. Captain von Schiller wrote, "Returning from a trip to Moscow one moonlit night, we heard clearly 10 or 12 rifle shots near the Russo-Latvian frontier. The Letts said it was the Russians and the Russians said it was the Letts, but that is not the point. There were no less than 10 bullet holes in our hull to be repaired with all dispatch." Nor was that limited to a bad section of the Russian border. "One day when Dad had the duty the German dirigible *Graf Zeppelin* arrived with a few bullet holes in it. It seems some 'red neck' did a little target practice as the ship flew in low over the New Jersey farmlands. This aroused a bit of attention among the news people who otherwise had only a 'ho hum' story on their hands."[427]

Happily, bullet-resistant safety and increased efficiency go together. C. P. Burgess also wrote, "The use of fuel gas, as in the '*Graf Zeppelin*,' shows a further opportunity for an important improvement in performance." Drawings in the National Archives show such a bladder system was considered for the 6 million cubic foot ships announced in 1925, sized to allow similar performance to the five million cubic foot ships the British were planning.[428] Karl Arnstein recognized a fuel-gas powered ship would have about a 15% advantage over the previous design of tying helium balloons to gasoline cans. In 1930 he wrote, "...it may be burned with an economy of about .4 lbs per HP, as contrasted to .45 for gasoline burned in the same engine. Its greatest advantage lies in the fact that it can be made to have such a density as will keep the ship in equilibrium even when a certain amount of liquid fuel is also to be burned. The weight of the fuel gas is hardly a factor, since we can carry it with no expenditure of weight... The actual volume of lifting gas required for inflation is materially diminished, since all that must be carried are the dead weights, the payload, and the service loads. An appreciable lowering of helium costs is possible."

Further, C. P. Burgess would write, "In fact, fuel gas is more advantageous for non-rigid than for rigid ships. It is comparatively easy to install ballonets for fuel gas inside the envelope in such a way as to eliminate the fire hazard by entirely surrounding the combustible gas with helium, whereas such an installation presents almost insurmountable difficulties in rigid airships of conventional design. Moreover, carrying fuel gas in ballonets, instead of liquid fuel in tanks inside the car, permits the designer to effect an important reduction in the size, weight and air resistance of the car." [429]

The US Navy's own creation, the non-rigid K-1, was the compromise result of Bu Air Design 88. The car was a Naval Aircraft Factory creation, and a fuel bladder of 51,700 cu. ft. capacity was slipped into a "replacement" envelope from Goodyear. (It could thus be quietly purchased

---

[427] Ken Ekelund, "Naval Aviation Southern Florida circa 1930," remembering his father LTJG Ekelund serving at NASL and his being quoted in New York papers, Florida Aviation Historical Society #155

[428] *Aviation*, January 5, 1925, pg.9. "This airship will be the largest in the world, excelling each of the two 5,000,000 cu. Ft. ships which have been ordered by the British Government for commercial use between England and Australia." Even the extra million cubic feet wouldn't allow the performance to match, and the ZRS was later enlarged to 6,500,000 cu. ft.

[429] C.P. Burgess, "Performance of a Nonrigid Airship of 320,000 cu. ft. Volume, Inflated with Helium and Fuel Gas," Design Memorandum No. 89, January, 1929

using only maintenance funds, getting around the need for Congress to vote money.) A single gaseous fuel bladder of about 4,000 pound capacity was suspended inside the envelope. Air could be pumped into the ballonets to compensate for the fuel bag's decreasing weight as fuel was consumed. The extra space in the small control car, thereby devoid of gasoline tanks for the first time since the 1919 D-ships, allowed room to include a winched spy basket and even bunks for the relief crew. This seemingly simple refinement delivered increased performance in comparison with a "conventional" design completed a couple years later, the TC-13. This is more of an apples-to-apples comparison. The Army ship was a dozen feet longer, helping add about 29,000 cubic feet of volume that should have lifted an additional 1740 pounds. Yet the fuel-bag equipped K-1 had 135-lb. *greater* useful load than the TC-13![430] Experiments continued during K-1's tumultuous eight-year career. The Naval Aircraft Factory tested a new AEL-2 carburetor for the R-975-D engine in 1932. Fed fuel gas from the interior bag "(1) in the cruising range 1.7 inches of water pressure gives approximately the best setting, whereas 1.2 inches water pressure gives approximately the best economy."[431] In either case, the ship's static condition was almost unaffected by running the fuel 'tank' dry, since the gaseous fuel had about the same weight as air. Sadly, the compromise lash-up design was not popular with her crews. Its 'enormous' size was thought to be too large for a non-rigid, and it was not a comfortable ship to fly. It did not have a particularly stellar record and there was an unearned reputation K-1 was somehow accused as not being as safe as other blimps.[432] As always, helium added more complication for K-1. When gaseous fuel leaked from the bladder into the helium, there was then no practical way of purging it clean - short of discarding the entire fill, improbable with expensive helium. This unexplainable hysteria over safer fuel gas was not limited to US Navy men, since the more efficient, scaled up fuel gas system for the LZ-128 was abandoned with that ship – in favor of diesel, following the diesel-powered R.101 fire! One can only imagine how much the more efficient LZ-128 might have accomplished.

Luckily the last two Army TC ships were designed with heavy takeoffs in mind, distributing their lower fin loads between two smaller surfaces spread apart to permit very high climb-out angles, precursor to the true X-fin configuration. Its 1935 design memo[433] called for what was seen as the next minor evolution, the follow-on K-2, to be equipped with a regulator so its fuel bladder didn't starve the engines when pressure was low, as had sometimes happened in K-1. There were to be measures taken to prevent bladder surging. Hopefully for reasons more meaningful than an undeserved reputation, but probably for fear of helium loss if compromised, the fuel bladder idea was not included. Our WWII airship crews accepted the lesser efficiency and the very heavy take-off (too often knocking off lower fins in the process) and the squadron-challenging terror of light landings. Eric Brothers found Patent 2,451,815 for a Goodyear airship, filed July 16, 1941 but not granted until Oct 19, 1948. It contains the first detailed X-fin controls, an elliptical, braced keel along the lower third of the bag to carry engines and tanks. The envelope also shows a provision for fuel gas ballonets.

The rolling-start takeoff technique evolved with successive Navy ships until a ZPG-3W (unwittingly) flew some 21,000 pounds heavy. Postwar missions were of ever increasing duration, but sooner or later that burned off fuel weight would have to be replaced. Working over

---

430 C.P. Burgess, "Comparison of Navy Airship K-1 and Army TC-13," Design Memo No. 129, courtesy Jeff Cook.

431 "Test of Engine for K-1 non-rigid Airship," Ser. No. AEL-LR-78, Aeronautical Engine Laboratory, Naval Aircraft Factory, Philadelphia, 8 NOV 32. Courtesy Hepburn Walker, Jr.

432 Author's conversations with airshipmen William Clarke and Hepburn Walker, Jr.

433 "Requirements for Proposed Airship K-2," Design Memorandum No.193, C. P. Burgess, Aer-La-1-MW, USN Bu Air, March, 1935

water, the postwar Navy ships could simply toss out a bag and haul up some water ballast with their winch. Yet missions don't always allow this luxury even carrying such equipment, nor are they always over water.

When the ability of fuel to lift its own weight is examined, natural gas and methane could be considered owing in part to their cost and universal availability. Hydrogen gas would be even more advantageous. It wasn't a new idea when, on the 1919 US Navy C-5 airship, a valve was added to an engine intake manifold to burn hydrogen from the bag in compensation for fuel weight consumed while flying. With no engine modifications, just hooking up a line from the bag to the intake manifold made no compensation for the hydrogen's faster burn rate, and the engine tended to overheat. (Indeed, running hydrogen right into a gasoline engine has been proposed as a method of faster warm-up, to allow the dirty-when-cold motors to meet strict emission standards.[434])

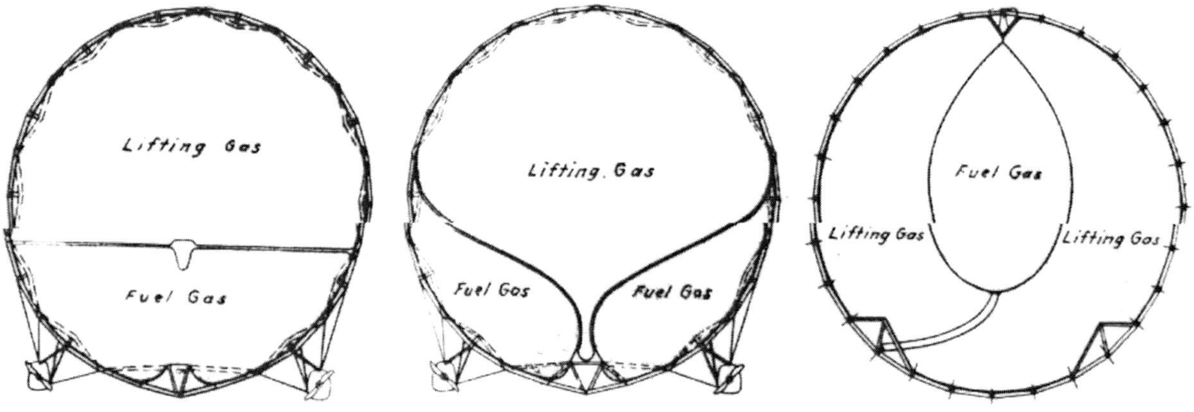

A Goodyear–Zeppelin draft press release circa 1927 noted the planned design of the LZ-127 saying, "As the natural gas is lighter than air... used for fuel could be carried in gas cells, these cells to be surrounded by helium cells for greater safety... heavy fuel gas is very much less explosive... and in fact brings the fire hazard even below that of gasoline... no concentrated heavy loads of any fuel tanks stress the structure in the corridor. " (Contrast with photos, pages 55, 93 and especially 98.) University of Akron Archives; NARA.

In November of 1925 C. P. Burgess explored this idea in his design memo #61, "Hydrogen and Coal Gas as Airship Fuel," in which he again stated the obvious, "Airships inflated principally with helium may be fitted with internal fuel ballonets surrounded by helium to minimize the danger from fire... in order to permit the gaining of a reasonable altitude without loss of helium, or the entire fuel load may be carried as a combustible gas." Burgess ran the numbers, which showed, "Combining this difference with the 34% advantage of hydrogen per unit volume, the total advantage is 104%. This means the hydrogen ship in which the hydrogen is burned as fuel in conjunction with the gasoline will have more than double the range and endurance of the exclusively helium ship." Burgess put forth one example of less than 3 million ft^3 in which a comparable helium-only ship would have to have eight times the volume. "In most cases the advantages of the hydrogen ship would be even greater... With the 34% caloric advantage of hydrogen per unit volume, its total advantage is 158%. On the assumption of the endurance varying as volume 1/3, the helium ship in this case would have to be 17 times larger than the hydrogen one of equal range!" Burgess urged immediate action before an upcoming mission, saying "It is recommended that a study be made for construction of combustible gas ballonets in the *Los Angeles* from a structural standpoint and engine operation. The gain in endurance appears to be precisely what is needed to make the Honolulu flight a safe instead of hazardous

[434] "The Hydrogen Start," *Popular Science*, June 1999, pg. 45

venture." Sadly, Burgess was not heeded and ZR-3 never got to scout for the fleet from Ewa; ZR-3 was never to cross the US. Almost ten years later C.P. was still trying to get the Navy to understand, with his design memo #185. "A great advantage of hydrogen over all other gases is its usefulness as 'anti-ballast,' and the possibility of using it to take a ship off the ground at or near 100% inflation, valving or burning it as altitude is gained. Another and perhaps even greater advantage is that a hydrogen ballonet in a helium gas cell should give very little trouble from surging, whereas serious difficulties from this cause are liable to occur with heavier fuel gases."

As discussed, building tanks to carry hydrogen fuel would have a built-in safety advantage since they would be so much more robust. A rule of thumb calls for LH_2 tankage to be about threc times as large as correspondingly energetic gasoline tanks. Liquid hydrogen fuel's minuscule weight means these larger tanks may be spread over a more catenary-friendly length of the envelope. The design benefits from LH_2's lesser weight by reducing the heavy structure and suspension necessary with ponderous petroleum fuels. Until we're conditioned to stop thinking of airship fire in association with hydrogen, we won't remember it does not matter what is *inside* the bag if the *bag itself* is flammable. Replacement envelope D-528, designed for a ZSG-4, caught fire *without gasoline* at Akron, just testing it with air inflation on 21 JAN 1955. Once hydrogen is not kept in a flammable container, the words of P. L. Teed make even more sense: "...in a modern suitably designed airship the possibility of it catching fire is so small as to be almost negligible... to obtain a hydrogen fire on an airship it is necessary that the hydrogen and air must be mixed together, and that in some portion of the mixture the temperature must be above red heat, a combination of circumstances which is improbable, for it presupposes a leak of hydrogen and fire occurring simultaneously in close proximity."

If we could set aside our revulsion to the idea of putting some hydrogen inside an airship *bag* again, the solution to the paradox is already here. It's a variation on the old idea that helium cargo airships should have an extra bag being inflated with cheap replaceable hydrogen. It lifts the cargo; once the cargo is placed at its destination, it's simply vented as the cargo is dropped, and the airship flies home under helium.

The disposable hydrogen would have lifted extra tonnage of gasoline, then could be vented to gain altitude, then vented some more when fuel weight was consumed. Air could have been added to regular ballonets to occupy the identical cubic area when descending. "Lately some thought has been given to using both hydrogen and helium in an airship. Hydrogen is about 14 times as light as air, while helium is only about 7 times as light as air... It has been proposed to carry a nominal percentage of the total inflation in hydrogen in special cells surrounded by helium. The combination would give a compromise of the fireproof qualities of helium and the freedom of valving associated with a hydrogen inflated ship. It would give a means of maintaining buoyancy equilibrium..." Yet its single-sortie limitations and weighty complexity overruled its construction even in the try-almost-anything experimental days of the war.[435]

Once we accept the idea, what is to prevent the hydrogen in the bag from being consumed as fuel, instead of just valving it off? Would it be so different than the fuel bladder in the K-1, with modifications to the suspension? The masters of the rigid airship did not have a problem with this idea when faced with less efficient helium lift. Bauer & Duggan: "A further possibility...

[435] It should be noted that the same wartime urgency also precluded development of the promising prewar experiments in sea anchoring and sea water ballast recovery following the accidental death of LT Clinton Rounds, their chief proponent. This is in spite of the fact both American and German rigids, even the LZ-129, had developed methods for recovering weight from water. Even the proven lessons of blimp rescue weren't relearned until 1943.

with the LZ-130 was the burning of hydrogen in the engines. This would allow hydrogen to be used as a lifting gas and also as a fuel, thus obviating the need to valve off gas in a wasteful fashion." In a paper to the Society of German Engineers (VDI), the story of the German effort was told. "In September of 1935 work was started with an old Maybach Airship Motor, Type 'M2. IVa' which had an output of 250 HP… it became possible to attain approximately one third this HP after long trial runs using hydrogen. The main characteristic feature of this motor was the single mixing valve employed at each cylinder. Further development and trials with larger motors of various designs and construction proved the influence and effects of the shape of the combusting chambers. The relative best results were achieved with a 'BMW 6' motor. The top performance of this motor by a 1:6 compression ratio using A. C. gas was 650 hp. With hydrogen, a continuous output of 200 HP was achieved by a 1500-RPM P/M. The consumption was 0.85 m^3/p sh. corresponding to 2040 K CAL/P sh. relative to a lower calorific value of hydrogen as 240 K cal/M^3. The exhaust cooled to 20ºC resulted in a yield of 0.7 L. water per M^3 of consumed hydrogen gas, by continuous operation, a 70% humidity and a 15ºC air temperature. With this engine a compensating weight factor for the LZ-129 could have been achieved, as follows: the hydrogen consumption amounted to 170 M3/h and the recovered amount of water from the exhaust gasses 120 Kg/h. Calculating 1 M^3 hydrogen – approx. 1 Kg lift resulting in a weight increase of 290 Kg/h.[436]

Burgess, again, was way ahead of us. In his memo #146, he stated, "The hydrogen ballonets in this proposed scheme will take up about 33% of the nominal gas volume of the ship. It is believed possible to provide this ballonet volume without sacrificing the principle of completely surrounding the hydrogen by helium." If the hydrogen ballonet was actually inside the helium chamber, we have a vehicle whose gaseous fuel is essentially fireproof. For the first time in history, we are free of the terrors of heavy petroleum fuels anxiously awaiting a chance to leak, spread, linger and find an ignition source. External suspension would allow more traditional fuel bladder with something like a balloon net to harness the lift. In the fuel system block diagram this might be labeled the "variable-volume intermediate fuel reservoir." With a nod to the master, C.P. Burgess, we could call this our *hydrogen ballonet.*

The plan would only work if consumption rate makes the hydrogen disappear at least as fast as we would like to increase altitude. Remember hydrogen's opponents complaining how it takes so much of it to yield any real energy? 1 gallon of gasoline is equal in energy to 380 SCF (in metric terms, the gasoline = 10 Nm3). Our fuel cells are taxed supplying the electrical needs of our power-hungry avionics suite, producing pure drinking water for the crew in the process. What water is not needed becomes eligible ballast. The ship's engines are burning large amounts of fuel to gain altitude, their exhausts helping heat the bag as its helium cools the engine and exhaust and helps condense the combustion product, water, into quick-release voids. More likely, this consumption would be greater than required by a desirable rate of climb. No problem, today's gas management hardware and software would find the simultaneous computation of keeping the right volume in the fuel bladder in check with barometer, thermometer and consumption rate a rather simple calculation to make.

To approach pressure height, the climb continues until the fuel bladder is pushed flat by the expanding helium. Once at mission altitude, fuel is drawn from the storage tank and bypasses or is fed into the bladder at the same rate as it is being consumed. As fuel weight and other consumables' weights disappear, most are matched by water recovered, owing to hydrogen's

---

[436] Von Dr. Ing. Fritz Sturm, VDI aand Dipl.lng. G. Molt, VDI, *Ballastwassergewinnung im Luftschiff LZ 130 „Graf Zeppelin"* dated 15 April 1939, translation by William Kramer, courtesy John Mellberg.

lopsided water production collecting oxygen weight from thin air. Eventually, the fuel bag's lift is gradually needed less, and its size is allowed to decrease. Air is pumped into the fore and aft ballonets to make up the cubic area and keep the bag pressure nominal. During the entire mission, regardless of temperature, superheat, barometer, and weight consumption, we have the ability to fly straight and level for the best economy. Descending, the process continues until on the deck, the fuel bladder could be flat. The fore and aft air ballonets would be used to push lift forward or aft to aid in landing, if necessary both as full as possible to kill the most lift. Heavy, pressing down on its wheels, the airship is more easily masted.

This should not be a major design problem center-of-buoyancy vs. center-of-thrust. In Navy airships with center ballonets, designers treated the two center bags as one; they shared a common filler duct and exhaust valves. Pilots working in the atmospheric ocean commonly let all the air out of the fore and aft bags on ascent, and simply "flew" the center bags. Most of the gasoline airships aloft today have but one, a center ballonet.

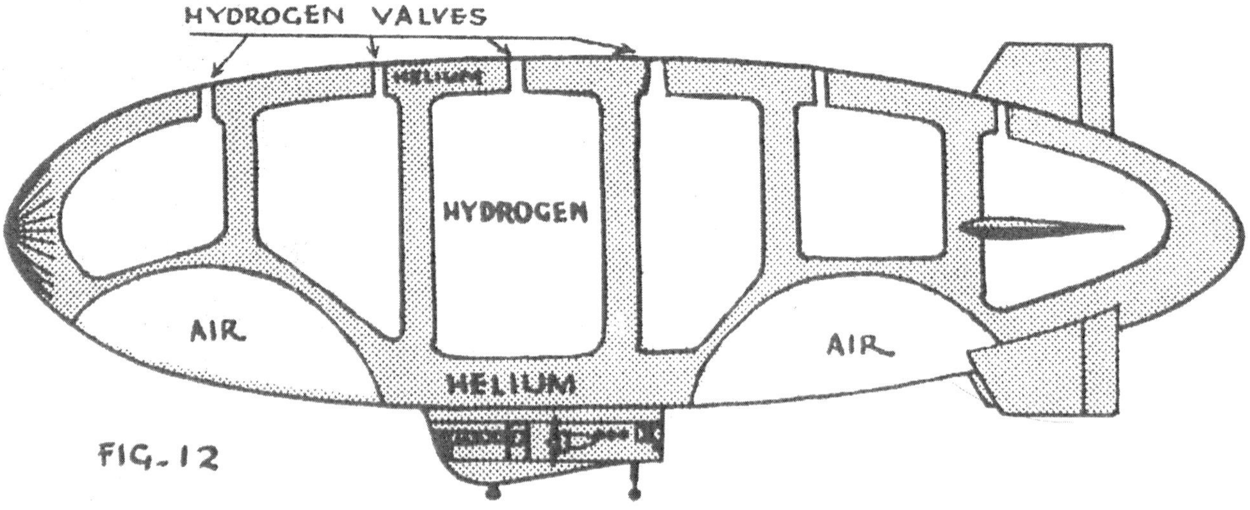

FIG.-12

This drawing from a 1943 Moffett Field training manual shows a proposal for fitting the TC-13 with hydrogen bladders.[437]

Another Burgess design memo discussed "Use of an Easily Liquifiable Fuel Gas for Buoyancy Control in Airships." In this memo, number 254, Burgess cited iso-butane as the most ideal for this purpose. Providing 21,000 Btu per pound, as a gas it weighs no more than the air it displaces. With onboard equipment condensing the gas as altitude increases, helium could expand to fill the void. In a *Macon*-sized airship, some 10,000 pounds of variable lift was seen as covering most situations, and this would require 20,000 pounds of iso-butane. Burgess reasoned the 6,000 pound penalty for the tanks and complicated apparatus would be an acceptable tradeoff for the performance gained. "The gain in lift by this method of take-off is 6150 lb., fully compensating for the weight of the apparatus, independently of the saving by flying in better trim." Again remembering we are constantly producing water and have heat to consume, we might look back to steam as at least a backup system. As T. J. Goodey wrote, "However, when an airship whose lift gas is steam is fitted with a steam engine, naturally the spent steam from the engine will be discharged into the envelope, and this eliminates the condenser as a separate unit. Probably for the first time in the history of the steam engine, the problem of condensation is

[437]AIRSHIP BUOYANCY courtesy Robert Koeberle, WWII Goodyear/Moffett Field airshipman. Donated to NMNA, another copy is on display by the Moffett Field Historical Society.

neutralized without penalty. Furthermore, the same boiler will be used for supplying steam both to the engine and directly into the envelope, and this is an effective synergy. An interesting further possibility, tangential to the main thrust of this paper, is that the lift provided by steam can be enhanced by mixing in a certain proportion of hydrogen, which of course must be kept within safe limits. A wealth of data on flammability of hydrogen-air-steam mixtures has been provided by work on the safety of nuclear plants. It has been established[438] that if an initial mixture of 10% H_2 / 90% H_20 (by volume) is diluted by addition of air, in any proportion, the resultant mixture is never flammable. However, as an initial mixture of 15% H_2 / 85% H_20 is progressively diluted by addition of air, at one stage the mixture becomes marginally flammable. We consider that an initial admixture of 12% hydrogen by volume to steam will be quite safe."

Sounds good, but in fact all this would only allow performance similar to that enjoyed by hydrogen airshipmen seven decades ago. Once down, a Zeppelin could not rise up again until gassed up. If we could put a man on the Moon in 1969, we can build a more capable airship than was flying in 1917. The trick is to use some of that expertise and technology perfected during the Apollo program. Once we accept the idea of hydrogen fuel, then come to terms with the idea of letting some of it expand into a bag before consuming it, we need only one more step. The main fuel supply will be stored in liquid form. With LH_2, we can construct an airship that offers *continuous selection of static condition,* over a wide range of challenges we'll encounter operating in our ocean of atmosphere.

The key to this writer's scheme of using liquid hydrogen as airship fuel and lift-on-demand is based on these three facts from your high school chemistry book.[439]

1) Hydrogen has the *most* Btu by *weight* of any fuel - a point ignored since few have seriously considered fueling an airship with it since May 6, 1937.

2) Hydrogen has the *least* Btu by *volume* of any fuel, an argument used against its adoption in compact automobiles.

3) LH_2 *expands* 851.33 times as it evaporates on its way to the combustion chamber or reaction bed.

Let's go over this again, because this is the entire basis for a new generation of airships. Hydrogen's high heat value per pound is 61,031 Btu. The next best fuel by weight, methane, is nearly *two-thirds less*, only 23,890 Btu per pound. On the flip side, hydrogen's high heat value by volume is a paltry 319.4 Btu/SCF.[440] The next weakest fuel energy by volume, again methane, is more than *three times as much*, 994.7 Btu/SCF.[441]

Toward the last of his life, a rigid airshipman suggested that, when such equipment would become available, alternately liquefying and evaporating the helium lifting gas to control static condition.[442] Sadly, as C.P. Burgess might have said, the liquefaction plant idea splits on the same rock as the compressor scheme. The final nail for helium, even if a liquifier was weightless and free, is that a gallon of LH_e only evaporates to 100 ft^3. LH_2 evaporates more than eight

---

[438] B. W. Marshall Jr., 'Hydrogen: Air: Steam Flammability Limits and Combustion Characteristics in the FITS Vessel,', Sandia Nat. Labs. Report SAND84-0383, Dec. 1986

[439] INTRODUCTORY CHEMISTRY: MODELS AND BASIC CONCEPTS, Amend, John R., John Wiley and Sons, 1977, page 296.

[440] Standard Cubic Foot of gas measured at 1 atmosphere of pressure, at 70°F.

[441] 1 gallon of LH_2 weights 0.59 pounds and evaporates to 113 ft^3 volume. 1 pound of GH_2 occupies 192 SCF.

[442] May 1975 *Nav Air News*

hundred fifty seven times its liquid volume, but the equipment needed to liquefy it would not be practical to fly. Suppose that we could achieve the desired results without taking all the heavy equipment along? The hydrogen generator and liquefaction plant can stay back at the base or support ship.

How the hydrogen is stored, versus how it is consumed, is the key. Difficult concept to grasp? You've been doing the same thing with your automobile. You don't carry the oil refinery. You have been storing the fuel in liquid form and atomizing it (to the best degree) for consumption in the engine chambers. So instead, we simply take on our hydrogen fuel in liquid form, and use its boiloff for the desired effect. Since hydrogen's displacement of air is most efficient at the least practical pressure, adding more hydrogen weight cuts down on the efficiency of the scheme. Therefore even with modern high-strength fabrics, the fuel bladder itself may not be practical to match the variables of liquid's boiloff vs. fuel demands. A smaller but slightly higher-pressure reservoir - accumulator may be necessary for the system's operational flexibility. Hydrogen's specific volume at STP is 191.96 cu. ft./lb. To determine tank size in a pressure vessel, where 1 bar = 14.696 PSI , we use cubic feet = X gals. (.1337). So tank capacity = [(X gallons) (0.1337)] (x bar). Therefore, a 100 gallon tank holds 133.7 SCF @ 10 bar (147 PSI).

Since initially this boiloff would be a slow process, we might still want to use a tube bank of compressed gas on the mast to give us the initial charge for the hydrogen ballonet. While not practical to fly, a ready supply of quickly expandable hydrogen would speed the liftoff process. Typical tube trailers hold up to 126,000 cubic feet of hydrogen under up to 2640 psi pressure. Tube modules carry up to 150,000 cu ft at 2400 psi. (Both are manufactured and maintained under DOT-3A/AA specifications, which test pressurize the cylinders five thirds their service pressure.) One could easily imagine a couple of tube modules built into a mobile mast, and a fuel tube leading up through the nose. Inside, the standard fore and aft ballonets are full of air, and the center fuel ballonet is empty. Next, instead of having to start with dead air occupying almost half of our available lift space, we can temporarily borrow fuel from our supply tank and inflate the fuel ballonet. We open a valve from the fuel supply, hydrogen expands into the fuel bag. This expands, mashing the helium, which presses all the air out of the fore and aft airbags. During our inflation of the interior fuel bladder, and the corresponding deflation of the air ballonets, the ship becomes increasingly light. As mission requirements and the conditions dictate, we could trim for equilibrium. At mission altitude, maintenance of pressure to accommodate various minor changes in altitude and conditions would be a simple task. As the hydrogen fuel leaves its contained form, it expands rapidly, inflating our fuel ballonet. The evaporation yields about an 851 to 1 expansion ratio. As it expands, it tends to compress the helium, which of course cannot escape. The air ballonet's valves, set for a low pressure, allow the air to escape as the helium presses them flat. In doing so we subtract the air's slight weight, but add the hydrogen's displacement of the same air volume, giving us the corresponding lift of the hydrogen. When we do not need the lift, we restrict this process and run our consuming engines and fuel cells to draw the bag down, then more direct from the tank.

The key to this scheme working where so many others have failed is not just that 851 to 1 ratio of liquid to gas we've always known about. Rather, it is the rate the newly released gases' consumption meets the airship's need for fuel while also managing its envelope pressure and lift. Discounting fuel cell draw for the moment, we must rely on automotive engine experience to show how this would work. Frank Lynch of Hydrogen Components Inc. has done a great deal of experimental work and testing of hydrogen-fueled engines and components from his shops in Denver. "Depending on how we burn the hydrogen, the engine may be a bit more or less efficient than the gasoline engine… it is best just to say the efficiency is similar," he writes. The lower heat value of hydrogen is 270 Btu per SCF. So, if an automobile goes a given distance on 1 gallon of gasoline, the same car will consume 433 SCF of H_2 to reach the same distance.

Dr. Mike Swain of the University of Miami has worked optimizing engine design for hydrogen consumption. Utilizing modified heads and other relatively simple modifications, his team has achieved efficiencies approaching the astonishing rate of 40%. With so little energy being consumed, one would wonder if the consumption rate would allow an acceptable rate of climb. Dr. Swain projects that at a high engine load, perhaps the equivalent of what HTA pilots called minimal acceptable takeoff power, a 300 horsepower engine would be consuming about 528 SCF of hydrogen a minute. With two engines, and with the option of loading the fuel cells to maximum consumption as well, it would still be possible to decrease the fuel bladder size about 1200 SCF per minute without wasting any fuel at all.

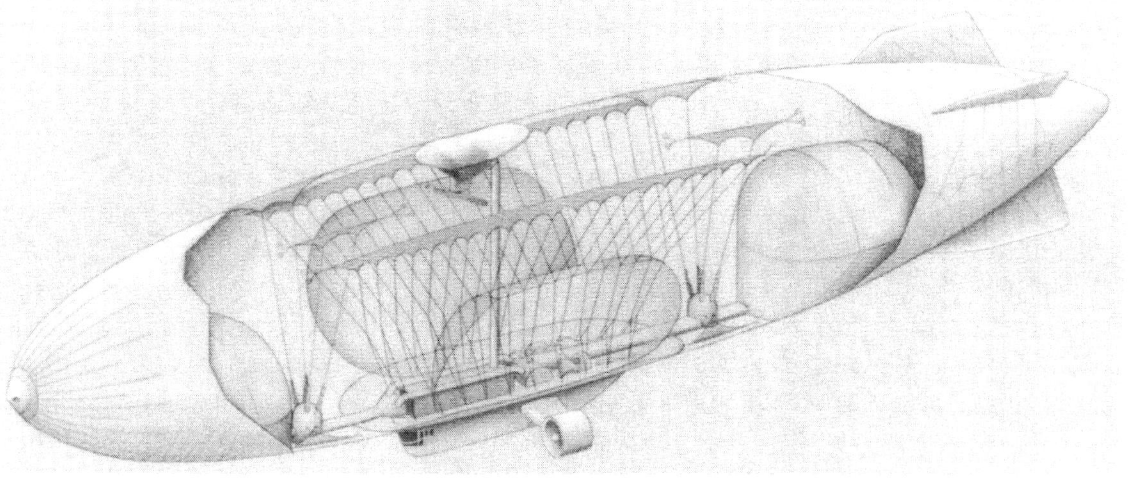

Proposal to update the ZPG-3W in the late 1980s still contained the separate suspension system to manage the tonnage of petroleum fuel bladders. Center split ballonet was flown and managed as a single center bag. If these center cells were converted to hydrogen, their huge exterior surface area could be used for Lockheed's proposed membrane fuel cells. GAC concept courtesy Larry Lanchonie.

The old rule of thumb shows that the lifting gas expands about 1/30th of its volume for every 1,000 feet of altitude gained. An airship could be designed with the hydrogen fuel ballonet sized for the consumers to use up, to make room for expanding helium on ascent. The airship design will be mature when the hardware allows liftoff at near equilibrium followed by climb to altitude at nearly the precise rate the fuel bag is being drawn down. The crew would then operate the ship at fuel-efficient equilibrium by balancing the lift, or lift-killing air ballonets, over a wide range of operational variables.

Converted to hydrogen, the otherwise unmodified ex-Navy L-19 reached 7,100 feet during its test flight.[443] While we would not enjoy the same efficiency, we would actually have wider parameters. If high altitude, lighter payload is our mission, we can start with all three interior bags at a high percentage and ascend until all the ballonets are flat. The engines and fuel cells would then be fed directly from the boiloff, without further inflating the fuel ballonet. If maximum payload is the mission, heavy takeoffs would be even more capable. Obtaining permission, we need only clear obstructions on our flight path, going no higher than a computed altitude where the fuel in the 3rd ballonet need not be burned off further. If we need to descend

[443] E-mail from pilot Lou Prost, instructor pilot, *Underberg*

right away, we reverse the procedure. When we want to descend later, after having consumed some fuel weight, we just pump air into the ballonets same as always, fore or aft for desired trim, while sucking the fuel bag down to a desired lift. This scheme would allow airship operations previously impossible even for hydrogen-filled ships. A normal launch and flight to a cargo pickup point could be carried out with nominal trim. We can then trim to come in heavy and arrange the lifting fixture around the payload. Inflating our hydrogen ballonet to maximize lift, we could still lift off in the ship's length. Getting rid of cargo in the middle of nowhere is the opposite procedure, just as capable. Extra performance could be rung out simply by adding additional hydrogen at heavy liftoff, and simply venting it away at load drop off to prevent the airship from being light. A tube trailer full of hydrogen could be delivered to the site to speed the return operation.

You could balance the hydrogen and air ballonets for unlimited adjustment across a wide range of payload, temperature and barometric conditions. Forget those shot bags, stop carrying so much parasitic emergency water ballast - you now have a powered vehicle with the buoyancy control of a hot air balloon. Further, for the first time in decades, the bag need not be angled into the air stream, up or down. The hydrogen-fueled ship would enjoy the fuel efficiency of level flight, made possible by trimming the ship into equilibrium at all times over a wide range of superheat, temperature inversions, humidity and barometer readings. The heavy lift cargo airship, out performing the largest conceivable helicopter, is being built as this book goes to press. Imagine how much more capable *Cargolifter* will become when improved with hydrogen fuel. The exchange of water ballast for payload weight in addition to the vast difference obtainable with a hydrogen ballonet set would greatly increase its capacity and operational parameters.

This lift-from-fuel scheme isn't just once a mission, like a 1917 Zeppelin. You can keep doing it all the day long, up and down, payload or empty, because you waste nothing. As long as you have fuel, you have virtually dial-in aerostatic control. By selectively using air or hydrogen, you have the ability to kill lift or add ascentive force at will. Purity at lower limit and the helium tanker doesn't show up on time? Purge it clean, add some more payload and inflate the hydrogen fuel bladder a few more percentage points. You would literally have a range of dial-in anti-gravity! It was if we had gone aboard the TV Jules Verne airship *Aurora* and had Paspartou demonstrate the "uplifticating" lever.

In the process, we have created a most survivable vehicle. The traditional airship advantage of altitude independent of forward speed, or even control surface operation, is retained. The fuel cell incorporates the previously unknown advantage of electrical power independent of engine operation. Now, the use of hydrogen fuel allows for certain options even in the event of *total envelope failure*, because we suddenly have a ready supply of lift on tap – and still have three bags to put it in. Instead of the impossible dream that helium could be moved into the ballonets during an envelope compromise emergency, the expanding fuel could drive out ballonet air and provide rudimentary back-up balloons. Now aerated and flammable, its containers would not be. Such a system would at least allow safer descent when there is a big hole in the main gasbag. No doubt critics would rather condemn future crews to the same fate as ZPG-3W Bu No 144242's men, of whom 18 drowned when the ship fell into the water following what a lawsuit alleged was a 135 foot tear in the top of the envelope. (Others maintain the bag ripped only after the ship hit the water largely due to crew and pilot error. Yet it is difficult for them to explain how so large an airship could accelerate so quickly without a massive loss of lift.) To this writer, even the seemingly abhorrent concept of running hydrogen into a fire-resistant air bag would be preferable to impacting so rapidly that crews are injured, or that the vehicle would sink so fast the crew cannot abandon ship. One might as well be in an airplane!

During the final decade of active pressure airship design, ever increasing capabilities came with larger sizes. Two million ft^3 "hot dog on a roll" airships, featuring external suspension for maximum utilization of the interior, seemed the ultimate. Pressure airship designs of just two million cu ft in the 1960s gave way to five million in the late 1990s. Hydrogen fuel frees us from the spiraling costs of ever-larger ships, unnecessary for typical missions.

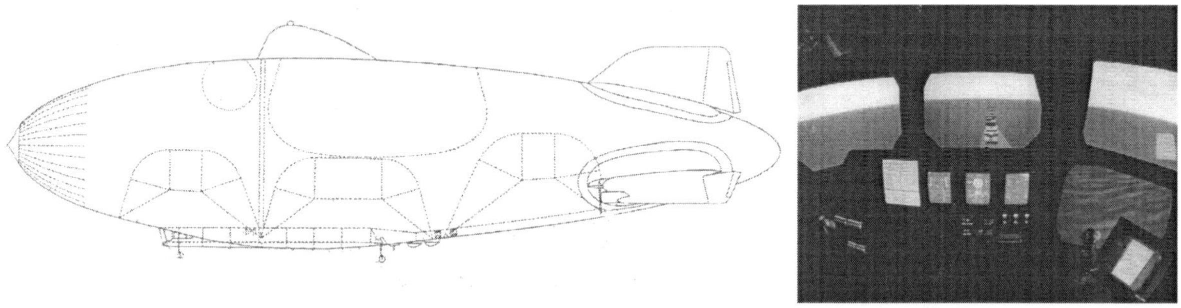

With external suspension, Goodyear's last designs offered hundreds of thousands of cubic feet of electronics-soothing helium in their stern-propulsion designs. The NASA Ames simulator (right) was built up and programmed to train pilots for the proposed YZE-2A, ironically not far from *Macon's* airdock. Courtesy Jeff Cook and NASA/Ames.

With an eye toward designing a versatile airship whose large interior volume could be configured for a variety of missions, we could consider the last Goodyear-Navy pressure airships as an excellent starting place to design a new hydrogen-ballonet equipped airship. The records set by the ZPG-2 are impressive; it is hard to imagine crews that could stand spending more than 93 continuous hours on anti-sub station, or more than 250 hours in the air.

ZPG-2 Bu No 141561's mission to the Arctic is legendary, but William Althoff's book ARCTIC MISSION keeps it in perspective. To carry the crew and support for this ambitious undertaking, more helium was pumped into the bag, running the ballonets nearly flat. When bad weather was encountered near New York, the airship couldn't climb over; it diverted to Lakehurst. Weather conspired to keep pressure height so low that in the unusual temperatures, fuel consumption was very high trying to keep the big bag aloft. They had no choice but to divert once again, to Akron. Finally some people were let off, then the weather cooled, and the flight made its way into history. It has been argued the larger ZPG-3W could have made all those record-breaking flights in operational trim, not stripped down for endurance. Allowing that modern materials will hopefully save about as much weight as modern safety and other necessities would add, the one and a half to two million cubic foot ship would seem capable of most any realistic mission today.

So, if we pick up where the US Navy left off, we have a ZPG-3[444] blimp with gross weight of 93,496 pounds designed for maximum altitude of 10,000 feet under optimal barometric conditions. Right away we could lose the gasoline bladder catenary system inside, the lighter but stronger LH_2 tanks being supported by the car's suspension system. The one and a half million cubic feet of helium was compressed by four big air bags totaling 383,000 cubic feet during ordinary barometric and mission conditions. Reworking the center ballonets and ductwork, the center ballonet becomes our expandable fuel bladder.

[444] A developed but unbuilt ASW variant of the ZPG-3W radar carrying airship.

The gaseous supply in the mast empties into the ship as we're ready to begin our mission. At full inflation, discounting the difference weight of the fuel and with little else changed, the ZPG-3 would now have roughly an additional 8,000 pounds - *four tons* - of *optional* lift. Making the fuel bladder a bit larger would allow even more air to be driven out of their ballonets. We were only counting on converting its center ballonet, about one third of the total, for fuel-lift use. If we go with Burgess' 33% for the fuel ballonet alone, we'd have a bladder of some 495,000 cubic feet. That would give us an option of adding (or subtracting) 31,185 pounds, or roughly *fifteen and one-half tons* of lift!

Let's not forget that 25% of total figure *rarely* meant pressure height actually would be the theoretical 10,000 feet in everyday conditions. Typical mission loads coupled with the real world temperature and barometer usually kept the ZPG-2s below 4000 feet and the ZPG-3Ws below 5000 feet. So increasing the fuel ballonet to Burgess' 33%, as well as having the option of taking up some potential expansion space with air in the fore and aft ballonets, means higher altitudes really would be possible. It's not that typical missions are going to demand altitudes that will require the crew to wear oxygen masks. (FAA rules oxygen for crew if at 12,500 feet for 30 minutes.) Oftentimes the difference between finding a wind going your way, or having to buck a gale going the wrong way, is just a few thousand feet. Having almost another mile of altitude in which to explore our atmospheric ocean will open a world of possibilities.

Using the compressed gas in the tube trailers on the mast, we rapidly fill the hydrogen bladder until full (or until the wheels start to lift off the deck). As we demast, a small crew (or perhaps even the pilot, via onboard instruments) could do the weigh-off – the ship would soon be at equilibrium. Starting take-off, rate of climb vs. fuel consumption is the question. Now the fact that hydrogen has only 10.7 kilojoues per liter, generally considered to be *three orders of magnitude too little energy* for practical applications, works *for* our airship.

The ZPG-3Ws' big nine-cylinder Wright R-1820-88 *Cyclones*, rated at a nominal 1,275 horsepower, could develop 1,575 horsepower at maximum takeoff power. At high RPM and very high manifold pressure they would consume about 75 to 80 gallons of avgas an hour, or for the two, about 2 ½ gallons of gasoline per minute. At that rate, the 3W was getting about 16 pounds lighter every 60 seconds. Scaling up the values of known engines to this 1500 hp figure, the two would be sucking down about 5,280 SCF of hydrogen per minute. The helium would be looking to expand at the rate of $1/30^{th}$ its volume for every 1000 feet of altitude gained. In the 1.5 million ft^3 ZPG-3, that meant about 49,500 ft^3 of air had to be let out per 1000 ft. of ascension. Even though the engine hydrogen is only being consumed at about 5,280 cu ft per minute, our airship could, if needed, climb at the rate of about 110 feet per minute with no reserve air capacity or wasting any gaseous fuel. Generating electrical power would consume more yet, allowing additional ascension speed if required. C. P. had his figures right; when we've burned off 459,000 ft^3 of hydrogen from the fuel ballonet, we can't go any higher – nor should we need to, because we're above 10,000 feet! With modern materials allowing higher envelope pressures, and more effective stern propulsion pushing aside thinner high altitude air, we will find it easier to overcome resistance like what Santos-Dumont called "pushing a candle through a brick wall."

During the mission, we are consuming fuel, making water, and recovering all the ballast water necessary as well. At the end of the mission, air is pumped back in the forward and aft ballonets as we descend to keep the envelope taut as we go back down into denser atmosphere. Balancing the fuel and air bladders keeps the ship in trim or slightly heavy right down to the deck. Recovery of the heavy ship would be possible with a smaller crew. Wouldn't the airship become much heavier as the lesser-lift helium expanded during climb replacing the greater-lift hydrogen? One remembers the optimistic Rogers' prediction that "The only apparent disadvantage of helium is the fact that it is about twice as heavy as hydrogen… this is of little practical importance, [with] the buoyancy, or lifting power, of helium being 93 per cent that of hydrogen." In this application, his statement is not misleading. The total, eventual lift difference would be less than ¾ ton, easily compensated for, and if below pressure height, less noticeable in the dynamics of flight and our ever-changing atmosphere.

A good place to start: like a modern USS *Patoka*, today's aerostat tenders could generate hydrogen from seawater. The large fuel ballonet would provide the aerostat an emergency backup lift cell, and membrane fuel cells could provide DC power for the electronics. Courtesy TCOM.

The logical debut of the hydrogen ballonet idea would be in an unmanned aerostat. (A "black" program may have already done so, if rumored altitude records discussed at the 1998 Airship Association Symposium in Bedford are true.) The performance gain in weather-beating altitude would be noticeable. Like an airship, the changeover would not involve complete redesign. If history is a guide, the changeover could be accomplished by modifying existing hardware and firmware. A small hydrogen-powered rotary engine could power an AC generator, or more modest loads could be handled by an inverter fed from membrane fuel cells along the sides. No more four-mile long, highly capacitive extension cord adding tons of weight to a tether. The tether could even carry a tiny fuel tube, and the system hardly even needs a pump!

Of course there will be more serious potential problems with the new idea. What if the higher-pressure fuel gas seeped out of the fuel ballonet into the helium chamber? Extracting trace amounts would be unnecessary, but would be possible using palladium filters. It's unlikely the lower pressure lifting gas would get into the fuel system, but 13% or less helium contamination in the hydrogen means it would still be useful as fuel without a catalyst.[445] "…helium is moving as an atomic gas H_e, and not as a molecule, like hydrogen H_2. In spite of its atomic weight (H_e) being twice as high as the hydrogen (H_2) molecule, the assumed diameter of the helium atom is smaller compared than the diameter of the molecule hydrogen (H_2). This is a remarkable characteristic of the inert gas helium. Therefore helium – regarded as a quantity of gas, enclosed

[445] C. W. Siebel quotes the more conservative figure of 10% hydrogen in helium as being the safe limit.

in a container (envelope) – is more volatile than hydrogen."[446] The fuel bladder therefore must be constructed to prevent helium seepage as well. Cost of the fuel would still be a consideration, so one would hope as the price of hydrogen drops in accordance with its improved generating methods, and the price of petroleum fuels rises with their real or contrived scarcity, at some point they would pass. Annual H_2 production in the 1990s reached 3 billion cubic feet. Most of the hydrogen consumed today is made at the same place it is consumed, averaging about 32 cents per pound. If it must be moved in gaseous form, as of this writing hydrogen was selling at about $14 per pound in single, standard K-bottles. Price goes down with energy density transported, of course, to the point where LH_2 delivered to the space shuttle launch pad at about $1.50 per pound. (There are at this writing about 25,000 shipments of 14,000-gallon tankers to over 500 locations in the US annually.) The extra lift in a cargo airship would already overcome many economic considerations, since too-big, too-heavy cargo shippers can pretty much name their own price any time. More costly fuel would endanger the bottom line to a smaller degree because it would be *carrying its own weight* – and then some.

At the beginning of the airship's war against the submarine in 1942, radar sets were being added to the K-ships. Few crews had trained operators, and pilots looked upon the sets as useless weight. Once demonstrated by trained operators, pilots quickly went from wanting to throw radar over the side, to refusing to fly without it. Likewise, no one was sure about the reliability of reversible propellers when first tested on an airship. Would they come apart and puncture the bag? Yet any pilot who transitioned from the old K-ship to the improved ZP2K modification would assure you they are great. Going back to landing without being able to apply reverse thrust would be like going back to the dark ages.[447]

Likewise, we can imagine a pilot trained to operate an airship that has a very wide range of control over its static condition would wonder how anyone ever operated a motorized gasbag that had fixed lift, no matter what. These new generation pilots would leave their bases at EQ, fly to a cargo pickup, lift the cargo away, fly it to altitude, lower it to a destination, regain EQ and fly home, landing slightly heavy, then become very heavy at the mast. (The wartime mission of the L-8 delivering special windows to the USS *Yorktown's* deck crowded with Doolittle's B-25s would seem all the more marvelous!) They would never forget the overwhelming advantages of an airship aloft. Instructor Anton F. Heinen wrote: "One more misleading conception with reference to airships very generally held is the opinion that it needs superengineering to provide at the same time, as seems to be correct at first superficial approach, extreme lightness in structure and an unheard of strength due to the idea (quite mistaken as has been shown), that it must continuously 'battle the elements'. The fundamental difference between the airship and the surface vessel, where the premised condition is very real and one of the most important factors in engineering procedure, lies in the fact that this very 'battle' is absent. A dramatic episode in the airship's history, which occurred during the *Shenandoah* operation at Lakehurst, (there have been numerous others before), could very well illustrate the point, If we succeed in breaking through the earth-bound reportorial clamor of the daily press, which saw superskill and even heroism where there was nothing of the kind, and failed to see the real significance of the happening in spite of earnest attempts by the participants to emphasize the other side of it.

---

[446] Supplement to *Gasbag Journal*, September, 1999, Burgess uses these figures: "Hydrogen is a diatomic gas with a molecular weight of 2.016. When chemically pure it has a specific weight under standard conditions of .00532 pounds per cubic foot and a specific gravity relative to air of .0696. Its ascensional force under standard conditions is therefore .07076 pounds per cubic foot…"

[447] "Propellers Used as Brakes on U S Navy's Largest Blimp," *Science News Letter*, 26 AUG 44 and "Reversible Propellers Make the New Navy Blimp More Maneuverable," *Scientific American*, JAN 45.

Training pilots and crews, of course, should be done with special care. Pilots are warned to stay out of clouds, because those silver linings might be airplanes going the opposite directions. LTA craft are inherently safer, but there is still no substitute for experience in their peculiar operation. Following the April 8, 2000 MV-22 *Osprey* accident that killed 19 Marines, it was discovered the lost pilot had 3,500 hours in KC-130s but only 80 in MV22s. "The post-crash briefing alluded that the pilot got into 'power-settling,' a situation that can occur in a rotary-wing aircraft when the rate of descent is too great and the forward airspeed too low. The instability and the aircraft's own rotor wash cause a loss of lift and the aircraft becomes difficult to control. Recovery procedures for this situation in rotary-wing aircraft involve *reducing power* and lowering the nose. This is a fundamentally different procedure that is used in fixed-wing aircraft (such as the KC-130), where the pilot, when confronted with a similar situation (a stall), is trained to *add* power and lower the nose."[448] This is surprisingly similar to the reaction that possibly contributed to the Navy's last blimp accident, and not very dissimilar to the year 2000 events that put the *Spirit of Akron* into the trees. These last two accidents involved dedicated, careful pilots with heavier-than-air experience but relatively little LTA time. Obviously, the operation of a hydrogen-fuel, variable lift airship would be a different animal altogether, and would require unique training. It would be worth the trouble. As Barkley and Buckley wrote, "...when sound operating procedure complements the inherent safety of the airship, no other means of transport can exceed its reliability."

Air-sailors were a disappearing breed, and the skilled crews that kept their gasbags in flying trim are also few and far between. The time to train pilots is right now. Old lessons will have to be re-learned, for no history book could teach all the experiences gained from 100 years of operations in our atmospheric oceans. One can imagine a crewman, after having topped up the envelope with the driest gas available, is secure in the knowledge his airship won't suffer water intrusion if rained on. Yet even in dry weather, he is puzzled to see water dripping out of the envelope's condensate drain following a mission – not realizing moisture in the gas will fog up and collect inside the bag as the airship ascends. Operation of a displacement new large vessel with greater 'depth' in our atmospheric ocean will be a learning experience.

We would be wise *not* to follow the late 1950s US Navy practice of simultaneous qualifications in fixed-wing, rotary-wing, and LTA. We need a new generation of air-sailors, men who fly the weather maps to ride the eddies and currents of our atmospheric ocean. As one wrote, "Therefore, instead of going against the current on the left side of the whirlpool, the *Hindenburg* sets her course on the opposite side of the storm-center. She becomes a part of the stream of wind, like a chip carried along on a stream of water... Dropping off the hurricane's edge on the western side last month, like a small boy who has stolen half a turn on a merry-go-round, the air clipper's navigator found he had made a land speed of 160 miles an hour while her air speed remained at eighty." Of course new technology will make the job easier, but there is no substitute for the raw skill that crossed the North Pole. "At a time when there was no radar, no radio beaconing, few other navigational aids, and almost no meteorological ground stations, it was vital the officers and men in the pilot cabin have enough accumulated experience to keep the airship on a steady course, at the most beneficial altitude from the standpoint of winds, and at the most favorable airspeed to make good forward progress yet conserve fuel."[449]

Roy McAlister wrote, "When hydrogen is produced in sufficient amounts to achieve the economies of scale it will be the cheapest renewable fuel... But it is often suggested that the

---

[448] S. D. Danyluk, "Epitaph for an Aviator: 'I Told Them I Needed Training,'" USNI *Proceedings*, Sep. 2000

[449] W. Cross, DISASTER AT THE POLE, Lyon Press, NY, 2000, courtesy Mr. Cross.

Hindenburg disaster ended the chance for practical applications of hydrogen." The sky is literally the limit when we finally shrug off the mental baggage of hydrogen's imagined dangers. Foremost among these is the perception that the LZ-129 was the ultimate expression of hydrogen's dangers. History has another example where the hydrogen *seemed* to be the problem. A British pioneer aeronaut found out the hard way when the first London-to-Paris flight, begun in November of 1910, suffered various delays, including deflation in trees. As Nick Walmsley explains, the *City of Cardiff* landed near Douai only to have the French authorities charge Willows with *illegal importation of gas*. Finally on December 31st, Willows made some short trips and circled the Eiffel Tower. Unfortunately the authorities hadn't forgotten about the gas duty - they warned him that he must pay 743 francs or leave France next day. He promptly deflated the airship. Had he had flown back he would have won a large prize purse by carrying back with him one M. Reynard, a French aeronautical enthusiast. With this and other historical lessons we see ignorant *people* are usually the problem, not the gas.

Likewise, there is still a great deal to learn about nature's most simple element. There are actually two kinds of molecules, ortho-hydrogen and para-hydrogen. The ordinary gas is composed of 25 % para- and 75% ortho-. We can't make pure ortho-; para-hydrogen is produced by absorbing ordinary H_2 in charcoal at about -225°C (-373°F), and has slightly shifted boiling and melting points. Some believe there is a useful variation in what is called "Brown's Gas", after its inventor Yul Brown, who made the gases' generators. It's claimed an ignited torch of the gas is all but cool to the touch, but its reactions with various metals generate very high temperatures. Thermal water splitting, using temperatures up to 3000°C, has been demonstrated. Hydrogen has a valence of one, giving it other unique properties. Only 1.23 volts are needed to split water at 25°C; commercial electrolysis systems that operate with 1.75 volts draw about 1 amp/cm^2. This produces about 46.8 kW-hr/kg, for an energy efficiency of some 70%.[450] Yet lowering the power consumed in the splitting process is an important area of research. A man has sold seminars describing what seems like a way to violate thermodynamic law by what sounds like the application of *resonance* with electrolysis. Every cavity has a matching electrical frequency at which electrical resistance is minimal and current flow can maximize. (Very high frequencies do not use wiring, but rather continuous resonant cavity-like-plumbing called waveguides. The resonant cavity of ordinary line transmission current would yield a waveguide something like the Holland tunnel.) His seminar seems to suggest he has built as enclosure where water's bonds can be broken with the most efficient frequency where electrical energy consumed would be minimal. One could easily believe many a determined technician went broke in pursuit of this dream, finding that perfect combination that yields more energy than is consumed, like a fusion reactor. But if a latter day Nick Tesla sees something imperceptible to the rest of us, it will not repeal the laws of aerostatics any more than fusion reactors actually defy thermodynamic laws. If cheaper and easier to produce, the process would make hydrogen more useful to airships.

Physicists are now speculating about making lightweight structures out of *metallic* hydrogen.[451] Experiments have suggested metallic hydrogen could be a room temperature superconductor! All-electric airships would certainly benefit. Others believe a secret Government program has taken the next logical step. "There is also documentation of a NASA-financed project that overcame the technical roadblocks that caused *Suntan* to stumble. In the early 1970s Gerald Rosen, a professor of physics at Philadelphia's Drexel University and one of the highest paid theoretical physicists in the United States, was contracted by NASA to determine whether it

[450] Electrolysis machine makers include Stuart Energy Systems of Toronto, Teledyne Brown in Maryland, and Proton energy in CT.
[451] W. Nellis, "Making Metallic Hydrogen," *Scientific American*, May 2000, pg. 84

would be possible to store hydrogen as individual atoms rather than as molecules. His calculations predicted that it was not only possible, but that so much fuel could be stored in a small space that the Apollo astronauts could have traveled to the moon in a rocket the size of a pickup truck. And so, official denials ignored, *Aurora* remains a lively topic for speculation."[452]

The concept of having such an inexhaustible fuel and variable lift supply would allow construction of a vehicle never imagined outside science fiction. Yet, as is often the case, yesterday's dream is tomorrow's hardware. None of these wonders will be available to us if we cannot get it through our thick skulls that the fire that destroyed the *Hindenburg* was not a hydrogen-driven fire.

By the turn of the century, visionaries had predicted 'air-ships' could simply follow the trade winds to carry commercial loads across the oceans. This theory was based on the wide range of operating altitudes offered by the hydrogen balloons of the day. The ability to find a wind going in your direction is an environmentally friendly, energy and cost effective method of movement lost with the limitations of today's airships. The airshipmen piloting the *Graf Zeppelin* could operate around the world's varied weather conditions in what today would be considered nothing more than a floating bomb. Yet hydrogen filled Zeppelin airships completed more than 175 Atlantic crossings, compared to none for helium filled rigids. These airshipmen were not insane; they simply had vision based on experience that we lack.

"The chemical compound dihydrogen monoxide, or DHMO, has been implicated in the deaths of thousands of Americans every year, mainly through accidental ingestion. In gaseous form, it can cause severe burns. And, according to a new chemical report, 'the dangers of this chemical do not end there.' The chemical is so caustic that it 'accelerates the corrosion and rusting of many metals…is a major component of acid rain, [and]… has been found in excised tumors of terminal cancer patients.' Symptoms of ingestion include 'excessive sweating and urination,' and 'for those who have developed a dependency on DHMO, complete withdrawal means certain death.' Yet the presence of the chemical has been confirmed in every river, stream, lake and reservoir in America."[453] It seems certain we must act quickly to ban this hazardous substance – until you realize they were talking about plain old H_20, ordinary water.

Today, in various forms in the US alone, more than 50 million pounds of hydrogen are consumed every day in electronics, food processing, glass, chemicals, oil refining, metals processing, and electric utilities to name a few. Air Products, Inc., and other hydrogen producer's experience base and safety standards are unparalleled with many other hazardous materials shipped every day. Once accepted by buoyant flyers, building a much safer flame-retardant airship with greater range, and survivability that allows some buoyancy and electric power even after total envelope *and* engine failure, is a desirable goal. The hardware is within our grasp. The only stumbling block is between our ears; we must get a grip on our fears.

If we realize our anti-hydrogen prejudice is based on bad press and misinformation, we should be able to overcome the massive inertia inherent in our aerospace industry – and the corporations that bankroll and insure them. This is a non-technical barrier. Airshipmen, already accustomed to keeping air away from their lifting gas, can adopt simple real-world safety precautions developed in over four decades of using hydrogen as fuel. Harnessing nature's most abundant element, we can open up a new universe of possibilities right here in our atmospheric ocean.

---

[452] "Skunk Works Magic," *Popular Mechanics*, September 1999
[453] James Glassman, *Washington Post* columnist, 28 OCT 1997

Not doing so will eventually mean the death of the airship. Adopting hydrogen fuel and proving its safety in public is the only way airships will survive after the helium is gone.

Safe adventures – and real profits - await us in the troposphere.

"Knowing is not enough; we must apply. Willing is not enough; we must do."
- Johann von Goethe

About the Authors

Dr. Addison Bain, Ph.D., was born in Kalispell, Montana, and joined NASA at Huntsville, Alabama on ground based hydrogen systems. Transferred to Kennedy Space Center, Florida, he represented KSC at the Aerospace Research Data Institute (ASRDI) and helped compose the NASA hydrogen safety handbook. In 1973, Bain help found the International Association For Hydrogen Energy (IAHE). Bain retired from KSC as the Director of Hydrogen Fuel Programs. In the photo above Bain (left) is seen with Albert Grotzinger, Zeppelin veteran, in 1996.

Richard Gerard Van Treuren, born in Ohio and raised near Cleveland, joined the US Navy in 1969. He served on the USS *Roosevelt* (CVA-42) with VA-176, and was ship's company on the USS *Kitty Hawk* (CV-63). Introduced by friends to the Space Shuttle as he was finishing his tour at Pt. Mugu, California, Van Treuren joined the Orbiter contractor team in 1979. Van Treuren married the former Deborah Pelzer in 1982. In between processing Orbiters as an Astronaut Crew Changeout technician (16-1), Van Treuren has produced books, articles and videos. In this Eric Brothers photo at Nordholz, Germany, the author holds a sample of gas cell membrane in front of a display showing bullet-damaged cell material.